Cardiovascular Molecular Morphogenesis

Books in the Series

Vascular Morphogenesis: in vivo, in vitro, in mente 0-8176-3920-9
C. D. Little, V. A. Mironov and E. H. Sage, editors

Living Morphogenesis of the Heart 0-8176-4037-1
M. V. de la Cruz and R. R. Markwald

Vascular Morphogenesis: In Vivo, In Vitro, In Mente

Charles D. Little, Vladimir Mironov and E. Helene Sage

Editors

Birkhäuser
Boston • Basel • Berlin

Editors:
Charles D. Little
Department of Cell Biology and Anatomy
Medical University of South Carolina
Charleston, SC 29425-2204

Vladimir Mironov
Department of Cell Biology and Anatomy
Medical University of South Carolina
Charleston, SC 29425-2204

E. Helene Sage
Department of Biological Structures
University of Washington School of Medicine
Seattle, WA 98195-7420

Library of Congress Cataloging-in-Publication Data

Vascular morphogenesis : in vivo, in vitro, in mente / Charles D.
Little, Vladimir Mironov, E. Helene Sage, editors.
p. cm. — (Molecular biology of cardiovascular development)
Includes bibliographical references and index.
ISBN 0-8176-3920-9. — ISBN 3-7643-3920-9
1. Blood-vessels—Growth. 2. Blood-vessels—Differentiation.
I. Little, Charles D., 1946– . II. Mironov, Vladimir, 1954– .
III. Sage, E. Helene, 1946– . IV. Series.
[DNLM: 1. Blood Vessels—embryology. 2. Cardiovascular System—embryology.]
QP101.V326 1998
573.1′8—dc21
Shared Cataloging for DNLM
Library of Congress 98-37689
CIP

ISBN 0-8176-3920-9
ISBN 3-7643-3920-9
Typeset by Braun-Brumfield, Inc.
Printed in the U.S.A.

9 8 7 6 5 4 3 2 1

Contents

List of Contributors

James B. Bassingthwaighte, Department of Bioengineering, Harris Hydraulics, Room 310, University of Washington, Seattle, WA 98195-7962

Daniel A. Beard, Department of Chemistry, Mail Code 5183, New York University, 31 Washington Place, Rm 1021, New York, NY 10003

Mark A.J. Chaplain, Department of Mathematics, University of Dundee, Dundee DD1 4HN, Scotland, UK

Bodo Christ, Anatomisches Institut, Abteilung Anatomie II, Albert-Ludwigs Universität Freiburg, Albertstrasse 17, 79104 Freiburg, GERMANY

Patricia D'Amore, Harvard Medical School and Children's Hospital, Laboratory for Surgical Research, 1050 Enders Building, 300 Longwood Avenue, Boston, MA 02115

Marco C. DeRuiter, Department of Anatomy and Embryology, Leiden University, PO Box 9602 2300 RC Leiden, The NETHERLANDS

Françoise Dieterlen-Lievre, Institute d'Embryologie Cellulaire et Moleculaire UMRC 9924, College de France, Centre National de la Recherche Scientifique, 49 bis avenue de la Belle Gabrielle, 94736 Nogent-sur-Marne Cedex-FRANCE

Christopher J. Drake, Department of Cell Biology and Anatomy, Medical University of South Carolina, 161 Ashley Avenue, Charleston, SC 29425

Adriana C. Gittenberger-de Groot, Department of Anatomy and Embryology, Leiden University, PO Box 9602, 2300 RC Leiden, The NETHERLANDS

Karen K. Hirschi, Children's Nutrition Research Center, Baylor College of Medicine, 1100 Bates Street, Houston, TX 77030

Haymo Kurz, Anatomisches Institut, Abteilung Anatomie II, Albert-Ludwigs Universität Freiburg, Albertstrasse 17, 79104 Freiburg, GERMANY

Zheng Li, Center for Biomedical Engineering, WD-12, Harris Hydraulics Lab, Room 310, University of Washington, Seattle, WA 98195

Charles D. Little, Department of Cell Biology and Anatomy, Medical University of South Carolina, 161 Ashley Avenue, Charleston, SC 29425

Sharon R. Lubkin, Biomathematics Program, Box 8203, North Carolina State University, Raleigh, NC 27695-8203

Daphne Manoussaki, NIH, NIDCD-LCB, Building 9, Room 1E116, Bethesda, MD 20892

Hans Meinhardt, Max-Planck-Institut für Entwicklungsbiologie, Spemannstrasse 35, 72076 Tübingen, GERMANY

Vladimir A. Mironov, Department of Cell Biology and Anatomy, Medical University of South Carolina, 161 Ashley Avenue, Charleston, SC 29425

Roberto Montesano, Université de Genève, Département de Morphologie, Faculté de Médecine, CMU, rue Michel-Servet, 1, 1211 Genève 4, SWITZERLAND

James D. Murray, Department of Applied Mathematics, University of Washington, Box 352420, Seattle, WA 98195-2420

Roberto F. Nicosia, Department of Pathology and Laboratory Medicine, Allegheny University of the Health Sciences, New College Building, Mail Stop #435, Broad & Vine, Philadelphia, PA 19102-1192

Su-Ja Oh, Department of Anatomy, Catholic University Medical College, Seoul 137-701, Korea

Michelle E. Orne, School of Mathematical Sciences, University of Bath, Bath BA2 7AY, UK

Luc Pardanaud, Institute d'Embryologie Cellulaire et Moleculaire UMRC 9924, College de France, Centre National de la Recherche Scientifique, 49 bis avenue de la Belle Gabrielle, 94736 Nogent-sur-Marne Cedex—FRANCE

Michael S. Pepper, Université de Genève, Département de Morphologie, Faculté de Médecine, CMU, rue Michel-Servet, 1, 1211 Genève 4, SWITZERLAND

Robert E. Poelmann, Department of Anatomy and Embryology, Leiden University, PO Box 9602, 2300 RC Leiden, The NETHERLANDS

E. Helen Sage, Department of Biological Structure, University of Washington, Box 357420, Seattle, WA 98195-1524

Konrad Sandau, Fachbereich Mathematik & Naturwissenschaften der Fachhochschule Darmstadt, Schoefferstr 3, Darmstadt, GERMANY

Chitra Suri, Regeneron Pharmaceuticals, Inc., 777 Old Saw Mill River Road, Tarrytown, NY 10591-6707

Robert Vernon, Department of Biological Structure, Health Sciences, University of Washington School of Medicine, Box 357420, Seattle, WA 98195-7420

Jörg Wilting, Anatomisches Institut, Abteilung Anatomie II, Albert-Ludwigs Universität Freiburg, Albertstrasse 17, 79104 Freiburg, GERMANY

George D. Yancopoulos, Regeneron Pharmaceuticals, Inc., 777 Old Saw Mill River Road, Tarrytown, NY 10591-6707

Tada Yipintsoi, Faculty of Medicine, Prince of Songkla University, Hat-yai, Songkla 90112, Thailand

Series Preface

The overall scope of this new series will be to evolve an understanding of the genetic basis of (1) how early mesoderm commits to cells of a heart lineage that progressively and irreversibly assemble into a segmented, primary heart tube that can be remodeled into a four-chambered organ, and (2) how blood vessels are derived and assembled both in the heart and in the body. Our central aim is to establish a four-dimensional, spatiotemporal foundation for the heart and blood vessels that can be genetically dissected for function and mechanism.

Since Robert DeHaan's seminal chapter "Morphogenesis of the Vertebrate Heart" published in Organogenesis (Holt Reinhart & Winston, NY) in 1965, there have been surprisingly few books devoted to the subject of cardiovascular morphogenesis, despite the enormous growth of interest that occurred nationally and internationally. Most writings on the subject have been scholarly compilations of the proceedings of major national or international symposia or multiauthored volumes, often without a specific theme. What is missing are the unifying concepts that can make sense out of a burgeoning database of facts. The Editorial Board of this new series believes the time has come for a book series dedicated to cardiovascular morphogenesis that will serve not only as an important archival and didactic reference source for those who have recently come into the field but also as a guide to the evolution of a field that is clearly coming of age. The advances in heart and vessel morphogenesis are not only serving to reveal general basic mechanisms of embryogenesis but are also now influencing clinical thinking in pediatric and adult cardiology.

Undoubtedly, the Human Genome Project and other genetic approaches will continue to reveal new genes or groups of genes that may be involved in heart development. A central goal of this series will be to extend the identification of these and other genes into their functional role at the molecular, cellular, and organ levels. The major issues in morphogenesis highlighted in the series will be the local (heart or vessel) regulation of cell growth and death, cell adhesion and migration, and gene expression responsible for the cardiovascular cellular phenotypes.

Specific topics will include the following:

- The roles of extracardiac populations of cells in heart development.
- Coronary angiogenesis.
- Vasculogenesis.
- Breaking symmetry, laterality genes, and patterning.
- Formation and integration of the conduction cell phenotypes.
- Growth factors and inductive interactions in cardiogenesis and vasculogenesis.
- Morphogenetic role of the extracellular matrix.
- Genetic regulation of heart development.
- Application of developmental principles to cardiovascular tissue engineering.

R.R. Markwald
Medical University of South Carolina

Foreword

Judah Folkman, M.D.

In the past three decades there has emerged a new field of vascular biology. It has evolved by hard won new knowledge contributed by dedicated researchers in many parts of the world. These are a most collegial group of scientists. It is not uncommon that advances in this field are greeted in competing laboratories with the same enthusiasm as in the laboratory that originally reported the findings.

Anyone who has worked long enough in this field has been witness to its steady progress. While in the 1960s, endothelium was considered to be simply a passive container for the blood, it is now known by the cytokines and mitogens it releases to be an active participant in inflammation, coagulation and repair. When the field of vascular biology was young its literature was mainly descriptive; the physiology of capillary flow and permeability was its quantitative frontier. Today, new regulatory proteins, new genes, and new methods are reported by the month in a multiplicity of journals representing many disciplines. Thirty years ago we had almost no idea of how arteries and veins form in the embryo, to say nothing of the development of whole vascular networks of tubes and branches. Today, however, the molecular cross-talk between endothelial cells and smooth muscle cells is being rapidly elucidated so that the formation of a vascular tube can be understood in terms of the proteins and their receptors which dictate its shape and its dimensions. Thirty years ago, tumor vessels were assumed to be simple dilated host vessels, i.e., a by-product of necrotic tumor cells. The idea that tumors recruited new vessels and depended on them for progressive growth was unpalatable. Today, it is widely accepted that tumors are angiogenesis-dependent; angiogenesis itself has become a major target of tumor therapy. A new field of angiogenesis research has branched from vascular biology. In fact, the various phenotypes of cancer, including tumor growth, invasion, metastasis, progression, tumor dormancy, and tumor cell apoptosis, are not autonomous, but are understood to be under the tight control of the microvascular endothelial cell. Progress in angiogenesis research itself is rapid, from the discovery of the molecular mediation of hypoxic induction of neovascularization to the beginning linkage of oncogenes and tumor suppressor genes to the angiogenic phenotype in cancer.

But, it was not until 1973 that it even became possible to passage endothelial cells in vitro. Because human umbilical vein endothelial cells were the first to be cultured, it became conventional wisdom that all endothelial cells were similar to these cells and to each other. This idea quickly faded in 1978 when it was shown that capillary endothelial cells had more fastidious requirements in cell culture than did umbilical vein endothelial cells. Today the heterogeneity of microvascular endothelium has been found to be so extensive that Auerbach (Auerbach 1991) can report on organ specific endothelial differences and Ruoshlati (Arap et al 1998) can address chemotherapeutic drugs to the endothelium of tumor vessels, while largely avoiding endothelial cells in normal tissues. The prevailing opinion that all angiogenesis is similar, is eroding. New tumor vessels appear to be relatively smooth and muscle-less, in contrast to new wound vessels which become coated with pericytes or smooth muscle relatively quickly. It remains to be seen whether these differences are regulated in part by the angiopoietins (Maisonpierre et al 1997).

While it has long been appreciated that hematopoietic cells and vascular endothelial cells arise from the same precursor cell, why are the detailed lineages of hematopoietic cells not found in vascular endothelial cells? This may be a temporary situation because early lineages are beginning to be uncovered in vascular endothelial cells.

The 'endothelial lineage problem' is just one of many fundamental and fascinating subjects that are discussed in this book, Vascular Morphogenesis: in vivo, in vitro, in mente. It is a compendium of our current knowledge of how the vascular system itself is formed. The editors, Charles Little, Vladimir Mironov, and Helene Sage have done an exemplary job in organizing the various cells, molecules, and processes which collaborate to form blood vessels. Newcomers to vascular biology will find this book invaluable as a basis for their own research, and it will also be a standard reference for clinicians who care for patients with vascular malformations of all types. Some of the most perplexing problems in medical practice are abnormalities of vascular morphogenesis. They afflict children and often persist into adulthood. Enlarging lymphangiomas and growing arteriovenous malformations destroy bone and may be life-threatening. Platelets are trapped in certain severe hemangiomas and venous malformations, resulting in bleeding due to thrombocytopenia. We have little or no idea of how these vascular abnormalities progress or cause such physical damage to the patient. Our therapies are still primitive. With the exception of hemangiomas we have virtually no pharmacological therapy for these vascular lesions. This outstanding book should be a starting point for the discoveries of the future which hopefully may ameliorate these diseases.

Finally, a knowledge of vascular morphology and morphogenesis is turning out to be essential to the understanding of diseases beyond vascular abnormalities. Recent experimental studies show that antiangiogenic therapy which specifically inhibits endothelial cell proliferation can regress large tumors in mice, maintain them in a microscopic dormant state, or eradicate them (Boehm et al 1997). This can be accomplished without the development of acquired drug resistance, principally because of the low mutation rate of the vascular endothelial cell compared to tumor cells. In other words, redirection of anti-cancer therapy to the vascular endothelial cell in the tumor bed and away from the tumor cell is more effective, and less toxic than conventional cytotoxic chemotherapy. However, this new paradigm of cancer therapy depends critically on the configuration of microvessels and tumor cells in a tumor. Certain subtleties may be overlooked if cancer is perceived only in terms of cancer cells in the absence of microvessels on which they are dependent. While there is little or no risk of acquired drug resistance when antiangiogenic therapy is administered with inhibitors such as angiostatin or endostatin which only target endothelial cells, i.e., "direct anti-angiogenic therapy" (Figure 1), "indirect" antiangiogenic therapy may suffer a high risk of drug resistance. Therefore, antiangiogenic therapy which depends, for example, upon neutralization of an angiogenic factor produced by tumor cells, or upon blockade of the receptor for this factor, may eventually select for tumor cells which employ a different angiogenic factor and thus induce drug resistance.

The examples discussed in this Foreword illustrate only a small fraction of the many facets of vascular morphogenesis which await inquiry by the vascular biologist, whether he or she is a basic scientist or a clinician. This book is the best guide that we have for our journey.

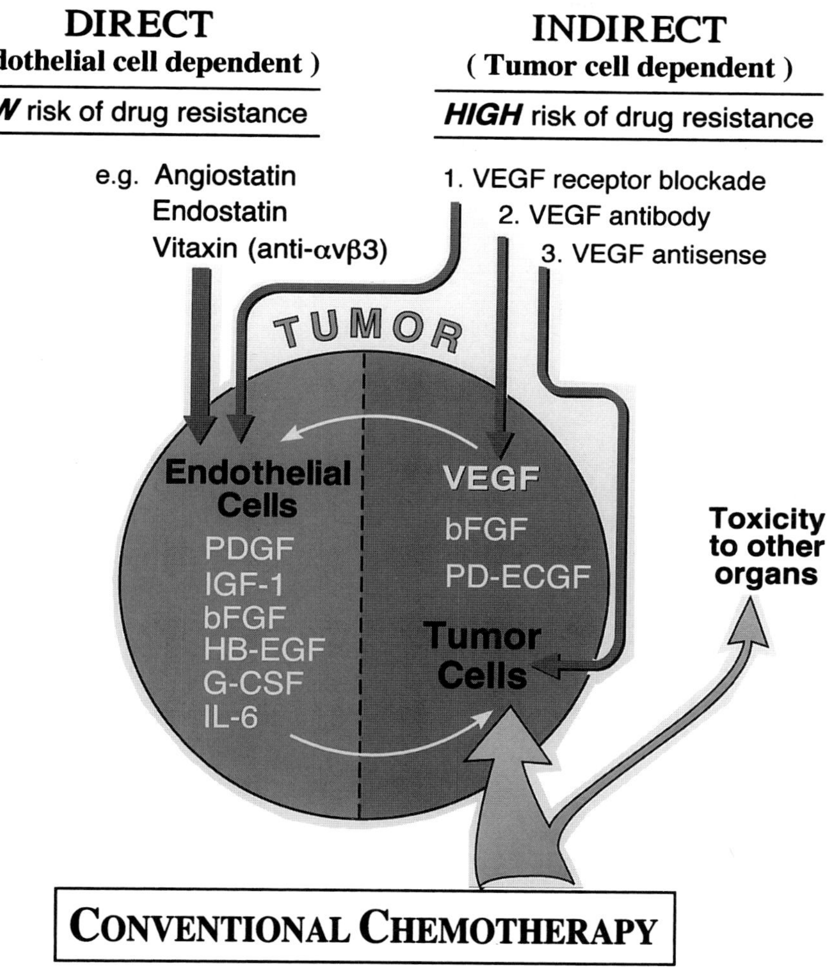

Fig. 1 Diagram to illustrate the critical role of the vascular endothelial cell as a determinant of whether anti-cancer therapy will induce acquired drug resistance. "Direct" antiangiogenic therapy which specifically targets endothelial cells has a low risk of inducing drug resistance. In contrast, "indirect" antiangiogenic therapy that depends upon blockade of tumor-derived angiogenic activity has a higher risk of inducing drug resistance, because tumor cells may eventually appear that release a different angiogenic factor. For example, a tumor that is producing mainly VEGF, may eventually generate a subpopulation of tumor cells that produce bFGF. Conventional chemotherapy which targets mainly tumor cells almost always induces drug resistance.

References

Arap, W. et al. 1998. Cancer treatment by targeted drug delivery to tumor vasculature in a mouse model. Science in press.

Auerbach, R. 1991. Vascular endothelial cell differentiation: Organ specificity and selective affinities as the basis for developing anticancer strategies. Int. J. Radiat. Biol. 60:1–10.

Boehm, T. et al. 1997. Antiangiogenic therapy of experimental cancer does not induce acquired drug resistance. Nature 390:404–407.

Maisonpierre, P.C. et al. 1997. Angiopoietin-2, a natural antagonist for Tie2 that disrupts in vivo angiogenesis. Science 277:55–60.

Part I

Vascular Morphogenesis In Vivo

Introduction: Understanding Blood Vessel Assembly in the Embryo

Charles D. Little

As we approach the end of the 1900s we are rediscovering the elegant embryology that flourished at the beginning of the century. Modern developmental biologists are beginning to place earlier embryological studies on a firm cellular and molecular basis. Sophisticated probes for specific molecules have stimulated a renaissance in the use of the light microscope. Fluorescent markers, digital image processing, and intravital labeling methods have had a major affect on understanding vascular morphogenesis. Targeted mutation of genes related to vascular development in mice bring the powerful tools of molecular biology to bear on understanding assembly of the vascular system. Recent progress is so rapid that keeping up with the latest discoveries is difficult.

Most of the molecules and several fundamental mechanisms that impinge on vascular development were first described in vitro. Arguably, the most important concept to emerge from this work is the idea that all vascular endothelial cells are able to manifest a planar, hexagon-like pattern. These polygonal arrays recapitulate in detail the quasiplanar primary vascular patterns of vertebrate embryos. Clearly, cultured endothelial cells will self-organize into patterned arrays with quasilumens; provided a suitable extracellular matrix and medium containing a serum supplement are present (see the chapters on in vitro studies).

Vascular biologists have recently discovered several major families of molecules that function during in vivo assembly of blood and lymph vessels (regarding lymphatic vessels, see the chapter by Wilting and colleagues): cell-surface-receptor tyrosine kinases and their inductive ligands, integrins and their ligands, cell–cell adhesion molecules, cytoplasmic regulators of cell behavior . . . the list goes on and on. Endothelial cell cultures were used to identify and characterize most of the important molecules that are known to affect vascular tube formation. Our debt to the workers who accomplished this in vitro work is great; now the challenge is to translate lessons learned in vitro into an understanding of endothelial cell function within the embryo.

Less is known, unfortunately, regarding the developmental biology of vascular smooth muscle and pericytes. For example, we do not understand smooth muscle behavior during the morphogenesis of the vessel wall. This large information gap exists despite enormous efforts centered on disease-based studies (hypertension, atherosclerosis); and also because smooth muscle cells exist along a continuum of phenotypes—there is no such thing as "the" smooth muscle cell. In addition, smooth muscle cells lose important physiological properties when grown using standard culture conditions—most notably the ability to contract. By focusing on embryological systems, our understanding of smooth muscle biology is advancing rapidly. Investigators are now employing the methods of cell and developmental biology to study directly the morphogenesis of blood vessel walls (for example, see the following articles, a list that is by no means complete—Sato et al., 1995; Waldo et al., 1996; Topouzis and Majesky, 1996; Mikawa and Gourdie, 1996; Suri et al., 1996; Hungerford et al., 1996, 1997; DeRuiter et al., 1997; Lee et al., 1997).

To understand blood vessel assembly, the most promising approach is to consider the vessel wall as a composite of two interacting cell populations—endothelial cells and (presumptive) smooth muscle. Fortunately, biologists are now addressing such interactions experimentally (see the chapter by Hirschi and D'Amore). The discovery of inductive molecules such as the angiopoietins and their TIE1 and TIE2 receptors (Sato et al., 1995), allows direct studies on endothelial cell:smooth muscle interactions. This work will revolutionize our understanding of vessel wall morphogenesis (see the chapters by Suri and Yancopolous and DeRuiter et al.).

Developmental biologists now comprehend in broad outline most cellular, and many molecular, mechanisms required for vascular morphogenesis. The task presently facing investigators is to understand how these mechanisms impinge on, and orchestrate, the assembly of an integrated vascular system.

References

DeRuiter, M.C., Poelmann, R.E., VanMunsteren, J.C., Mironov, V., Markwald, R.R., and Gittenberger-de Groot. Embryonic Endothelial Cells Transdifferentiate into Mesenchymal Cells Expressing Smooth Muscle Actins In Vivo and In Vitro. Circ. Res. 80:444–451, 1997.

Hungerford, J.E., Owens, G.K., Argraves, W.S., and Little, C.D. Development of the Vessel Wall as Defined by Vascular Smooth Muscle and ECM Markers. Dev. Biol. 178:375, 1996.

Hungerford, J.E., Hoeffler, J.P., Bowers, C.W., Dahm, L.M., Falchetto, R., Shabanowitz, J., Hunt, D.F., and Little, C.D. Identification of a Novel Marker for Pirmordial Smooth Muscle and Its Differential Expression Pattern in Contractile Versus Noncontractile Cells. J. Cell Biol. 137:925–937, 1997.

Lee, S.H., Hungerford, J.E., Little, C.D., and Iruela-Arispe, M.L. Proliferation and Differentiation of Smooth Muscle Cell Precursors Occurs Simultaneously During Development of the Vessel Wall. In press.

Mikawa, T., and Gourdie, R.G. Pericardial Mesoderm Generates a Population of Coronary Smooth Muscle Cells Migrating into the Heart Along with Ingrowth of the Epicardial Organ. Dev. Biol. 174:221–232, 1996.

Sato, T.N., Tozawa, Y., Deutsch, U., Wolburg-Buchholz, K., Fujiwara, Y., Gendron-Maguire, M., Gridley, T., Wolburg, H., Risau, W., and Qin, Y. Nature 376(6535):70–74, 1995.

Suri, C., Jones, P.F., Patan, S., Bartunkova, S., Maisonpierre, P.C., Davis, S., Sato, T.N., and Yancopoulos, G.D. Requisite Role of Angiopoietin-1, A Ligand for the TIE2 Receptor, During Embryonic Angiogenesis. Cell 87:1171–1180, 1996.

Topouzis, S., and Majesky, M.W. Smooth Muscle Lineage Diversity in the Chick Embryo. Dev. Biol. 178:430–445, 1996.

Waldo K.L., Kumiski D., and Kirby M.L. Cardiac Neural Crest is Essential for the Persistence Rather than the Formation of an Arch Artery. Dev. Dyn. 205:281–292, 1996.

1.1

The Morphogenesis of Primordial Vascular Networks

Christopher J. Drake and Charles D. Little

This chapter will address the morphogenesis of the first blood vessels formed in the vertebrate embryo and several mechanisms regulating this process. The focus will be on the development of vessels and vascular networks that arise within the embryo proper. We have chosen to study the formation of the intraembryonic vessels (stage 7–10) because their morphogenesis is both well characterized and is free of associated hematopoiesis. The intraembryonic vessels arise from isolated mesodermal cells, angioblasts, and rapidly organize into a network of endothelial tubes arranged in a reproducible spatial pattern. We refer to these networks as primary vascular networks. The physical characteristics of primary networks, coupled with the fact that they can be studied in vivo, make them an important tool in understanding general principals of vessel formation. We will, in the following pages: (1) describe the early events of vasculogenesis that lead to the formation of the first vessels and vascular networks; (2) review our experimental efforts to understand the formation of this vasculature; and (3) discuss some of the interesting unanswered questions regarding early vessel morphogenesis.

Primary Vascular Networks

The morphogenesis of the first embryonic blood vessels begins with the establishment of a mesodermal cell subpopulation that will give rise to endothelial cells. How this subpopulation is generated remains unclear. The most compelling data regarding the establishment of the endothelial cell lineage suggest a pivotal role for three cytokines: TGFβ, bFGF, and VEGF. Two of these cytokines, TGFβ and bFGF, are known to function very early in the pathway as inducers of mesoderm, and may have additional roles specifically related to the endothelial lineage (for review, see Risau and Flamme 1995; Wilting and Christ 1996). A role for VEGF in the genesis of endothelial cells is indicated by studies where the genes for either VEGF or the VEGF receptor (Flk-1/KDR, VEGFR-2) were abrogated in mice (Carmeliet et al., 1996; Ferrara et al., 1996). The effects of disrupting VEGF expression are dramatic in that the absence of a single VEGF allele caused embryonic lethality. Although the vasculature of both VEGF+/– and VEGF–/– embryos were abnormal, it is clear that an angioblastic population was present. The effects of abrogating the VEGF receptor FLK-1 appeared to have a more direct effect on events related to the endothe-

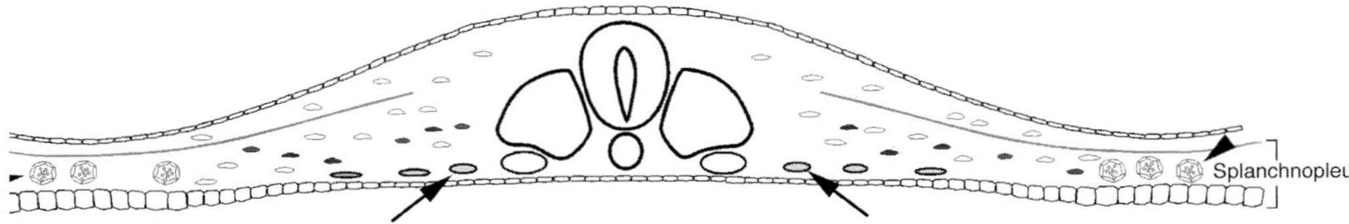

Fig. 1 A cross-sectional, diagramatic representation of an early avian embryo. The primary axial structures (neural tube and notochord) are avascular at stages when intraembryonic (arrow) and extraembryonic vessels (arrowhead) first appear in the splanchnopleural layer. Angioblastic cells are first detected as randomly distributed cells in the splanchnic mesoderm (black ovals). The first intraembryonic vessels form at a position ventral to the splanchnic mesoderm (large shaded ovals). Extraembryonic blood islands contain hematopoietic centers (arrowheads).

lial lineage. FLK-1–/– embryos had no organized blood vessels and few cells representing the angioblastic population (Shalaby et al., 1995). It is clear from these studies that VEGF plays a fundamental role either in the maintenance of the endothelial cell lineage or in events immediately proceeding endothelial cell specification.

The cellular events required for de novo formation of blood vessels are collectively referred to as vasculogenesis. Initial vasculogenic activity is confined to the embryonic splanchnopleure (SP; Fig. 1). Using our optical assay we are able to visualize vasculogenesis in whole-mounted quail embryos. This method involves using the laser scanning confocal microscope along with the mouse monoclonal antibody QH1 which recognizes a epitope present on the surface of quail endothelial cells (Pardanaud et al., 1987). When QH1 immunolabeled embryos are examined en face the complete vasculogenic process, from the genesis of angioblasts to the formation of vessel segments, vascular networks, and larger diameter vessels is apparent in a single embryo (Fig. 2). The process can be depicted as having distinct phases, which are summarized in Figure 3 (panels A–C).

Our previous studies and work in progress, using the expression of the transcription factor TAL1/SCL as novel marker of angioblasts, suggest that the first step in vessel formation is the "aggregation" of angioblasts. We refer to the cells that comprise such nonlumenized aggregates, as primordial endothelial cells. In a process that is poorly understood, aggregates of primordial endothelial cells polarize and transform into a vascular epithelium surrounding a lumen. These initial vessel segments are short blind-ended tubes (unpublished observations and Drake and Jacobson, 1988). Next, vascular networks are formed by processes that entail both the generation of additional segments from primordial endothelial cells, and the expansion of existing segments.

Primary vascular networks exhibit the following characteristics: (1) they are derived directly from mesoderm; (2) they form very quickly (hours in avians); and (3) they are free of smooth muscle cells, pericytes or other associated cells. Additionally, the primary vascular networks that we study are not subject to the mechanical forces of blood flow, since the heart is not as yet functional. A developing primary vascular network is seen in Figure 4. This rudimentary vasculature exhibits a number of characteristic features; the most obvious of which is the network-like polygonal organization. An important, but unappreciated, aspect of the pattern of networks is the avascular spaces. It is unclear whether the establishment and maintenance of avascular areas is an active or passive process. Our data on the formation of the first networks suggest that they arise by the interconnection of numerous short vessel segments. Once rudimentary networks are established they appear to be expanded by cell division and the extension of cell processes (see Fig. 4). Both of these endothelial cell activities likely contribute to the formation of new vessels and thus to the expansion of the network.

Experimental Studies of Primary Vascular Networks and Vascular Morphogenesis

To test the hypothesis that cell matrix interactions are required for vascular pattern formation, we designed an experimental system that would allow us to examine vessel morphogenesis in whole avian embryos. Our strategy was to use a microinjection approach. The usefulness of in vivo microinjection in the study of developmental processes in the avian embryo is well documented (Bronner-Fraser, 1985; Jacob et al., 1991). Preliminary microinjection studies determined that the precise delivery of reagents to the splanchnopleure, the site of initial intraembryonic vasculogenesis, was not feasible using embryos in situ (i.e., on the yolk). To circumvent this problem an ex ovo culture method, based on the work of New, was devised (New, 1955). Using paper rings, embryos attached to their vitelline membrane are removed from the yolk. This preparation, and the use of transmitted light optics, permits accurate and precise delivery of experimental reagents into the splanchnopleural ECM, that fills the extracellular space between the endoderm and splanchnic mesoderm, at stages 6–10. Injected embryos are then returned to culture on yolk/agar beds for specific intervals (Drake et al., 1992).

Analysis of the experimental results is accomplished by light microscopic analysis using a variety of imaging devices. The most important reagent for monitoring vascular morphogenesis in the quail embryo is the mouse monoclonal antibody QH1 (Pardanaud et al., 1987). Analysis of vascular development in the 2–9-somite stage quail is also aided by the fact that: (1) initial vessel formation occurs to a large extent in a single plane; (2) the process is rapid: at the 2-somite stage, blood vessels are absent, 10 hours later at the 9-somite stage, the embryo has an intact circulatory network; and (3) morphogenesis occurs in the absence of systemic variables such as blood flow and the presence of smooth muscle or pericytes.

Experimental Perturbation of Integrin Adhesion Receptors

A recent in vitro study by Vernon and colleagues examined the need for a malleable ECM in the process of vascular patterning (Vernon et al., 1994). It was determined that bovine aortic endothelial clones that secrete collagen I are competent to form vascular patterns, while clones that do not produce this collagen do not engage in such morphogenesis. Thus, in this culture system there is a strict correlation between the production of an ECM constituent and the ability of clones to engage in vascular pattern formation. In fact, it is only after the cells are elevated on a collagenous ECM, that cell shape changes and directed cell motility occurs. When the vascular patterns, derived from the collagen-producing clones, are compared with those of a 6-somite quail embryos, the similarities in scale and design are immediately apparent. This study, and many similar studies over the last decade (see Part I of this volume), have led to the hypothesis that direct, active cell-ECM adhesive activity is required for vascular morphogenesis.

We chose to perturb such interactions at the level of the adhesion receptors in order to evaluate the role of cell–matrix interactions in network formation. In recent years a number of transmembrane receptors for extracellular matrix molecules have been identified. Of these, the integrins are perhaps the most thoroughly studied and most diverse class of ECM receptors. Inte-

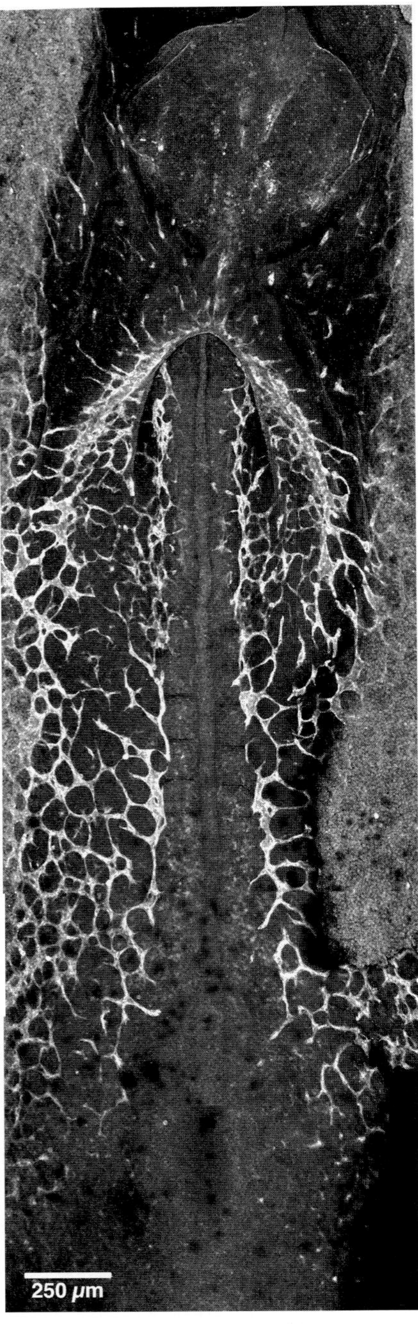

Fig. 2 A frontal view of a whole-mounted 6-somite quail embryo. The embryo at this stage of development is about 3 to 4-mm long. The perspective is from above looking down, as if the embryo were still on the yolk. The endothelial cells are stained with the QH1 antibody. This micrograph is constructed by projecting several individual image planes onto one "collapsed" virtual plane. This process yields a roadmap-like view of the developing embryonic vasculature. Bar = 250 μm.

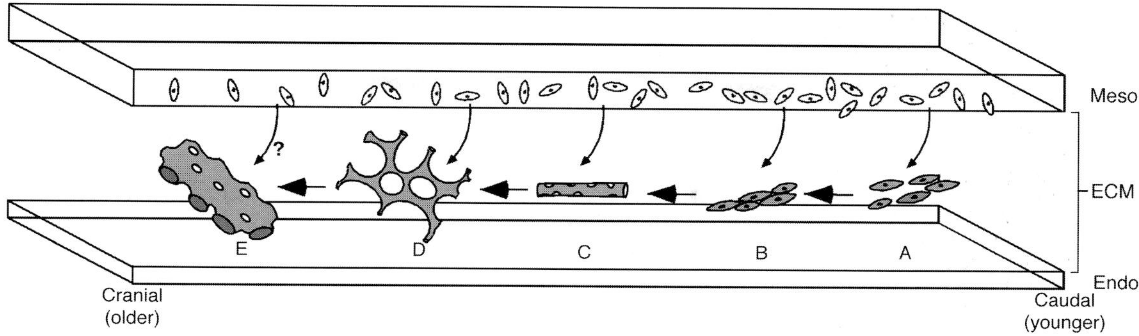

Fig. 3 A model depicting the process of vessel assembly within the splanchnopleural ECM. Five steps are defined along the cranial-to-caudal axis of a hypothetical embryo (A–E). The most primitive stage depicted is the appearance of endothelial precursor cells or angioblasts (white cells), in the splanchnopleural ECM. Current evidence suggests, as indicated by the arrows, that a ventral movement of angioblasts from the mesoderm is associated with the appearance of cells within this ECM (A). Further we speculate that a dorsal-to-ventral gradient of vasculogenesis may contribute to later vascular morphogenesis (curved arrows, A–D). The expression of the QH1 epitope, a definitive marker of endothelial cells, is first detected on isolated cellular aggregates (B). Presumably, cell–cell adhesion molecules are first expressed at this stage. The cellular aggregates at position "B" do not have lumen. We refer to the cells that comprise such structures as primordial endothelial cells. The first vascular structures with a lumen appear as isolated vessel segments (C). The endpoint for most vasculogenic activity in the embryo is the formation of organized polygonal arrays of small caliber vessels comprising a vascular network (D). However, as represented in (E), we believe that vascular networks can contribute to the formation of large caliber vessels, such as the dorsal aorta or the endocardium by the process of fusion. The question mark at position "E" denotes our ignorance as to when mesodermally-derived angioblasts no longer contribute to vascular development. Meso = splanchnic mesoderm; endo = endoderm.

grins are heterodimeric molecules composed of α and β subunits (Ruoslahti, 1991; Hynes, 1992). These integral membrane glycoproteins act for the most part as cellular receptors for the ECM (Juliano and Haskill, 1993; Yamada and Miyamoto, 1995). Their interaction may be specific, such that one heterodimeric receptor interacts with only one ECM molecule; or a heterodimer may be promiscuous in its interactions, binding a number of different ECM molecules. Pertinent to this chapter is the fact that 16 of 22 known integrins are relevant to vascular biology, including seven out of ten $\beta 1$ family members and two members of the integrin $\beta 3$ family. Using the powerful technique of targeted mutations to integrins, many studies have confirmed that these receptors are functional during vascular morphogenesis (for review, see Hynes and Bader, 1997). Interestingly, studies showed that one integrin ($\alpha v\beta 3$) can compensate functionally for the absence of a different integrin, i.e., $\alpha 5\beta 1$ (Yang and Hynes, 1996).

The importance of integrin-mediated cell–matrix interaction has also been demonstrated in a number of developing systems using integrin antagonists (Brooks et al., 1994; Friedlander et al., 1995). Two function-inhibiting monoclonal antibodies, CSAT and LM609, have played a

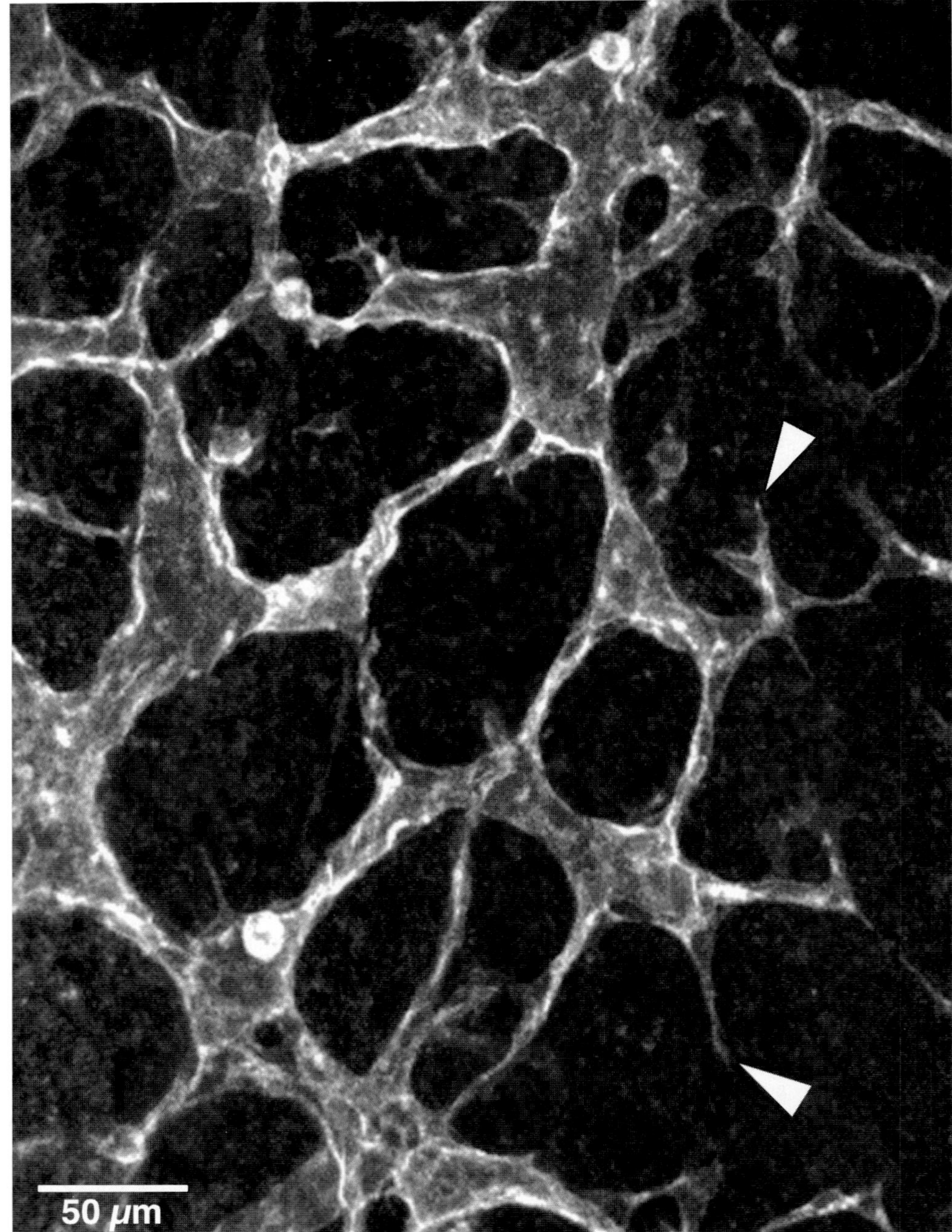

Fig. 4 A typical primary vascular network comprised of hexagonally arranged fine vessels surrounding avascular zones. Cellular protrusions appear to be in a position to "close" partially-formed polygons (arrowheads). This primitive vascular bed is relatively planar. Blood circulation will not commence for several hours and no smooth muscle-like cells are present. The bright round cells are engaged in mitosis. Bar = 50 μm.

prominent role in our studies. All members of the avian β1 integrin family are recognized by the monoclonal antibody CSAT; a reagent that interferes with the ligand binding site of the heterodimer and neutralizes the ligation of all β1 hetrodimers. To date, the list of β1 ligands includes fibronectin, laminin, vitronectin, thrombospondin, and collagens type I and IV, among other proteins (Hynes, 1992). In vitro studies in avian embryos using CSAT (Neff et al., 1982), or a similar antibody JG-22 (Greve and Gottlieb, 1982) have demonstrated that β1 integrins mediate important cellular processes, most notably cell migration (Bronner-Fraser, 1985; Jaffredo et al., 1988).

Our studies examining the role of integrins in vascular development began with CSAT, and thus was an evaluation of the entire β1 integrin family. We injected CSAT antibodies into early-stage embryos that were in an active period of vessel formation. We expected that the appearance of primordial endothelial cells at correct positions would be influenced by CSAT, as microinjection of CSAT antibodies had previously been shown to effect the migration of other cell types. This prediction, however, proved to be incorrect. While overtly malformed vessels were observed in CSAT injected embryos, the position of the vessels were correct. This observation had two possible interpretations: (1) CSAT did not effect the migration of endothelial cells; or (2) endothelial cells did not undergo extensive migration. Although CSAT did not effect the relative position of the vessels, it was clear that the morphology of the endothelial assemblies was entirely abnormal (Drake et al., 1992). Instead of vessels with lumens, only solid cords of endothelial cells were present. The cords appeared to be condensations of cells that were able to form cell-to-cell adhesions but were not able to interact with the ECM scaffold. The organization of cordlike assemblies of endothelial cells is observed in normal embryos; however, such assemblies are transient. The persistence of such assemblies in CSAT-injected embryos is interpreted to reflect a failure of endothelial cells to engage in the cell-ECM adhesions that must precede the thinning and flattening of endothelial cells observed in normal vessels with lumens.

While these results established a direct link between a family of adhesion receptors and vascular morphogenesis; the fact that a number of different integrins could be responsible was a limitation. With this in mind, we next entered into collaboration by which we could test whether integrin αvβ3 participated in early vascular pattern formation. This integrin was chosen for several reasons, chiefly the availability of a function-inhibiting antibody, LM609, and the fact that this integrin plays a pivotal role in angiogenesis (Brooks et al., 1994).

Our initial efforts established that the integrin αvβ3 was expressed by quail endothelial cells engaged in vasculogenesis. Next we introduced the antibody LM609 into embryos to evaluate αvβ3 function. Microinjected embryos developed a vasculature that was abnormal. The polygonal arrays of interconnected vessels characteristic of normal primary vascular networks were poorly formed in these embryos. Endothelial cells comprising these malformed networks lacked normal cellular protrusions. Also evident in injected embryos were abnormal clusters of endothelial cells. The formation of these structures and the malformations in the network were attributed to the failure of endothelial cells to engage in normal protrusive activity (Drake et al., 1995). The results suggested that avβ3-mediated adhesion is required for the formation of primary vascular networks. These findings establish a morphological function (extension of cell protrusions) to a specific integrin during formation of the primary vascular network.

Many of the β1 integrins and the αvβ3 integrin, recognize the triplet amino acid sequence Arg-Gly-Asp, or RGD; moreover, a large number of other ECM glycoproteins contain these sequences (Hynes, 1992). Taking advantage of the fact that RGD is an integrin recognition sequence, a number of experimental reagents that mimic the RGD motif have become available. Both linear and cyclized RGD-containing peptides have been shown to influence a number of cell adhesion processes (Cheresh and Spiro, 1987; Aumailley et al., 1991). Pertinent to this chapter is work in which RGD peptides were shown to inhibit vascular development both in in vitro studies using cultured rat aortic explants, and in in vivo studies using the quail chorioallantoic membrane (Nicosia and Bonanno, 1991; Britsch et al., 1989).

When we injected high concentration aliquotes (50 mg/ml) of linear GRGDS in the vasculogenic mesoderm of early quails, no effects were noted (Drake et al., 1992); however, when much lower concentrations (5 mg/ml) of an RGD peptide in a cyclized form were injected at 2–8-somite stages, distinct vascular and endocardial abnormalities were observed. The abnormalities included interrupted vascular patterns and poorly extended endothelial cell clusters similar to the LM609 results (unpublished observations). The cyclized peptide was a kind gift of Dr. Simon Goodman & E. Merck, (Darmstadt, Germany).

Vascular Endothelial Growth Factor

To identify some of the upstream regulatory events that govern vascular morphogenesis, we recently turned our attention to cytokines. Although many cytokines are likely to be important in the vascular development we chose to investigate the role of the vascular endothelial growth factor (VEGF). There are overwhelming data suggesting that VEGF plays a fundamental role in vessel formation. Work previous to our study showed that: (1) the effects of VEGF are both direct and specific to endothelial cells (Leung et al., 1989); (2) knockout of VEGF receptors in mice leads to profound effects on vascular development (Fong et al., 1995; Shalaby et al., 1995); (3) VEGF protein is likely to be present in the quail embryo, considering that its mRNA has been detected by in situ hybridization studies at stages relevant to this study (Flamme et al., 1995); and (4) recombinant human VEGF165 (rhVEGF165) is bioactive in avian embryos (Leung et al., 1989; Wilting et al., 1992). Collectively, these facts emphasize the relevance of VEGF in mediating processes associated with vascular network formation. It should be noted that in studies conducted subsequent to ours, abrogation of the VEGF gene further demonstrated the importance of VEGF to vasculogenesis (Carmeliet et al., 1996; Ferrara et al., 1996).

The experimental method we chose to evaluate VEGF function during vasculogenesis is analogous to protein overexpression at a specific developmental stage. To produce abnormally high levels of VEGF, recombinant human protein was introduced by a single-bolous microinjection. Upon delivery, elevated levels of rhVEGF165, induced profound malformations in the development of individual vessels, and perturbed formation of vascular networks.

Abnormal vessel development and alteration of vascular patterning was attributable to two specific endothelial cell behaviors: (1) the inappropriate formation of vessels in areas that would normally be avascular, and (2) the unregulated fusion of vessels. The latter process we refer to as vascular hyperfusion. The earliest manifestations of a VEGF effect was an alteration in the polygonal pattern of the vessels. Both irregularities in the size of the polygons forming the network and the inappropriate formation of vessels in normally avascular areas contributed to the loss of pattern. These alterations were followed by a progressive, inappropriate, fusion of vessels which led to the loss of individual vessel identity. Thus, instead of a well-defined network of vessels, the embryos exhibited enlarged vascular sinuses (Drake and Little, 1995). We speculate that endogenous VEGF has similar fusion-promoting activity and that in the tightly regulated environment of the embryo, normal "fusion" activity contributes to the formation of larger vessels such as dorsal aortae and endocardium. We postulate that these embryonic vessels are formed in part by localized fusion of network primordia under the regulation of VEGF. A corollary to this speculation, having implications for normal vascular patterning, is that in areas where the activity of VEGF is attenuated, avascular zones are maintained. Avascular zones or null zones are a critical component of proper vascular morphogenesis.

The above data and reasoning lead us to hypothesize that the expression and bioavailability of the endogenous VEGF must be tightly controlled during vascular development. Preliminary data in quail embryos suggest that endogenous VEGF is immobilized on ECM fibers in vasculogenic regions (unpublished observations). Thus it is possible that similar to other cytokines, the ECM may also act to regulate the activity of VEGF (Hogan, 1995; Vlodavsky et al., 1991). Cell transfection studies have shown that VEGF isoforms exhibit very different solubility properties (Park et al., 1993). These data suggest that VEGF, like members of other cytokine families such as FGFs and TGFβs, is sequestered in the ECM, perhaps by virtue of its heparin-binding domain. This property is supported by data suggesting that heparin mediates the binding of VEGF165 to its receptor (Gitay-Goren et al., 1992; Tessler et al., 1994). Microinjection of rhVEGF into the quail embryos may have circumvented this extracellular regulation mechanism (binding), by exposing cells to abnormally elevated levels of VEGF.

One explanation for the abnormalities observed in VEGF-injected embryos is increased mitogenic activity of VEGF (Ferrara and Henzel, 1989; Wilting et al., 1993). While mitogenic activity may play a role in VEGF-mediated hyperfusion, our studies have led us to believe that VEGF affects cell behavior in ways that are independent of mitogenic activity. Two such activities are the regulation of cell adhesion, and the recruitment of cells to the endothelial lineage.

The potential for VEGF affecting cell–matrix adhesions is supported by recent findings that VEGF can induce, in dermal microvascular endothelial cells, the expression of mRNAs encoding the αv and β3 integrin subunits (Senger et al., 1996). We speculate that in the embryo, elevated levels of VEGF could act to stimulate primordial endothelial cells to increase integrin-mediated protrusive activity and that the resulting altered adhesive state of these cells contributes to vascular fusion. That the adhesive state of endothelial cells can effect vascular patterning is consistent with our study, described above, utilizing the αvβ3 antagonist, LM609, in which we showed that vascular network formation is decreased when ligation of αvβ3 integrin is inhibited.

The second, nonmitogentic, activity of VEGF, which may relate to excessive vascular fusion, is its demonstrated ability to influence the fate of primordial endothelial cells. It is reasonable to speculate that excessive VEGF may potentiate lineage induction, since genetic abrogation of VEGF (Carmeliet et al., 1996; Ferrara et al., 1996), or the VEGF receptor FLK-1 (Shalaby et al., 1995), leads to deficiencies in the endothelial cell lineage. In this scenario, injected exogenous VEGF would act to recruit an elevated number of mesodermal cells to the endothelial lineage and thus contribute to fusion. Efforts are currently underway to evaluate this latter possibility using a novel maker of angioblasts (work in progress; also see text below). Whatever the mechanisms by which VEGF induces vascular hyperfusion, the process must be rapid, as evidence of excessive vascular fusions is observable within 5 hours postinjection.

The Endothelial Lineage

The experimental VEGF data just described piqued our interest in vascular biology's earliest event—the generation of the endothelial lineage. In this regard we sought to identify factors expressed by endothelial precursor cells/angioblasts that would define this mesodermal subpopulation. A number of antigenic determinants (e.g., CD34, CD31, Tie 2) are expressed in both the endothelial and the hematopoietic lineages. We chose to investigate whether the transcription factor TAL1/SCL is expressed in angioblasts. TAL1/SCL is best known as the factor critical to induction of the hematopoietic lineage, and is expressed in blood islands (extraembryonic mesoderm).

In addition, recent studies demonstrated that TAL1 is expressed in mouse endothelial cells (Kallianpur et al., 1994). We therefore sought to determine if avian endothelial cells also expressed TAL1; and, if so, to investigate whether their progenitors, angioblasts, also express this transcription factor.

Using double immunofluorescence confocal microscopy we observed: (1) that TAL1/SCL is expressed by both quail and chicken endothelial cells; (2) that expression of TAL1/SCL proceeds QH1 expression (the definitive marker of quail endothelial cells); (3) that TAL1/SCL, but not QH1, expression defines a subpopulation of cells within the substance of the splanchnic mesoderm; (4) that repositioning of the angioblast-like cells into an ECM defines a dorsal-to-ventral component of vasculogenesis; and (5) that TAL1/SCL-positive/QH1-negative angioblasts participate directly in vessel formation (work in progress; see Fig. 5).

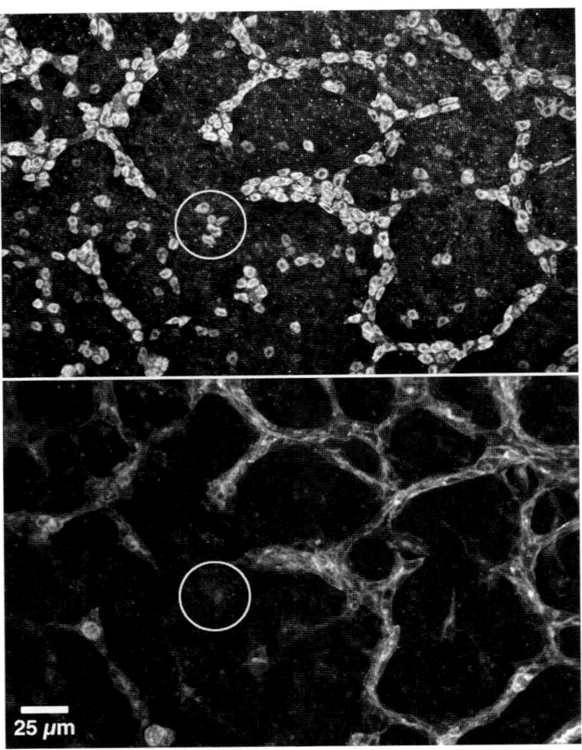

Fig. 5 A region of the developing vascular network double-immunostained using antibodies to TAL1 and the QH1 epitope. This figure establishes that the immunostaining patterns of TAL1 and QH1 are coincident, except at positions containing the most immature vascular elements. The upper panel shows nuclear TAL1 immunofluorescence. A general polygonal pattern can be discerned based on the expression pattern of TAL1. When the same field of cells is examined using a separate excitation wavelength for the QH1 endothelial marker, a slightly different pattern is observed (lower panel). Comparison of the two images shows that there are TAL1-positive cells that are devoid of QH1 staining (circled areas). Such TAL1-positive/QH1-negative cells are not affiliated with polygons. Thus, the most primitive vascular progenitor cells are immunoreactive for the TAL1 transcription factor but not for the cell surface endothelial marker, QH1. Bar = 25 μm.

The ability to observe the earliest events of vasculogenesis has led to new insights into initial vessel assembly. Successful vasculogenesis, in our view, has at least four requirements: (1) formation of the endothelial lineage; (2) ventral motility of angioblasts; (3) the presence of an instructive ECM; and (4) the expression of cell–cell adhesion molecules. These observations are delineated in an integrated model (Fig. 6). The drawing is intended to depict sequential steps during vascular morphogenesis. Each panel represents vascular development at a fixed position along the cranio-caudal axis viewed at four time points. The first panel (Step 1) corresponds to mesoderm at the angioblastic stage described above. This stage is characterized by the presence of TAL1 angioblasts (white cells) resting in the dorsally-positioned splanchnic mesoderm. Angioblasts are sparse, and are separated from one another by intervening splanchnic mesoderm. A few cells have moved ventrally and entered the splanchnopleural ECM (shaded areas); while others, organized in loose aggregates, have begun to express the QH1 epitope (shaded cells).

The second panel (Step 2) shows that angioblasts are still detectable in the splanchnic mesoderm. Within the splanchnopleural ECM, multiple aggregates of QH1-positive cells are now detectable at "centers" where traction forces are being exerted on the ECM. The tension may initially be exerted in a radial fashion, but as adjacent aggregates begin to focus the lines of tension, the individual clusters then begin to pull against each other (dashed line). Note that angioblasts are still thought to be seeding ventrally from the dorsally-positioned splanchnic mesoderm.

In Step 3, the continued assembly of vascular polygons is hypothesized to be derived by two separate mechanisms: (a) by continued seeding of angioblasts from splanchnic mesoderm; and (b) by protrusive activity on the part of endothelial cells. The protrusive mechanism is thought to be similar to in vitro vessel assembly on extracellular matrix-coated culture surfaces (Vernon et al., 1994). In this system, endothelial cells appear to detect the traction forces exerted by nearby aggregates and to respond to this stimulus by aligning with, and extending protrusions along, the lines of tension.

The final panel (Step 4) depicts the outcome of this vasculogenic activity, vascular networks, composed of patent endothelial tubes (dark shading). It is possible that even at this stage, angioblasts may continue to move ventrally and contribute to preexisting endothelial tubes. A fundamental aspect of the model depicted is that assembly of vessels occurs within the splanchnopleural ECM; a matrix that is hypothesized to contain both instructive and permissive molecular cues. Some of the ECM-derived bioactive factors may be produced by the ventral-most cell layer, the primitive endoderm (Sugi and Markwald, 1996; Flamme, 1989).

This oversimplified model attempts to depict major morphological steps, as they relate to physical position in the embryo, to sequential timing events and to cell–matrix interactions. For clarity, critically important molecular mechanisms such as cell–cell interactions and cytokine stimulation are not indicated.

Perhaps the most important aspect of this model is the modular progression of vasculogenesis. Thus, it is predicted that the nascent polygonal arrays, which are fundamental to the formation of vascular networks, are established over time by: (1) small cellular movements in a dorsal-to-ventral dimension (angioblast repositioning); and (2) cell-ECM protrusive activity in a plane parallel to the embryonic plate (medio-lateral and cranio-caudal).

Summary, Perspective, and Speculation

Is vascular development autonomous from surrounding embryonic structures? The early vascular patterning of the heart, dorsal aortae, and lateral networks appears to be temporally independent of events regulating the formation of axial structure formation (notochord, somites, and

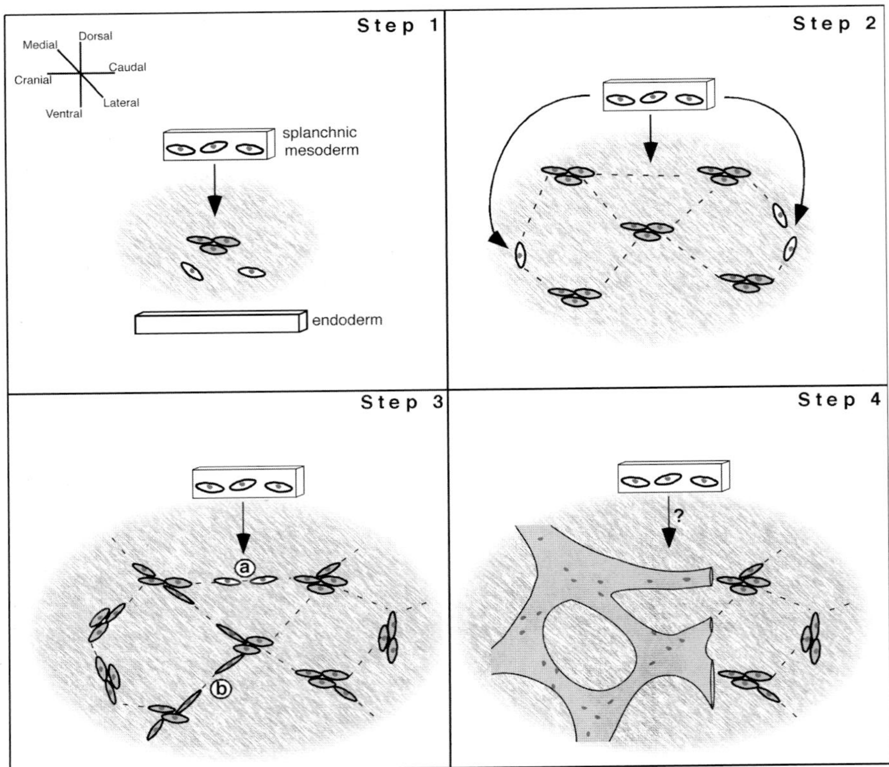

Fig. 6 A hypothetical model for the de novo formation of a polygonal vascular network. Step 1 shows that during formation of a primary vascular network angioblasts (white cells) are residents of the splanchnic mesoderm. Tube formation, however, does not occur in the plane of the mesoderm. Instead, mesodermally derived angioblasts (white cells) come to lie within an quasi-planar extracellular matrix (bracket, ecm; lightly shaded). The extracellular matrix is ventral to the mesoderm; therefore either vascular cells move ventrally a short distance (20–30 μm), or they somehow become surrounded by this ECM. Regardless, once in association with the splanchnopleural ECM, primordial endothelial cells form loose aggregates (shaded cells) that appear to be randomly, but homogeneously, distributed (Step 2). Next, we propose that primordial endothelial cells exert traction forces on ECM fibers, similar to the behavior of cultured endothelial cells. This activity results in the coalescence of cells into positions that roughly predict the polygonal boundries. Once traction centers are established a more uniform pattern emerges (Step 3). During this phase we speculate that the primary vascular networks can be expanded by two mechanisms. One involves the continued accretion of mesodermal progenitors (a), and the other mechanism involves directed protrusive activity (b). We hypothesize that both growth mechanisms entail cells sensing and aligning along matrix fibers that are under traction. The final stage depicted (Step 4) suggests that establishment of a traction-generated vascular pattern precedes convergence of cells into chords and conversion into a polarized epithelium surrounding a lumen. It is possible that even after primitive tubes have formed; angioblasts continue to contribute to nascent vessels (Step 4); however, we have no evidence for or against this possibility. It is important to note that null zones, or avascular zones, are an important feature of primary vascular networks.

neural tube). Specifically, the progress of vascular network formation cannot be strictly related to, for example, somite number. Of course, at some point (the 9–10-somite stage), all the separate cardiovascular rudiments must join to form a complete circulatory system. This observation suggests that hypothetical positional control genes, symmetry control genes, and polarity determination genes that regulate arterial, venous, and capillary morphogenesis may be distinct from similar genes that regulate axial patterning (neural tube, notochord, and somites). Recent work in heart segmentation supports this possibility (Yamamura et al., 1997). If this assertion is true, then presumably there exists a set of specific downstream effectors of vascular morphogenesis with respect to pattern. The recent exciting work on VEGF, VEGF receptors, and angiopoeitin I and its receptor, TIE2/Tek, are excellent examples of such effectors (Maisonpierre et al., 1997; Carmeliet et al., 1996; Ferrara et al., 1996; Shalaby et al., 1995; Fong et al., 1995; Suri et al., 1996; Sato et al., 1995; Dumont et al., 1994; for review see Hanahan, 1996; Folkman and D'Amore 1996).

Based on TAL/SCL protein expression, commitment to a vascular fate begins within the substance of the splanchnic mesoderm and culminates with endothelial tubes at a ventral position within the splanchnopleural ECM. The data also show that TAL1 continues to be expressed in overtly identifiable endothelial cells; this was established by demonstrating TAL1 coexpression with the QH1 epitope. It is important to note that QH1-negative/TAL1-positive cells are only observed within the splanchnic mesoderm proper. In contrast, both markers are detected on cells protruding into, or lying within, the splanchnopleural ECM. Thus, QH1-expression appears to correlate with the movement of cells into the splanchnopleural ECM and/or the establishment of cell–cell contact (Fig. 2). This line of evidence is consistent with our earlier observations that only upon entering the planar, relatively thick ECM do angioblasts undergo conversion to an epithelium, form lumens, and begin tube formation (Drake and Jacobson, 1988; Drake et al., 1990).

Our observations also suggest that avian angioblasts are distributed throughout the mesoderm in a manner that requires little, if any, cranio-caudal or medio-lateral migration; thus, the only significant movements are local rearrangements on the order of one-to-two cell diameters; this includes the critically important ventrally-directed repositioning into the splanchnopleural ECM. In our view, the small degree of movement required to form polygons does not justify the term migration. This is in contrast to other regions of the embryo where significant primordial endothelial cell motility is reported to occur, for example, during the formation of the aortic arches and cranial vessels (Noden, 1989; Poole and Coffin, 1989).

Based on the embryological studies and the wealth of in vitro studies discussed above, we speculate that upon entry into a relatively cell-free ECM, the angioblast rapidly differentiates into an endothelial cell. Concomitant with differentiation, these cells engage the ECM and commence the tractional structuring that leads to the classical vascular pattern of intersecting polygonal and anastomotic networks. This hypothesis fits very well with the in vitro work on vascular morphogenesis and the requirement for a malleable ECM.

If supported by subsequent studies, TAL1 will be the first definitive angioblastic marker. It must be noted however that TAL1 is only useful as an angioblast marker in intraembryonic regions. This is due to the fact that this transcription factor is also expressed in erythropoieic lineage. Therefore, in extraembryonic regions it is not possible to distinguish between erythroblast and angioblasts in blood islands on the basis of TAL1 expression.

To end our discussion of vessel morphogenesis, it is appropriate to make the point that nascent vascular endothelial tubes do not constitute a blood vessel. New work by investigators at Regeneron, Inc. (Davis et al., 1996; Suri et al., 1996; Maisonpierre et al., 1997) and others (Vikkula et al., 1996) provides a fascinating new insight into understanding vessel morphogenesis, especially development of the vessel wall (see the chapter by Suri and Yancopoulos). The implications of these studies are beyond the scope of this chapter; nevertheless, it is clear that the interplay between nascent endothelial tubes, the ECM, and presumptive smooth muscle cells is critical to normal vessel morphogenesis.

References

Aumailley, M., Gurrath, M., Muller, G., Calvete, J., Timpl, R., and Kessler, H. (1991). Arg-Gly-Asp constrained within cyclic pentapeptides. Strong and selective inhibitors of cell adhesion to vitronectin and a laminin fragment P1. Fed. Eur. Biochem. Soc. 291:50–54.

Britsch, S., Christ, B., and Jacob, H.J. (1989). The influence of cell–matrix interactions on the development of quail chorioallantoic vascular system. Anat. Embryol. 180:479–484.

Bronner-Fraser, M. (1985). Alterations in neural crest migration by a monoclonal antibody that affects cell adhesion. J. Cell Biol. 101:610–617.

Brooks, P.C., Clark, R.A., and Cheresh, D.A. (1994). Requirement of vascular integrin alpha v beta 3 for angiogenesis. Science 264:569–571.

Carmeliet, P., Ferreira, V., Brier, G., Pollefeyt, S., Kieckens, L., Gertsenstein, M., Fahrig, M., Vandenhoeck, A., Harpa, H., Eberhardt, C., Declercq, C., Pawling, J., Moons, L., Collen, D., Risau, W., and Nagy, A. (1996). Abnormal blood vessel development and lethality in embryos lacking a single VEGF allele. Nature 380(April): 435–439.

Cheresh, D.A., and Spiro, R.C. (1987). Biosynthetic and functional properties of an Arg-Gly-Asp-directed receptor involved in human melanoma cell attachment of vitronectin, fibrinogen, and von Willebrand factor. J. Biol. Chem. 262:17703–17711.

Davis, S., Aldrich, T.H., Jones, P.F., Acheson, A., Compton, D.L., Jain, V., Ryan, T.E., Bruno, J., Radziejewski, C., Maisonpierre, P.C., and Yancopoulos, G.D. (1996). Isolation of angiopoietin-1, a ligand for the Tie2 receptor, by secretion-trap expression cloning. Cell 887:1161–1169.

Drake, C.J., Cheresh, D.A., and Little, C.D. (1995). An antagonist of integrin avb3 prevents maturation of blood vessels during embryonic neovascularization. J. Cell. Sci. 108:2655–2661.

Drake, C.J., Davis, L.A., Hungerford, J.E., and Little, C.D. (1992). Perturbation of beta-1 integrin-mediated adhesions results in altered somite cell shape and behavior. Dev. Biol. 149: 327–338.

Drake, C.J., Davis, L.A., and Little, C.D. (1992). Antibodies to beta 1 integrins cause alterations of aortic vasculogenesis, in vivo. Dev. Dyn. 193:83–91.

Drake, C.J., Davis, L.A., Walters, L., and Little, C.D. (1990). Avian vasculogenesis and the distribution of collagens I, IV. laminin, and fibronectin in the heart primordia. J. Exp. Zool. 255:418–421.

Drake, C.J., and Jacobson, A.G. (1988). A survey by scanning electron microscopy of the extracellular matrix and endothelial components of the primordial chick heart. Anat. Rec. 222: 391–400.

Drake, C.J., and Little, C.D. (1995). Exogenous vascular endothelial growth factor induces malformed and hyperfused vessels during embryonic neovascularization. Proc. Natl. Acad. Sci. USA 92:7657–7661.

Dumont, D.J., Gradwohl, G., Fong, G.-H., Puri, M.C., Geertsenstein, M., Auerbach, A., and Breitman, M.L. (1994). Dominant-negative and targeted null mutations in the endothelial receptor tyrosine kinase, tek, reveal a critical role in vasculogenesis of the embryo. Genes Dev. 8:18977–1909.

Ferrara, N., Carver-Moore, K., Chen, H., Dowd, M., Lu, L., O'Shea, K.S., Powell-Braxton, L., Hillan, K.J., and Moore, M.W. (1996). Heterozygous embryoic lethality induced by targeted inactivation of the VEGF gene. Nature 380:439–442.

Ferrara, N., and Henzel, W.J. (1989). Pituitary follicular cells secrete a novel heparin-binding growth factor specific for vascular endothelial cells. Biochem. Biophys. Res. Commun. 161:851.

Flamme, I. (1989). Is extraembryonic angiogenesis in the chick embryo controlled by the endoderm? A morphological study. Anat. Embryol. 180:259–272.

Flamme, I., Breier, G., and Risau, W. (1995). Vascular endothelial growth factor (VEGF) and VEGF receptor 2 (flk-1) are expressed during vasculogenesis and vascular differentiation in the quail embryo. Dev. Biol. 169:699–712.

Folkman, J., and D'Amore, P.A. (1996). Blood vessel formation: What is its molecular basis? Cell 87:1153–1155.

Fong, G.-H., Rossant, J., Gertsenstein, M., and Breitman, M.L. (1995). Role of the Flt-1 receptor tyrosine kinase in regulating the assembly of vascular endothelium. Nature 376: 66–70.

Friedlander, M., Brooks, P.C., Shaffer, R.W., Kincaid, C.M., Varner, J.A., and Cheresh, D.A. (1995). Definition of two angiogenic pathways by distinct av integrins. Science 270: 1500–1502.

Gitay-Goren, H., Sokert, S., Vlodavsky, I., and Neufeld, G. (1992). The binding of vascular endothelial growth factor to its receptors is dependent on cell surface-associated heparin-like molecules. J. Biol. Chem. 267:6093–6098.

Greve, J.M., and Gottlieb, D.I. (1982). Monoclonal antibodies which alter the morphology of cultured chick myogenic cells. J Cell. Biochem. 18:221–229.

Hanahan, D. (1996). Signaling vascular morphogenesis and maintenance. Cell Biol. 277: 48–50.

Hogan, B.L.M. (1995). The TGF-B-related signalling system in mouse development. Sem. Dev. Biol. 6:257–265.

Hynes, R.O. (1992). Integrins: versatility, modulation, and signaling in cell adhesion. Cell 69: 11–25.

Hynes, R.O., and Bader, B.L. (1997). Targeted mutations in integrins and their ligands: their implications for vascular biology. Thromb. Haemostas. 78:83–87.

Jacob, M., Christ, B., Jacob, H.J., and Poelmann, R.E. (1991). The role of fibronectin and laminin in development and migration of the avian Wolffian duct with reference to somitogenesis. Anat. Embryol.

Jaffredo, T., Horwitz, A.F., Buck, C.A., Rong, P.M., and Dieteerlen-Lievre, F. (1988). Myoblast migration specifically inhibited in the chick embryo by grafted CSAT hybridoma cell secreting an anti-integrin antibody. Development 103:431–446.

Juliano, R.L., and Haskill, S. (1993). Signal transduction from the extracellular matrix. J. Cell. Biol. 120:577–585.

Kallianpur, A.R., Jordan, J.E., and Brandt, S.J. (1994). The SCL/TAL-1 gene is expressed in progenitors of both the hematopoietic and vascular systems during embryogenesis. Blood 83: 1200–1208.

Leung, D.W., Cachianes, G., Kuang, W.-J., Goddel, D.V., and Ferrara, N. (1989). Vascular endothelial growth factor is a secreted angiogenic mitogen. Science 246:1306–1309.

Maisonpierre, P.C., Suri, C., Jones, P.F., Bartunkova, S., Wiegand, S.J., Compton, D., McClain, J., Aldrich, T.H., Papadopoulos, N., Daly, T.J., Davis, S., Sato, T.N., and Yancopoulos, G.D. (1997). Angiopoietin-2, a natural antagonist for Tie2 that disrupts in vivo angiogenesis. Science 277:55–60.

Neff, N.T., Lowrey, C., Decker, C., Tovar, A., Damsky, C., Buck, C., and Horwitz, A.F. (1982). A monoclonal antibody detaches embryonic skeletal muscle from extracellular matrices. J. Cell Biol. 95:654–666.

New, D., (1955). A new technique for the cultivation of the chick embryo in vitro. J. Exp. Morphol. 3:320–331.

Nicosia, R.F., and Bonanno, E. (1991). Inhibition of angiogenesis in vitory by Arg-Gky-Asp-containing synthetic peptide. Am. J. Pathol. 138:829–833.

Noden, D. (1989). Embryonic origins and assembly of blood vessels. Am. Rev. Respir. Dis. 140: 1097–1103.

Pardanaud, L., Altmann, C., Kitos, P., Dieterien-Lievre, F., and Buck, C.A. (1987). Vasculogenesis in the early quail blastodisc as studied with a monoclonal antibody recognizing endothelial cells. Development 100:339–349.

Park, J.E., Keller, G.-A., and Ferrara, N. (1993). The vascular endothelial growth factor (VEGF) isoforms: differential deposition into the subepithelial extracellular matrix and bioactivity of extracellular matrix-bound VEGF. Mol. Biol. Cell 4:1317–1326.

Poole, T.J., and Coffin, J.D. (1989). Vasculogenesis and angiogenesis: two distinct morphogenetic mechanisms establish embryonic vascular pattern. J. Exp. Zool. 251:224–231.

Risau, W., and Flamme, I. (1995). Vasculogenesis. Annu. Rev. Cell Dev. Biol. 11:73–91.

Ruoslahti, E. (1991). Integrins. J. Clin. Invest. 87:1–5.

Sato, T.N., Tozawa, Y., Deutsch, U., Wolburg-Buchholz, K., Fujiwara, Y., Gendron-Maguire, M., Gridley, T., Wolburg, H., Risau, W., and Qin, Y. (1995). Distinct roles of the receptor tyrosine kinases Tie-1 and Tie-2 in blood vessel formation. Nature 376:70–74.

Senger, D.R., Ledbetter, S.R., Claffey, K.P., Papadopoulos-Sergiou, A., Perruzzi, C., and Detmar, M. (1996). Stimulation of endothelial cell migration by vascular permeability factor/vascular endothelial growth factor through cooperative mechanisms involving the $\alpha v\beta 3$ integrin, osteopontin, and thrombin. Am. J. Pathol. 149:293–305.

Shalaby, F., Rossant, J., Yamaguchi, T.P., Gertsenstein, M., Wu, X.-F., Breitman, M.L., and Schuh, A.C. (1995). Failure of blood-island and vasculogenesis in Flk-1-deficient mice. Nature 376:62–66.

Sugi, Y., and Markwald, R.R. (1996). Formation and early morphogenesis of endocardial endothelial precursor cells and the role of the endoderm. Dev. Biol. 175:66–83.

Suri, C., Jones, P.F., Patan, S., Bartunkova, S., Maisonpierre, P.C., Davis, S., Sato, T.N., and Yancopoulos, G.D. (1996). Requisite role of angiopoietin-1, a ligand for the Tie2 receptor, during embryonic angiogenesis. Cell 87:1172–1190.

Tessler, S., Rockwell, P., Hicklin, D., Cohen, T., Levi, B.-Z., Witte, L., Lemischka, I.R., and Neufeld, G. (1994). Heparin modulates the interaction of VEGF165 with soluble and cell associated flk-1 receptors. J. Biol. Chem. 269(17): 12456–12461.

Vernon, R.B., Lara, S.L., Drake, C.J., Iruela-Arispe, M.L., Angello, J.C., Little, C.D., Wight, T.N., and Sage, E.H. (1994). Organized type I collagen influences endothelial patterns during "spontaneous angiogenesis in vitro": planar cultures as models of vascular development. In Vitro Cell Dev. Biol. 107:2690–2695.

Vikkula, M., Boon, L.M., Carraway, K.L.I., Calvert, J.T., Diamonti, A.J., Goumnerov, B., Pasyk, K.A., Marchuk, D.A., Warman, M.L., Cantley, L.C., Mulliken, J.B., and Olsen, B.R. (1996). Vascular dysmorphogenesis caused by an activating mutation in the receptor tyrosine kinase TIE2. Cell 87:1181–1190.

Vlodavsky, I., Fuks, Z., Ishai-Michaeli, R., Bashkin, P., Levi, E., Korner, G., Bar-Shavit, R., and Klagsbrun, M. (1991). Extracellular matrix-resident basic fibroblast growth factor: implication for the control of angiogenesis. J Cell. Biochem. 45:167–176.

Wilting, J., and Christ, B. (1996). Embryonic angiogenesis: a review. Naturwissenschaften 153–164.

Wilting, J., Christ, B., Bokeloh, M., and Weich, H.A. (1993). In vivo effects of vascular endothelial growth factor on the chicken chorioallantoic membrane. Cell Tissue Res. 274: 163–172.

Wilting, J., Christ, B., and Weich, H.A. (1992). The effects of growth factors on the day 13 chorioallantoic membrane (CAM): a study of $VEGF_{165}$ and PDGF-BB. Anat. Embryol. 186:251–257.

Yamada, K.M., and Miyamoto, S. (1995). Integrin transmembrane signaling and cytoskeletal control. Curr. Opin. Cell Biol. 7:681–689.

Yamamura, H., Zhang, M., Markwald, R.R., and Mjaatvedt, C.H. (1997). A heart segmental defect in the anterior–posterior axis of a transgenic mutant mouse. Dev. Biol. 186:58–72.

Yang, J.T., and Hynes, R.O. (1996). Fibronectin receptor functions in embryonic cells deficient in alpha 5 beta 1 integrin can be replaced by alpha v integrins. Mol. Biol. Cell 7: 1737–1748.

1.2

Angiogenesis and Lymphangiogenesis: Analogous Mechanisms and Homologous Growth Factors

Jörg Wilting, Haymo Kurz, Su-Ja Oh, and Bodo Christ

Introduction

Angiogenesis, the development of blood vessels, has attained considerable attention during the last years. This is due in part to the detection of angiogenesis-dependent diseases (for review, see Folkman, 1985, 1987; Blood and Zetter, 1990; Plate et al., 1994). Furthermore, the early functioning of the circulatory system is of greatest importance for the development of the embryo (for review, see Benninghoff et al., 1930; Wagner, 1980; Welt et al., 1990; Wilting et al., 1995b; Wilting and Christ, 1996). In contrast, lymphangiogenesis, the development of lymphatic vessels, has only rarely been studied. This is mainly due to the fact that specific markers for lymphatic endothelial cells are not available during development. In this article we compare angiogenesis and lymphangiogenesis. We survey the mechanisms involved in embryonic development of both types of vessels. Furthermore, we are asking the question whether growth factors such as vascular endothelial growth factor (VEGF), its homologues, and their receptors, are involved in the specification of blood vascular and lymphatic endothelial cells.

Angiogenesis

Since the original introduction of the term angioblast by His (1900), the terms angiogenesis and vasculogenesis (von Schulte, 1914) have been synonymous. Both terms have been used in the general meaning of blood vessel development. Recently they have been newly defined, but, unfortunately, the definitions are not uniformly used. For example, angiogenesis has been used to describe immigration of angioblasts (Pardanaud et al., 1996), development of endothelial cells from pre-existing ones (Pardanaud et al., 1989), or sprouting of endothelial cells (Risau et al., 1988). Vasculogenesis has been defined as in situ differentiation of angioblasts (Risau et al., 1988; Paradanaud et al., 1989). As we have shown recently (Christ et al., 1990; Wilms et al., 1991; Wilting et al., 1995a, b; Wilting and Christ, 1996), the development of blood vessels is a complex process that cannot be divided into only two mechanisms. Therefore, incomplete definitions obscure the complexity of mechanisms involved in blood vessel development. For that reason, we will stick to the historically established meaning of the terms angiogenesis/vasculogenesis as synonymous.

There is a need to show, in detail, the development and behavior of endothelial cells and their precursors (angioblast/hemangioblasts), and the large variety of angiogenic mechanisms. We have recently reviewed the process of embryonic angiogenesis (Wilting et al., 1995b; Wilting and Christ, 1996). Here, we will provide a short outline of the relevant mechanisms.

Sabin (1917) proposed that embryonic angiogenesis can be subdivided into three phases. First, endothelial cells are derived from precursor cells. These seem to be either unipotential angioblasts or bipotential hemangioblasts (Sabin, 1920; Stainier et al., 1995; Pardanaud et al., 1996). The emergence of these cells does not depend on the normal formation of the mesoderm and can even be observed when gastrulation is prevented (Azar and Eyal-Giladi, 1979; Christ et al., 1991). During subsequent developmental stages, endothelial cells are derived from precursor cells and, simultaneously, from endothelial cells that are already integrated into the primary vascular plexus of the embryo (Sabin, 1917; Flamme, 1989). In later stages, new endothelial cells seem to emerge solely from already existing ones. Notably, cells displaying endothelial characteristics have been isolated from peripheral blood of adult humans (Asahara et al., 1997), but it remains to be studied whether these are angioblasts, endothelial cells that were scaled off, or a specific subpopulation of leukocytes. The data show that there are two sources of blood vascular endothelial cells: mesenchymal precursor cells (angioblasts/hemangioblasts) and, later, pre-existing endothelial cells. Angiogblasts possess high migratory potential (Noden, 1988; Wilms et al., 1991; Wilting et al., 1995a; Kurz et al., 1996). The cells aggregate and flatten, and some data suggest that the intercellular space becomes the lumen of the vessel (Hirakow and Hiruma, 1983); however, intracellular lumen formation by vacuolation has also been described (Benninghoff et al., 1930; Wolff and Bär, 1972).

Almost the complete intraembryonic mesoderm contains angioblasts (Noden, 1989; Wilms et al., 1991). These cells are also able to join previously formed vessels (Wilting et al., 1995a). In contrast, vascular sprouts and single endothelial cells are able to leave the vascular lining (Martinoff, 1907; Brand-Saberi et al., 1995). This is most obvious in the formation of endocardial cushions (Fig. 1; Markwald et al., 1975; Eichmann et al., 1993). During the 2nd to 3rd day of avian development, the emergence of endothelial cells from both angioblasts and pre-existing endothelial cells can be observed in almost all areas of the embryo. Studies on chick-quail chimeras indicate that there is practically no blood vessel which has been derived from only one of the two sources (Wilting et al., 1995a). During later stages, endothelial cells seem to be solely derived from already existing ones. Thereby, at least three mechanisms are involved in growth and expansion of the vascular bed: sprouting, intussusceptive microvascular growth, and intercalated growth (Wilting and Christ, 1996).

Sprouting can be subdivided into sequential steps (Folkman, 1985). Endothelial cells at the tips of sprouts form bulbous or several filamentous processes (Fig. 2). These contact each other and some of these contacts are remodelled into capillary loops. Cellular interactions mediated by cell surface receptors such as tie 1 and tie 2 (Sato et al., 1993) seem to be of great importance in this process. Mice deficient in the tie 2 receptor are characterized by the lack of capillary sprouts into the neural tube (Sato et al., 1995). The formation of capillary loops can also be the result of another mechanism that has been called intussusceptive microvascular growth (Caduff et al., 1986) or growth by sinusoids (Minot, 1900). This mechanism seems to be prevalent in respiratory and parenchymatous organs. By inserting a transluminal post into a huge sinusoidal capillary, this sinusoid becomes subdivided and remodelled into a capillary loop. This kind of growth allows almost constant perfusion of the vessel. Endothelial cell proliferation may also result in intercalated growth of vessels in order to increase their diameter and length. This has also been called nonsprouting angiogenesis (Folkman, 1987), and may give rise to highly tortuous vessels.

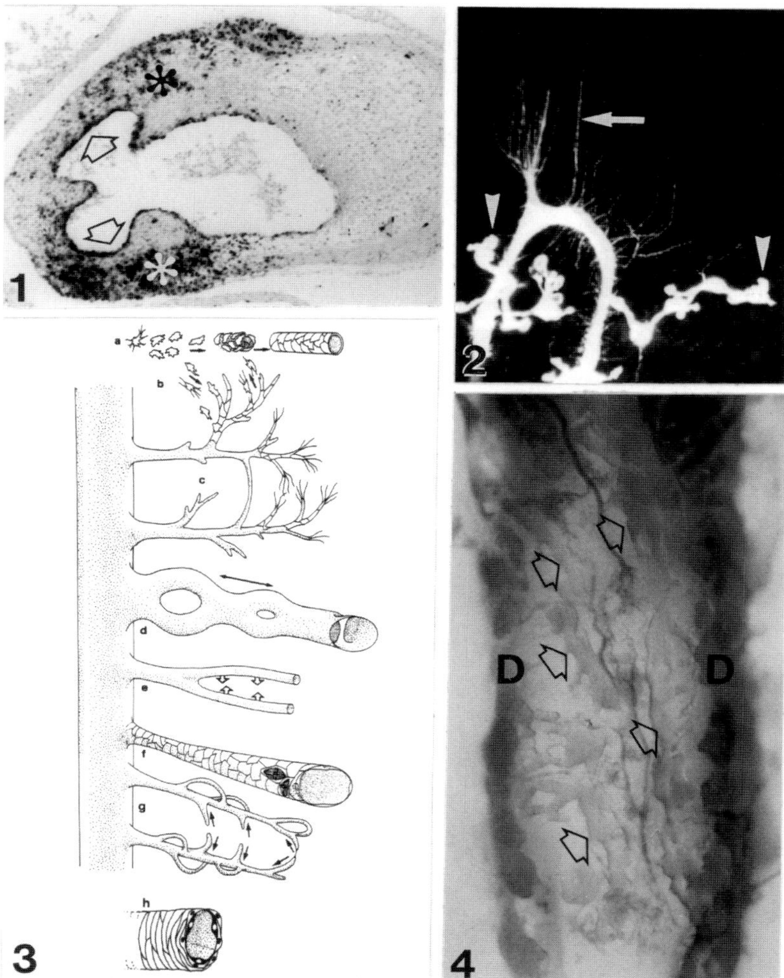

Fig. 1 In situ hybridization of the heart of a 4-day quail embryo with a Quek1/VEGFR-2 probe. Note expression in endocardial cells (arrows) and in mesenchymal cells of the endocardial cushions (asterisks).

Fig. 2 Confocal laser-scanning picture of QH1-stained endothelial cells in the optic tectum of a 4.5-day quail embryo. Note the bulbous processes (arrowheads) and the up to 60-μm long filamentous processes (arrow).

Fig. 3 Schematic illustration of a variety of angiogenetic modes. a Local angioblasts aggregate and form lumenized vessels. b Angioblasts become integrated into preexisting vessels, or detach from them. c Endothelial cells form sprouts, which anastomose into capillary loops. d Sinusoidal capillaries form transluminal pillars, thereby generating capillary loops. In the opposite process the pillars vanish, whereby a huge vessel is formed. e Fusion of blood vessels. f Intercalated growth of vessels to increase diameter and length, as indicated by mitotic spindles. g Regression of blood vessels. h Investment.

Fig. 4 Paired thoracoabdominal trunk (thoracic duct, D) and periaortal lymphatic plexus (arrows) demonstrated by intralymphatic injection of Mercox-resin.

During organogenesis, the primary vascular plexus becomes remodeled and the organotypical vessel pattern is formed. This process is characterized by the fusion of blood vessels in many regions, e.g., fusion of the paired dorsal aorta and fusion (homing) of coronary vessels with the aorta (Bogers et al., 1989). There seems to be a function of the tie 1 receptor in these later stages of angiogenesis (Partanen et al., 1996). Tie 1 deficient mice appear to be normal until E13 but die of hemorrhage around E15 (Sato et al., 1995; Puri et al., 1995). In some regions, such as the prechondrogenic zones of the limbs (Latker et al., 1986), regression of blood vessels takes place. Besides apoptosis and transdifferentiation, the most likely mechanism for regression of vessels seems to be emigration of endothelial cells. Vascular endothelial cells seem to be able to migrate within the endothelial lining preferably against the bloodstream (Christ et al., 1990; Wilms et al., 1991).

Succcessively, the vascular tree becomes invested by mesenchymal cells that give rise to pericytes, different types of smooth muscle cells, and fibrocytes. These cells originate from mesenchyme such as the neural crest and all mesodermal compartments of the embryo (Clark and Clark, 1925; Le Douarin, 1982; Wilting et al., 1995b). The interaction of tie 2, a highly endothelial cell specific receptor tyrosine kinase, and its ligand, angiopoietin-1, seems to be of major importance for the development of the vascular wall (Sato et al., 1993; Davis et al., 1996; Suri et al., 1996). Malfunction of this interaction results in failure of organotypic vascular remodeling and in venous malformations (Sato et al., 1995; Vikkula et al., 1996). The development of the blood vascular tree is schematically illustrated in Figure 3.

Differentiation of endothelial cells, that is, the development of organo-typical characteristics such as fenestrations, cell junctions, enzymes, or carrier systems, depends on interactions with local cells of the various organs. This has most clearly been shown in quail-chick chimeras (Stewart and Wiley, 1981; Wilting and Christ, 1989).

Lymphangiogenesis

In contrast to angiogenesis, lymphangiogenesis has received little attention. The main reason seems to be the absence of specific markers for developing lymphatic endothelial cells. In adult tissues, 5′-nucleotidase activity has been used to detect lymphatics (Werner and Schünke, 1989). However, our knowledge of lymphatics has been mostly derived from studies by injection methods and serial sections. Budge (1880, 1882, 1887) and Sala (1900) were the first to describe the developing lymphatic system of chick embryos. In his first study, Budge (1880) postulated that the first lymphatic vessels develop in the yolk sac of 3-day embryos. However, it was soon found that his injection technique had only demonstrated extracellular spaces (Sabin, 1909). Nevertheless, in follow-up studies, Budge (1882, 1887) clearly demonstrated the avian lymph hearts and lymphatic vessels such as those surrounding the chorioallantoic arteries and the aorta (Fig. 4).

The main difference between blood vascular and lymphatic endothelial cells resides in the fact that lymphatics develop much later. In the chick, the deep lymphatic system is first detectable during day 4–5 of incubation (Clark and Clark, 1920), whereas the first blood vessels are detectable after 1 day of incubation (Pardanaud et al., 1987). In the human, lymphatic primordia have been found in 6- to 7-week-old embryos of 10–14 mm total length (van der Putte, 1975). This is about 3–4 weeks after the development of the first blood vessels.

The basic question about lymphangiogenesis remains, as yet, unresolved. Are lymphatics derived by sprouting from veins (Sabin, 1909), from lymphangioblasts in the mesenchyme (Huntington, 1908; Kampmeier, 1912), or by both mechanisms (van der Jagt, 1932)? It will not be possible to answer this question in quail embryos with the help of the QH1 antibody alone, since this antibody marks both blood vascular and lymphatic endothelial cells (Fig. 5a). The first anlagen of the lymphatics, the jugular, posterior and retroperitoneal lymph plexuses, are always located immediately adjacent to veins, and are connected to them (Sabin, 1909; Clark 1912; van der Putte, 1975). Therefore, it is most probable that the first lymphatics are derived from venous endothelium.

Growth of the lymphatics has then been found to proceed centrifugally by sprouting (Sabin, 1909). With regard to the process of sprouting, blood vascular and lymphatic endothelial cells seem to behave in a similar manner. Nevertheless, it cannot completely be ruled out that lymphangioblast in the mesenchyme become integrated into the growing lymphatic system. However, we have to be aware of the fact that lymphatics develop very late during embryogenesis. We therefore favor the view of Sabin (1909) and van der Putte (1975), who suggest that lymphatic endothelial cells only develop from pre-existing ones. In this respect, therefore, embryonic angiogenesis and lymphangiogenesis are different.

The early lymphatic plexuses are then remodeled into lymph sacs which later will give rise to primary lymph nodes (Miller, 1913; Lewis, 1905; von Gaudecker, 1990). The formation of lymph sacs from lymphatic plexuses involves fusion of vessels in a mechanism similar to that observed in the blood vascular system. The early lymph sacs are filled with stagnant blood which is removed from the vessels as soon as the lymphatics take up function. In the chick, the non-functioning period lasts for about 1 day (Clark, 1912). The formation of the sinuses of lymph nodes resembles the process of intussusceptive microvascular growth. The lymph sac is invaded by lymphatic endothelial cells and connective tissue, and its lumen thereby becomes divided into sinuses (von Gaudecker, 1990; Wilting et al., 1997). During further growth of the lymphatic system, it is probable that intercalated growth increases diameter and length of the vessels, but evidence is lacking.

As noted above, the early lymph sacs are connected to veins. For example, the deep pelvic lymph sac is connected with five intersegmental coccygeal veins (Clark, 1912). Whereas the jugular lymph sacs form secondary connections with the jugular veins, the communication of the other lymph sacs with their veins becomes lost (Sabin, 1909). This process seems to be similar to the regression of blood vessels. During regeneration, however, it is possible that new lymphovenous anastomoses are formed (Berens von Rautenfeld and Drenckhahn, 1994). This has been studied experimentally by ligation of the thoracic duct (Lee, 1922).

In contrast to the most blood vascular capillaries, lymphatic capillaries do not develop a continuous basement membrane (Casley-Smith, 1980; Oh et al., 1997). The absence of laminin and type IV collagen has been used to distinguish between both types of vessels (Barksy et al., 1983; Ezaki et al., 1990). However, a basement membrane is present in collecting lymphatic vessels. Additionally, these vessels possess a contractile media and an adventitia (Berens von Rautenfeld and Drenckhahn, 1994). Investment of collecting lymphatics seems to be comparable to that of blood vessels. Thus, it is obvious that angiogenesis and lymphangiogenesis have many aspects in common, but differ in others.

In vitro, mouse thoracic duct endothelial cells and blood vascular endothelial cells can be stained with antibodies against angiostensin-converting enzyme. In contrast, acLDL binding has been observed in blood vascular endothelial cells, but is missing in those of the thoracic duct (Gumkowski et al., 1987).

Two further aspects about lymphangiogenesis are worthy of comment: Although the first lymphatics accompany veins, they later also grow along arteries. However, fusion of arterial and lymphatic endothelial cells has to be prevented. The mechanisms controlling this process are unknown. Furthermore, there are several organs, such as the central nervous system and the bone marrow, that are strongly vascularized but do not possess any lymphatic vessels. This may be due

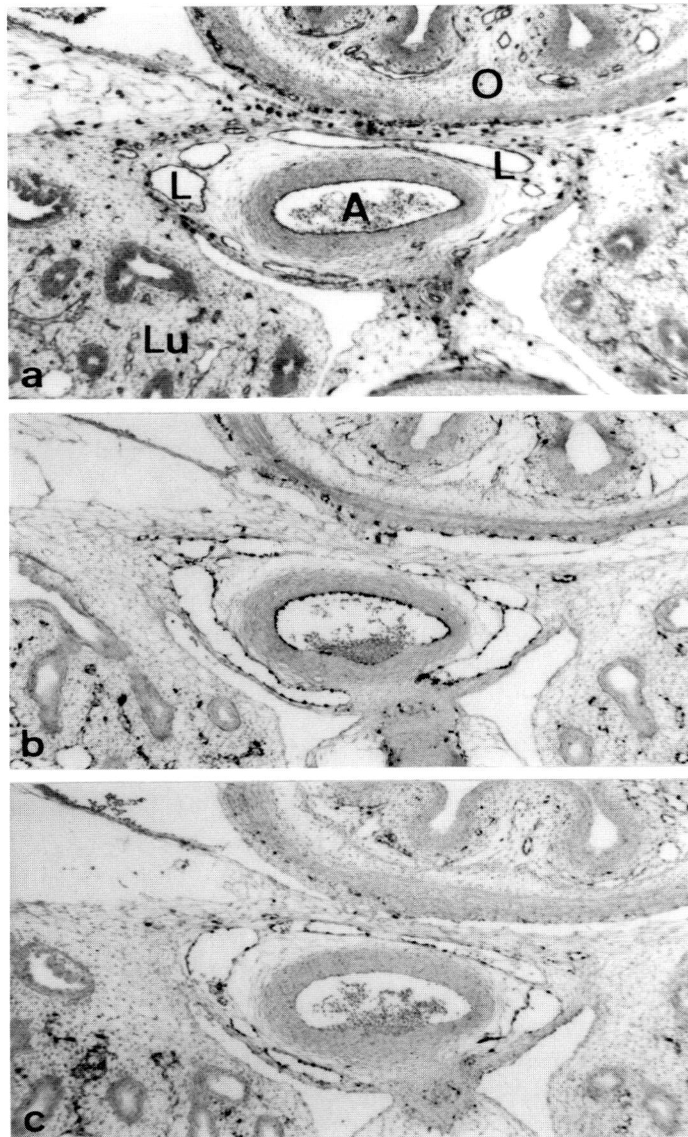

Fig. 5 Consecutive sections showing the aorta (A), periaortal lymphatic plexus (L), oesophagus (O), and lung (Lu) of a 9-day quail embryo. a QH1 staining. Note that blood vascular and lymphatic endothelial cells are stained. b In situ hybridization with a Quek1/VEGFR-2 probe. Note expression in blood vascular and lymphatic endothelial cells. c In situ hybridization with a Quek2/VEGFR-3 probe. Note expression in lymphatic but not aortic endothelial cells.

to unknown antilymphangiogenic factors or to different receptor expression in blood vascular and lymphatic endothelial cells (Kaipainen et al., 1995; Wilting et al., 1996, 1997).

Angiogenic and Lymphangiogenic Growth Factors

There is a family of five homologous growth factors that seems to be of major significance for vascular development and growth. These growth factors are vascular endothelial growth factor (VEGF), VEGF-B, VEGF-C, placenta growth factor (P1GF), and platelet-derived growth factor (PDGF).

VEGF has first been described as vascular permeability factor (Senger et al., 1983). It is a secreted homo- or heterodimeric glycoprotein which in the human has been found in four alternatively spliced forms with 206, 189, 165, and 121 amino acids (Leung et al., 1989; Houck et al., 1991; Tischer et al., 1989; Birkenhäger et al., 1996). Except for $VEGF_{121}$, VEGF is a heparin-binding factor. The two largest forms possess a basic domain that causes the factors to remain cell-associated after secretion. VEGF exerts its effects via two high affinity receptors, VEGFR-1 (flt1) and VEGFR-2 (flk1, KDR, Quek1) (Terman et al., 1992; Millauer et al., 1993; Eichmann et al., 1993). With only a few exceptions (Clauss et al., 1990; Wilting et al., 1997), VEGF receptors are expressed exclusively in endothelial cells (Fig. 5b). VEGF is a highly specific endothelial cell mitogen in vitro (Connolly et al., 1989), and from many factors studied, it is the only factor that induces angiogenesis in the differentiated avian chorioallantoic membrane. VEGF homo- and heterodimers, and also VEGF/P1GF heterodimers, induce development of new capillaries (Wilting et al., 1991, 1992, 1993, 1996; Kurz et al., 1995; Birkenhäger et al., 1996). The effect is most clearly visible in the precapillary region, whereas on the postcapillary side dilatations of vessels have been found. It seems to be of great practical use that capillary formation is not induced on the venous side. The arteries of the chorioallantoic membrane are accompanied by lymphatic vessels (Budge, 1887, Wilting et al., 1996). Data demonstrate that VEGF homo- and heterodimers do not induce lymphangiogenesis (Wilting et al., 1996; Oh et al., 1997), although VEGFR-2 is expressed in lymphatic endothelial cells (Wilting et al., 1996, 1997). The expression of VEGFR-1 has not been studied in these cells. One may speculate that this receptor is missing in lymphatic endothelial cells, but its expression has been demonstrated in blood vascular endothelium (Fong et al., 1995). Angiogenesis seems to be induced only by a combined activation of both VEGFR-1 and VEGFR-2.

VEGF-B has only recently been detected (Olofsson et al., 1996). It is a secreted factor and possesses about 43% identity to VEGF, 30% to P1GF, and 20% to PDGF. VEGF-B is predominantly expressed in heart, skeletal muscle, and pancreas. It forms homodimers but also heterodimers with VEGF, and stimulates DNA synthesis in endothelial cells. Receptor-binding and in vivo effects have not been studied so far.

VEGF-C has been cloned from human prostatic carcinoma cells (Joukov et al., 1996). The homologous portions of VEGF-C are about 30% identical to VEGF, 27% to VEGF-B, 25% to P1GF-1, and 23% to PDGF. VEGF-C is expressed in human adult tissues such as heart, muscle, ovary, and small intestine, but only weakly in brain, liver, and thymus. It binds with high affinity to VEGFR-2 and VEGFR-3 (flt4) and induces migration of capillary endothelial cells in vitro. VEGFR-2 is expressed in both blood vascular and lymphatic endothelial cells (Wilting et al., 1996, 1997). During early development, VEGFR-3 is expressed in all endothelial cells, but becomes restricted to lymphatic endothelial cells in later stages (Fig. 5c) (Kaipainen et al., 1995; Wilting et al., 1997). VEGFR-3 expression has also been found in human lymphangioma endothelial cells (Kaipainen et al., 1995). The lymphangiogenic potency of VEGF-C has recently been demonstrated in avian and murine embryos. Overexpression of VEGF-C in skin of transgenic mice results in lymphatic endothelial proliferation and vessel enlargement (Jeltsch et al., 1997). Application of VEGF-C on the differentiated avian chorioallantoic membrane induces development of new lymphatics, accompanied by proliferation and migration of lymphatic endothelial cells (Oh et al., 1997).

P1GF exists in two different forms, P1GF-1 and P1GF-2, with 132 and 153 amino acids, respectively (Maglione et al., 1991; Hauser and Weich, 1993; Park et al., 1994). It possesses 40% amino acid identity with VEGF and 20% with PDGF. P1GF is almost exclusively expressed in the placenta, but also in human umbilical vein endothelial cells. It binds with high affinity only to VEGFR-1 (Park et al., 1994). P1GF homodimers, in contrast to P1GF/VEGF heterodimers, are not angiogenic (Birkenhäger et al., 1996; Oh et al., 1997). This again indicates that angiogenesis is induced by a combined activation of both VEGFR-1 and VEGFR-2. The function of P1GF in the placenta remains to be elucidated.

PDGF is a homo- and heterodimeric factor consisting of A and B chains (Johnson et al., 1982; Ross et al., 1986). It is about 20% identical to VEGF and P1GF. PDGF is bound by α- and β-receptors with high affinity (Heldin et al., 1988). It is not angiogenic in the differentiated chorioallantoic membrane (Wilting et al., 1992). The factor is produced by endothelial cells (Collins et al., 1985), and stimulates proliferation and chemotaxis of smooth muscle cells. It has therefore been suggested that PDGF is involved in the investment of blood vessels with smooth muscle cells (Thayer et al., 1995). This may also apply to the collecting lymphatics. Further evidence for this hypothesis comes from studies on the PDGFRα deleted Patch mouse. In this mouse, the vascular wall is poorly developed (Schatteman et al., 1992). Furthermore, PDGF expression is upregulated in endothelial cells in response to shear stress (Resnick et al., 1993). This may account for local differences of the vascular wall.

Conclusions and Perspectives

In general, blood vessels and lymphatic vessels possess a common structure. Both are made up of intima, media, and adventitia. Angiogenesis and lymphangiogenesis seem to be comparable in many aspects, but differ in others. The main difference resides in the fact that lymphangiogenesis starts much later during development. Lymphatic endothelial cells seem to be derived from the venous endothelium, and new lymphatic endothelial cells are obviously only formed from pre-existing ones. Otherwise, mechanisms such as sprouting, intussusceptive microvascular growth, intercalated growth, fusion, regression, and investment of vessels has been described for both angiogenesis and lymphangiogenesis. It is highly probable and has partially been shown that homologous growth factors and their receptors control the development of both types of vessels. VEGF has been shown to be a highly potent angiogenic factor. VEGF-C is the first growth factor inducing lymphangiogenesis. Future studies will need to examine the activity of the different growth factors in different types of vessels and in different organs.

Acknowledgments

We are grateful to Dr. A. Eichmann for providing us with the probes for VEGFR-2 and 3. We thank Mrs. L. Koschny, Mrs. U. Pein, Mrs. M. Schüttoff, and Mr. G. Frank for their excellent technical assistance, Mrs. Ch. Micucci for the photographic work, and Mrs. U. Uhl for typing the manuscript. The QH1 antibody was obtained from the Developmental Studies Hybridoma Bank, maintained by the Department of Pharmacology and Molecular Sciences, Johns Hopkins University School of Medicine, Baltimore, MD, and the Department of Biological Sciences, University of Iowa, Iowa City, IA, under contract N01-HD-6-2915 from the NICHD. Our studies were supported by a grant (Wi 1452/1-1) from the Deutsche Forschungsgemeinschaft.

References

Asahara T., Murohara T., Sullivan A., Silver M., van der Zee R., Li T., Witzenbichler, B., Schatteman G., Isner J.M. (1997) Isolation of putative progenitor endothelial cells for angiogenesis. Science 275:964–967.

Azar Y., Eyal-Giladi H. (1979) Marginal zone cells—the primitive streak-inducing component of the primary hypoblast in the chick. J Embryol Exp Morphol 52:79–88.

Barsky S.H., Baker A., Siegal G.P., Togo S., Liotta L.A. (1983) Use of anti-basement membrane antibodies to distinguish blood vessel capillaries from lymphatic capillaries. Am J Surg Pathol 7:667–677.

Benninghoff A., Hartmann A., Hellmann T. (1930) Blutgefäß- und Lymphgefäßapparat, Atmungsapparat und Innersekretorische Drüsen. In: Handbuch der Mikroskopischen Anatomie des Menschen, von Möllendorff W, ed. Berlin: Springer, 1–160.

Berens von Rautenfeld D., Drenckhahn D. (1994) Bau der Lymphgefäße. In: Benninghoff, Alfred: Anatomie: Makroskopische Anatomie, Embryologie und Histologie des Menschen, Drenckhahn D., Zenker W., eds. München: Urban & Schwarzenberg, S. 756–761.

Birkenhäger R., Schneppe B., Röckl W., Wilting J., Weich H.A., McCarthy J.E.G. (1996) Synthesis and physiological activity of heterodimers comprising different splice-forms of vascular endothelial growth factor and placenta growth factor. Biochem J (in press).

Blood C.H., Zetter B.R. (1990) Tumor interactions with the vasculature: angiogenesis and tumor metastasis. Biochem Biophys Acta 1032:89–118.

Bogers A.J.J.C., Gittenberger-de Groot A.C., Poelmann R.E., Péault B.M., Huysmans H.A. (1989) Development of the origin of the coronary arteries, a matter of ingrowth or outgrowth? Anat Embryol 180:437–441.

Brand-Saberi B., Seifert R., Grim M., Wilting J., Kühlewein M., Christ B. (1995) Blood vessel formation in the avian limb bud involves angioblastic and angiotrophic growth. Dev Dynam 202:181–194.

Budge A. (1880) Über ein Kanalsystem im Mesoderm von Hühnerembryonen. Arch Anat Entwickl-Gesch 320–328.

Budge A. (1882) Über Lymphherzen bei Hühnerembryonen. Arch Anat Entwickl-Gesch 350–359.

Budge A. (1887) Untersuchungen über die Entwicklung des Lymphsystems beim Hühnerembryo. Arch Anat Entwickl-Gesch 59–89.

Caduff J.H., Fischer L.C., Burri P.H. (1986) Scanning electron microscope study of the developing microvasculature in the postnatal rat lung. Anat Rec 216:154–164.

Casley-Smith J.R. (1980) The fine structure and functioning of tissue channels and lymphatics. Lymphology 12:177–183.

Christ B., Poelmann R.E., Mentink M.M.T., Gittenberger-de Groot A.C. (1990) Vascular endothelial cells migrate centripetally within embryonic arteries. Anat Embryol 181:333–339.

Christ B., Grim M., Wilting J., v. Kirschhofer K., Wachtler F. (1991) Differentiation of endothelial cells in avian embryos does not depend on gastrulation. Acta Histochem 91:193–199.

Clark E.L. (1912) General observations on early superficial lymphatics in living chick embryos. Anat Rec 6:247–251.

Clark E.R., Clark E.L. (1920) On the origin and early development of the lymphatic system of the chick. Contrib Embryol 9:447–482.

Clark E.R., Clark E.L. (1925) The development of adventitial (Rouget) cells on the blood capillaries of amphibian larvae. Am J Anat 35:239–264.

Clauss M., Gerlach M., Gerlach H., Brett J., Wang F., Familetti P.C., Pan E., Oleander J.V., Connolly D.T., Stern D. (1990) Vascular permeability factor: a tumor-derived polypeptide that induces endothelial cell and monocyte procoagulant activity, and promotes monocyte migration. J Exp Med 172:1535–1545.

Collins T., Ginsburg D., Boss J.M., Orkin S.H., Prober J.S. (1985) Cultured human endothelial cells express platelet-derived growth factor B chain: cDNA cloning and structural analysis. Nature 316:748–750.

Connolly D.T., Heuvelman D., Nelson R., Olander J.V., Eppler B.L., Delfino J.J., Siegel N.R., Leimgruber R.M., Feder J. (1989) Tumor vascular permeability factor stimulates endothelial cell growth and angiogenesis. J Clin Invest 84:1470–1478.

Davis S., Aldrich T.H., Jones P.F., Acheson A., Compton D.L., Jain V., Ryan T.E., Bruno J., Radziejewski C., Maisonpierre P.C., Yancopoulos G.D. (1996) Isolation of angiopoietin-1, a ligand for the TIE2 receptor, by secretion-trap expression cloning. Cell 87:1161–1169.

Eichmann A., Marcelle C., Brént C., Le Douarin N.M. (1993) Two molecules related to the VEGF receptor are expressed in early endothelial cells during avian embryonic development. Mech Dev 42:33–48.

Ezaki T., Matsuno K., Fujii H., Hayashi N., Miyakawa K., Ohmori J., Kotani M. (1990) A new approach for identification of rat lymphatic capillaries using a monoclonal antibody. Arch Histol Cytol Suppl 53:77–86.

Flamme I. (1989) Is extraembryonic angiogenesis in the chick embryo controlled by the endoderm? Anat Embryol 180:259–272.

Fong G.H., Rossant J., Gertsenstein M., Breitman M.L. (1995) Role of the Flt-1 receptor tyrosine kinase in regulating the assembly of vascular endothelium. Nature 376:66–70.

Folkman J. (1985) Tumor angiogenesis. Adv Canc Res 43:175–203.

Folkman J. (1987) Angiogenesis. In: Thrombosis and Haemostasis, Verstraete M., Vermylen J., Lijnen R., Arnout J., eds. Leuven, Netherlands: Leuven University Press, 583–596.

Gumkowski F., Kaminska G., Kaminski M., Morrissey L.W., Auerbach R. (1987) Heterogeneity of mouse vascular endothelium: in vitro studies of lymphatic, large blood vessel and microvascular endothelial cells. Blood Vessels 24:11–23.

Hauser S., Weich H.A. (1993) A heparin-binding form of placenta growth factor (P1GF-2) is expressed in human umbilical vein endothelial cells and in placenta. Growth Factors 9:259–268.

Heldin C.H., Backstrom G., Ostman A., Hammacher A., Ronnstrand L., Rubin K., Nister M., Westermark B. (1988) Binding of different dimeric forms of PDGF to human fibroblasts: evidence for two separate receptor types. EMBO J 7:1387–1393.

Hirakow R., Hiruma T. (1983) TEM-studies on development and canalization of the dorsal aorta in the chick embryo. Anat Embryol 166:307–315.

His W. (1900) Lecithoblast und Angioblast der Wirbeltiere. Abhandl. math.-naturwiss. Kl. sächs. Akad Wiss (Wien) 26:171–328.

Houck K.A., Ferrara N., Winer J., Cachianes G., Li B., Leung D.W. (1991) The vascular endothelial growth factor family: identification of a fourth molecular species and characterization of alternative splicing of RNA. Mol Endocrinol 5:1806–1814.

Huntington G.S. (1908) The genetic interpretation of the development of the mammalian lymphatic system. Anat Rec 2:19–46.

Jeltsch M.M., Kaipainen A., Joukov V., Meng X., Lakso M., Rauvala H., Swartz M., Fukumura D., Jain R.K., Alitalo K. (1997) Hyperplasia of lymphatic vessels in VEGF-C transgenic mice. Science 276:1423–1425.

Johnson A., Heldin C.H., Westermark B., Wasteson A. (1982) Platelet-derived growth factor: identification of constituent polypeptide chains. Biochem Biophys Res Commun 104: 66–74.

Joukov V., Pajusola K., Kaipainen A., Chilov D., Lahtinen I., Kukk E., Saksela O., Kalkkinen N., Alitalo K. (1996) A novel vascular endothelial growth factor, VEGF-C, is a ligand for the Flt4 (VEGFR-3) and KDR (VEGFR-2) receptor tyrosine kinases. EMBO J 15:290–298.

Kaipainen A., Korhonen J., Mustonen T., van Hinsberg V.W., Fang G.H., Dumont D., Breitman M., Alitalo K. (1995) Expression of the fms-like tyrosine kinase 4 gene becomes restricted to lymphatic endothelium during development. Proc Natl Acad Sci USA 92:3566–3570.

Kampmeier O.F. (1912) The value of the injection method in the study of the lymphatic development. Anat Rec 6:223–233.

Kurz H., Ambrosy S., Wilting J., Marmé D., Christ B. (1995) The proliferation pattern of capillary endothelial cells in chorioallantoic membrane development indicates local growth control, which is counteracted by vascular endothelial growth factor application. Dev Dynam 203:174–186.

Kurz H., Gärtner T., Eggli P.S., Christ B. (1996) First blood vessels in the avian neural tube are formed by a combination of dorsal angioblast immigration and ventral sprouting of endothelial cells. Dev Biol 173:133–147.

Latker C.H., Feinberg R.N., Beebe D.C. (1986) Localized vascular regression during limb morphogenesis in the chicken embryo: II Morphological changes in the vasculature. Anat Rec 214:410–417.

Le Douarin N.M. (1982) The Neural Crest. London New York: Cambridge University Press.

Lee F.C. (1922) Establishment of collateral circulation following ligation of the thoracic duct. Bull Johns Hopkins Hosp 33:21–32.

Leung D.W., Cachianes G., Kuang W.J., Goeddel D.V., Ferrara N. (1989) Vascular endothelial growth factor is a secreted angiogenic mitogen. Science 246:1306–1309.

Lewis F.T. (1905) The development of the lymphatic system in rabbits. Am J Anat 5:95–121.

Maglione D., Guerriero V., Viglietto G., Delli-Bovi P., Persico M.G. (1991) Isolation of a human placenta cDNA coding for a protein related to the vascular permeability factor. Proc Natl Acad Sci USA 88:9267–9271.

Markwald R., Fitzharis T., Adams Smith W.N. (1975) Structural analysis of endocardial cytodifferentiation. Dev Biol 42:160–180.

Martinoff V. (1907) Zur Frage der sog. Gefäßsegmente des großen Netzes bei neugeborenen Säugetieren. Int Monatsschrift Anat Physiol 24:281–291.

Millauer B., Wizigmann-Voos S., Schnürch H., Martinez R., Moller N.P.H., Risau W., Ullrich A. (1993) High affinity VEGF binding and developmental expression suggest flk-1 as a major regulator of vasculogenesis and angiogenesis. Cell 72:835–846.

Miller A.M. (1913) Histogenesis and morphogenesis of the thoracic duct in the chick; development of blood cells and their passage to the blood stream via the thoracic duct. Am J Anat 15:131–198.

Minot C.S. (1900) On a hitherto unrecognized form of blood circulation without capillaries, in the organs of vertebrates. Proc Boston Soc Nat Hist 29:185–215.

Noden D.M. (1988) Interactions and fates of avian craniofacial mesenchyme. Development 103 (Suppl.) 121–140.

Noden D.M. (1989) Embryonic origins and assembly of blood vessels. Am Rev Respir Dis 140:1097–1103.

Oh S.J., Jeltsch M.M., Birkenhäger R., McCarthy J.E.G., Weich H.A., Christ B., Alitalo K., Wilting J. (1997) VEGF and VEGF-C: specific induction of angiogenesis and lymphangiogenesis in the differentiated avian chorioallantoic membrane. Dev Biol 188:96–109.

Olofsson B., Pajusola K., Kaipainen A., von Euler G., Joukov V., Saksela O., Orpana A., Pettersson R.F., Alitalo K., Eriksson U. (1995) Vascular endothelial growth factor B, a novel growth factor for endothelial cells. Proc Natl Acad Sci USA 93:2576–2581.

Pardanaud L., Altmann C., Kitos P., Dieterlen-Liévre F., Buck C.A. (1987) Vasculogenesis in the early quail blastodisc as studied with a monoclonal antibody recognizing endothelial cells. Development 100:339–349.

Pardanaud L., Yassine F., Dieterlen-Lièvre F. (1989) Relationship between vasculogenesis, angiogenesis and haemopoiesis during avian ontogeny. Development 105:473–485.

Pardanaud L., Luton D., Prigent M., Bourcheix L.M., Catala M., Dieterlen-Lièvre F. (1996) Two distinct endothelial lineages in ontogeny, one of them related to hemopoiesis. Development 122:1363–1371.

Park J.E., Chen H.H., Winer J., Houck K.A., Ferrara N. (1994) Placenta growth factor. J Biol Chem 269:25646–25654.

Partanen J., Puri M.C., Schwartz L., Fischer K.D., Bernstein A., Rossant J. (1996) Cell autonomous functions of the receptor tyrosine kinase TIE in a late phase of angiogenic capillary growth and endothelial cell survival during murine development. Development 122: 3013–3021.

Plate K.H., Breier G., Risau W. (1994) Molecular mechanisms of developmental and tumor angiogenesis. Brain Pathol 4:207–218.

Puri M.C., Rossant J., Alitalo K., Berstein A., Partanen J. (1995) The receptor tyrosine kinase TIE is required for integrity and survival of vascular endothelial cells. EMBO J 14:5884–5891.

Resnick N., Collins T., Atkinson W., Bonthron D.T., Dewey C.F., Gimbrone M.A. (1993) Platelet-derived growth factor-b chain promoter contains a cis-acting fluid shear stress responsive element. Proc Natl Acad Sci USA 90:4591–4595.

Risau W., Sariola H., Zerwes H.G., Sasse J., Ekblom P., Kemler R., Doetschman T. (1988) Vasculogenesis and angiogenesis in embryonic-system-cell-derived embryoid bodies. Development 102:471–478.

Ross R., Raines E.W., Bowen-Pope D.F. (1986) The biology of platelet-derived growth factor. Cell 46:155–169.

Sabin F.R. (1909) The lymphatic system in human embryos, with a consideration of the morphology of the system as a whole. Am J Anat 9:43–91.

Sabin F.R. (1917) Origin and development of the primitive vessels of the chick and of the pig. Contrib Embryol Carnegie Inst Publ (Washington) 6:61–124.

Sabin F.R. (1920) Studies on the origin of blood vessels and of red blood corpuscles as seen in the living blastoderm of chicks during the second day of incubation. Contrib Embryol Carnegic Inst Publ (Washington) 9:213–259.

Sala L. (1900) Sullo sviluppo dei cuori linfatici e dei dotti torici nell' embryone di pollo. Ric Lab Anat Norm Univ Roma 7:899–1000.

Sato T.N, Qin Y., Kozak C.A., Audus K.L. (1993) Tie-1 and tie-2 define another class of putative receptor tyrosine kinase genes expressed in early embryonic vascular system. Proc Natl Acad Sci USA 90:9355–9358.

Sato T.N., Tozawa Y., Deutsch U., Wolburg-Buchholz K., Fujiwara Y., Gendron-Maguire M., Gridley T., Wolburg H., Risau W., Qin Y. (1995) Distinct roles of the receptor tyrosine kinases Tie-1 and Tie-2 in blood vessel formation. Nature 376:70–74.

Schatteman G.C., Morrison-Graham K., Van Koppen A., Weston J.A., Bowen-Pope D.F. (1992) Regulation and role of PDGF receptor-subunit expression during embryogenesis. Development 115:123–131.

Senger D.R., Galli S.J., Dvorak A.M., Perruzzi C.A., Harvey V.S., Dvorak H.F. (1983) Tumor cells secrete a vascular permeability factor that promotes accumulation of ascites fluid. Science 219:983–985.

Stainier D.Y.R., Weinstein B.M., Detrich H.W., Zon L.I., Fishman M.C. (1995) Cloche, an early acting zebrafish gene, is required by both the endothelial and hematopoietic lineages. Development 121:3141–3150.

Stewart P.A., Wiley M.J. (1981) Developing nervous tissue induces formation of blood–brain barrier characteristics in invading endothelial cells. Dev Biol 84:183–192.

Suri C., Jones P.F., Patan S., Bartunkova S., Maisonpierre P.C., Davis S., Sato T.N., Yancopoulos G.D. (1996) Requisite role of angiopoietin-1, a ligand for the TIE2 receptor, during embryonic angiogenesis. Cell 87:1171–1180.

Terman B.I., Dougher-Vermazen M., Carrion M.E., Dimitrov D., Armellino D.C., Gospodarowicz D., Böhlen P. (1992) Identification of the KDR tyrosine kinease as a receptor for vascular endothelial cell growth factor. Biochem Biophys Res Commun 187:1579–1586.

Thayer J.M., Meyers K., Giachelli C.M., Schwartz S.M. (1995) Formation of the arterial media during vascular development. Cell Mol Biol Res 41:251–262.

Tischer E., Gospodarowicz D., Mitchell R., Silva M., Schilling J., Lau K., Crisp T., Fiddes J.C., Abraham J.A. (1989) Vascular endothelial growth factor: A new member of the platelet-derived growth factor gene family. Biochem Biophys Res Commun 165:1198–1206.

van der Jagt E.R. (1932) The origin and development of the anterior lymph sacs in the sea turtle (Thalassochelys caretta). Quart J Microsc Sci 75:151–165.

van der Putte S.C.J. (1975) The development of the lymphatic system in man. Adv Anat Embryol Cell Biol 51:1–60.

von Gaudecker B. (1990) Lymphatische Organe. In: Humanembryologie, Hinrichsen K.V., ed. Berlin: Springer Berlin.

von Schulte H.W. (1914) Early stages of vasculogenesis in the cat (Felis domestica) with especial reference to the mesenchymal origin of endothelium. Anat Rec 8:78–80.

Vikkula M., Boon L.M., Carraway K.L., Calvert J.T., Diamonti A.J., Goumnerov B., Pasyk K.A., Marchuk D.A., Warman M.L., Cantley L.C., Mulliken J.B., Olsen B.R. (1996) Vascular dysmorphogenesis caused by an activating mutation in the receptor tyrosine kinase TIE2. Cell 87:1181–1190.

Wagner R.C. (1980) Endothelial cell embryology and growth. Adv Microcirc (Basel) 9:45–75.

Welt K., Schippel K., Mironov V.A., Alimov G.A., Bobrik J.J., Banin V.V., Karagonov J.L. (1990) Gefäßendothel (Übersicht). I. Allgemeine Morphologie. 2a: Histogenese des Gefäßendothels. Gegenbaurs Morphol Jahrb (Leipzig) 136:163–199.

Werner J.A., Schünke M. (1989) Cerium-induced light-microscopic demonstration of 5′-nucleotidase activity in the lymphatic capillaries of the proximal oesophagus of the rat. Acta Histochem 85:15–21.

Wilms P., Christ B., Wilting J., Wachtler F. (1991) Distribution and migration of angiogenic cells from grafted avascular intraembryonic mesoderm. Anat Embryol 183:371–377.

Wilting J., Christ B. (1989) An experimental and ultrastructural study on the development of the avian choroid plexus. Cell Tissue Res 255:487–494.

Wilting J., Christ B. (1996) Embryonic angiogenesis: a review. Naturwissenschaften 83:153–164.

Wilting J., Christ B., Bokeloh M. (1991) A modified chorioallantoic membrane (CAM) assay for qualitative and quantitative study of growth factors. Anat Embryol 183:259–271.

Wilting J., Christ B., Weich H.A. (1992) The effects of growth factors on the day 13 chorio-allantoic membrane (CAM): a study of $VEGF_{165}$ and PDGF-BB. Anat Embryol 186:251–257.

Wilting J., Christ B., Bokeloh M., Weich H.A. (1993) In vivo effects of vascular endothelial growth factor on the chicken chorioallantoic membrane. Cell Tissue Res 274:163–172.

Wilting J., Brand-Saberi B., Huang R., Zhi Q., Köntges G., Ordahl C.P., Christ B. (1995a) Angiogenic potential of the avian somite. Dev Dynam 202:165–171.

Wilting J., Brand-Saberi B., Kurz H., Christ B. (1995b) Development of the embryonic vascular system. Cell Mol Biol Res 41:219–232.

Wilting J., Birkenhäger R., Eichmann A., Kurz H., Martiny-Baron G., Marmé D., McCarthy J.E.G., Christ B., Weich H.A. (1996) $VEGF_{121}$ induces proliferation of vascular endothelial cells and expression of flk-1 without affecting lymphatic vessels of the chorioallantoic membrane. Dev Biol 176:76–85.

Wilting J., Eichmann A., Christ B. (1997) Expression of the avian VEGF receptor homologues Quek1 and Quek2 in blood-vascular and lymphatic endothelial and non-endothelial cells during quail embryonic development. Cell Tissue Res 288:207–223.

Wolff J.R., Bär T. (1972) Nahtlose cerebrale Capillarendothelien während der Cortexentwicklung der Ratte. Brain Res 41:17–24.

1.3

Ontogeny of the Endothelial Network Analyzed in the Avian Model

Françoise Dieterlen-Lievre and Luc Pardanaud

The avian embryo offers privileged approaches to experimental analyses of development, because of its accessibility, comparatively large size, deployment of the blastodisc in a plane, and last but not least, due to the "quail-chicken" marker system (Le Douarin, 1973). The latter approach consists in transplanting cells, rudiments, or territories from one species to the other, either in a heterotopic or orthotopic location, and tracing the cells from the transplant during their migration and differentiation. Tracing is accomplished either by identifying quail-specific, nucleolus-associated heterochromatin, or by applying antibodies that recognize all cells or a cell lineage from one of the two species; a variety of monoclonal antibodies (mAb) with such properties are now available. One mAb (MB1/QH1) has been particularly useful in the field of endothelial and blood cell ontogeny because it has affinity for the so-called "hemangioblastic" lineage in the quail. While mouse embryos may be engineered so as to reveal the critical gene activities involved in developmental processes, the unique amenability of bird embryos to microsurgery has made it possible to pin down the origins and migrations of precursors that build the vascular network.

Most endothelial cells (EC) in adult organisms divide very rarely. When they do, in response to physiological stimuli such as wound healing and cycling of female reproductive organs, or to pathological stimuli like in tumor growth, they participate in a process termed angiogenesis (Folkman, 1974). Angiogenesis involves exit from the quiescent state, breaking of the basement membrane, mitosis, and migration. This process of dedifferentiation and reactivation of already mature cells is actually the only recognized mechanism through which new EC differentiate in the adult. In the embryo, endothelial precursors have to appear de novo, a process known as vasculogenesis (Risau and Lemmon, 1988). It has long been recognized that the whole blood forming system is normally a mesodermal derivative (reviews in Romanoff, 1960; Risau and Flamme, 1995). The generation of blood vessels was first described through careful observation of living amphibian or avian embryos under the microscope (Clark, 1918; Sabin, 1917, 1920; Auerbach et al., 1974). These descriptions relied on the visualization of blood vessels through the presence of circulating red cells. A large part of the events involved in the formation of blood vessels has already occurred at that step. The analysis of how the endothelial network builds up in the embryo really began when monoclonal antibodies, capable of recognizing EC at a very early stage of their commitment, became available. These mAbs detect EC before they interconnect into a chain and eventually a tube.

Vasculogenesis and angiogenesis have been reported to occur during ontogeny. We wish to point out that some specialists in the field discuss the propriety of these two terms, while others introduce further distinctions like angioblastic recruitment and angiotrophic influence (Brand-Saberi et al., 1995). A process of "intussusceptive growth" has also been described during the branching of the vascular network (Patan et al., 1993). It is not our purpose to discuss these notions here and we will employ the dual angiogenesis/vasculogenesis terminology, which is appropriate to describe our views regarding the formation of the primary endothelial network in the early embryo.

This chapter is based mainly on our own work that made use of various transplantation strategies between quail and chicken embryos and used endothelial specific markers. We thus arrived at a coherent interpretation according to which distinct complementary modes of endothelial development occur in different mesodermal subsets and complement each other to yield an orderly endothelial tree. Among markers, we mainly relied on hemangioblastic specific mAb QH1. Valuable data on EC origin in hematopoietic organs were obtained from early transplantation studies in which Feulgen-Rossenbeck DNA staining served to distinguish quail from chick cell nuclei (Jotereau and Le Douarin, 1978). Two of these classical studies concerning the thymus and the bone marrow bring about information about the development of blood vessels in these rudiments. Transplantation of the thymus rudiment (Metcalf and Moore, 1971) or, better, the third and fourth endodermal branchial pouches (Le Douarin and Jotereau, 1973) had previously demonstrated that all lymphoid progenitors were extrinsic colonizers. Connective tissue of the thymus, on the other hand, was shown to derive from the neural crest, by means of orthotopic transplantations of quail neural tube in a chicken host at the rhombencephalic level (Le Lièvre and Le Douarin, 1975). The neural crest clearly appeared devoid of endothelial potential: in these chimeras, EC all came from the chicken host head paraxial mesoderm (Couly et al., 1995). In the case of the bone marrow, both extrinsic EC precursors and hematopoietic stem cells (HSC) colonized this tissue in transplanted limb buds (Jotereau and Le Douarin, 1978). It is also clear from these studies that pericytes and EC have distinct origins. In this review we will consider successively which markers are available to identify EC and their precursors and the morphogenic process that emerged when these markers were applied to the analysis of normal and experimental embryos.

Immunodetection of the Hemangioblastic Lineage

The hemangioblast is a putative bipotential cell that is hypothesized to give rise to both endothelial and hematopoietic precursors. This cell was hypothesized to exist based on cytological aspects (His, 1886; Murray, 1932), though no hard-core experimental evidence has yet been obtained to confirm the existence of hemangioblasts. The idea arose because of the close anatomical links, that are common between endothelial and hematopoietic cells in the early embryo. These links are first observed in the blood islands of the extraembryonic area that are aggregates of identical cells, from which peripheral cells become endothelial while central cells become hematopoietic (Romanoff, 1960). A little later the embryonic aorta of birds (Sabin, 1920; Dieterlen-Lièvre and Martin, 1981) and mammals (Emmel, 1916; Jordan, 1916; Houser et al., 1961; Garcia-Porrero et al., 1995; Tavian et al., 1996) harbours clusters of hematopoietic cells typically aggregated onto its ventral endothelium.

Apart from these readily observed anatomical associations, it is intriguing that the two cell lineages share many characteristics in both birds and mammals. In the case of avian cells, mAb MB1, that recognizes endothelial and hematopoietic cells in the quail at all stages of embryonic and adult life, was raised by immunizing the mice with the μ chain of quail immunoglobulin (Péault et al., 1983). Thorough electrophoretic and biochemical analyses indicated that this mAb recognizes a number of acidic glycoproteins with Mr ranging between 80 and 240 kd, one of which is the immunoglobulin μ chain, another α2 macroglobulin, while the others are unidentified (Labastie, 1989). These unidentified bands most likely represent different degrees of glycosylation of the same protein. MB1 has been used to study chick thymic colonization by quail lymphoid progenitors (Coltey et al., 1987) and to analyze the early determination and differentiation of extraembryonic blood islands in the quail (Péault et al., 1988). Another mAb, QH1 (Pardanaud et al., 1987), raised against a completely different immunogen, namely bone marrow cells from E13 quail embryos, displays affinities very similar if not identical regarding cell lineages as well as molecules (Pardanaud et al., 1987). QH1 is available from a hybridoma bank (Developmental Studies Hybridoma Bank, University of Iowa, Iowa City, IA 52242, USA) and has been widely used by investigators studying avian endothelial cell differentiation (Pardanaud et al., 1987, 1989, 1996; Pardanaud and Dieterlen-Lièvre, 1993a, 1995; Noden, 1991; Coffin and Poole, 1988; Drake et al., 1992, 1995; Flamme and Risau, 1992; Wilting et al., 1995; Roncali et al., 1996), somite differentiation (Schramm and Solursh, 1990), heart development (Inakagi et al., 1993; Garcia-Martinez and Schwoenwolf, 1993; Linask and Lash, 1993; Sugi and Markwald, 1996), or hematopoietic cells (Dieterlen-Lièvre, 1994; Cuadros et al., 1994). In the hematopoietic lineage, QH1 recognizes cells from all sublineages from the emergence of early progenitors to the most differentiated stages. In the erythroid lineage however, mature erythrocytes lose affinity for QH1.

In toto Immunostaining of the Early Quail Blastodisc: Emergence of the Endothelial Network

Early studies of the vascularization process relied on microscopic observations of embryos, alive (Sabin, 1920), fixed, and stained with classical dyes, injected with India ink (Evans, 1909; Flamme, 1989), or prepared for scanning and transmission electronic microscopy (Hirakow and Hiruma, 1981, 1983). The identification of EC was based on their integration in tubular structures, the first EC were identified around the 4–5-somite stage. The advent of MB1 and QH1 mAbs permitted to study earlier events of vascular development.

In order to obtain a three-dimensional view of vascularization and observe the dynamics of this process, we initiated an in toto immunostaining approach in which whole quail blastodiscs ranging between early primitive streak stage to the 13-somite stage were treated with QH1 revealed by a fluorescent label (Pardanaud et al., 1987). The architecture of embryos was visible because of a general background of endogenous fluorescence emitted by the embryonic tissues. The specimens were observed ventral side up because most vessels develop in ventralmost mesoderm close to the endoderm, as first recognized by Sabin (1920).

The first EC appear as single cells at the level of the first somite, in the area opaca (extraembryonic area) at the head-process stage and in the area pellucida (embryonic area) at the 1-somite stage. These cells progressively interconnect especially at the level of the anterior intestinal portal and along the somites, thus building up the profiles of the heart and dorsal aortae. The emergence and connectivity of new QH1$^+$ cells then rapidly progress in both rostral and caudal directions so that, by the 5-somite stage, numerous short QH1$^+$ processes outline the beginning of vascular organization (Fig. 1a). These cellular processes are observed from the head to the somite level; in particular, QH1 cells assemble in a network along the ventral and lateral side of the somites in the region of the future dorsal aortae. During the next step (7-somite stage), the aortic rudiments begin to form cords (Fig. 1b); at this stage, the heart becomes interconnected

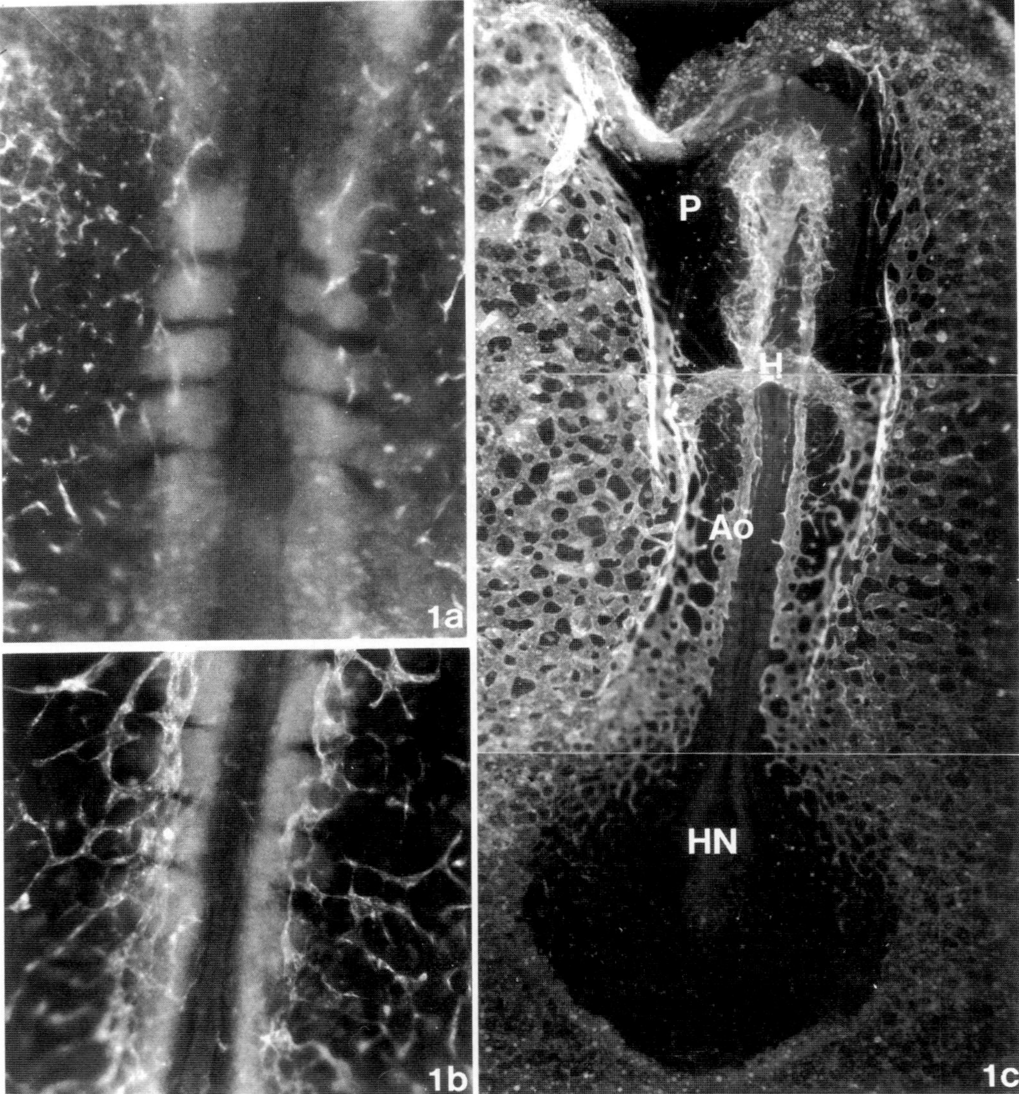

Fig. 1 Whole mounts of QH1-immunofluorescence treated quail embryos. (a) 4-somite stage. Ventral view. Single or interconnected $QH1^+$ cells are numerous in the area pellucida. A few are aligned along the somites. × 110. (b) 7-somite stage. $QH1^+$ cells have connected into the rudiment of the aortae on each side of the somites. × 110. (c) 13-somite stage. The aortae (Ao) are now tubes and the cardiac rudiment (H) is beating. The only unendothelialized territories are the proamnion (P), which is always devoid of mesoderm, and Hensen's node (HN). × 35.

with the extraembryonic arborization, the blood islands become visible caudally in the area opaca. Next, the anterior horns of the vascular network move up in front of the brain vesicles. A rostro-caudal gradient of vascular development is particularly evident, with about a 2-hour delay in differentiation of the vessels between head and tail. At this stage, the extraembryonic and intraembryonic vascular plexus mingle and no limit between the area opaca and the area pellucida can be observed in immunostained wholemounts of blastodiscs. At the 13-somite stage (Fig. 1c), the vascular tree covers the whole embryonic and extraembryonic territories, with the exception of the proamnion and Hensen's node region where gastrulation is still underway. The edge of the area vasculosa becomes entirely circumscribed by the QH1$^+$ sinus marginalis.

Further descriptive studies, using QH1 immunostained wholeamounts, followed our initial observations. The organization of the primitive tubular network into large and small vessels was analyzed (Coffin and Poole, 1988; Poole and Coffin, 1989). Drake et al. (1990) added immunodetection of the extracellular matrix proteins (collagens I, IV, laminin, and fibronectin) to QH1 staining. Angioblast emergence and the morphogenesis of the endothelium network have also been studied in mouse embryo wholemounts immunostained for Von Willebrand Factor (VWF) (Coffin et al., 1991). In this species, the aorta endothelium assembles from single positive cells while VWF-negative precursors give rise to other vessels, for instance intersegmental arteries.

A Reminder About Mesoderm Evolution

The endothelial network differentiates according to distinct modes in the different mesodermal subsets. In order to make our findings clear, it is useful to recapitulate the evolution of the mesoderm. During avian gastrulation, the mesodermal germ layer is generated by cells invaginating through the primitive streak. At first this mesodermal germ layer is uniform. As develop-

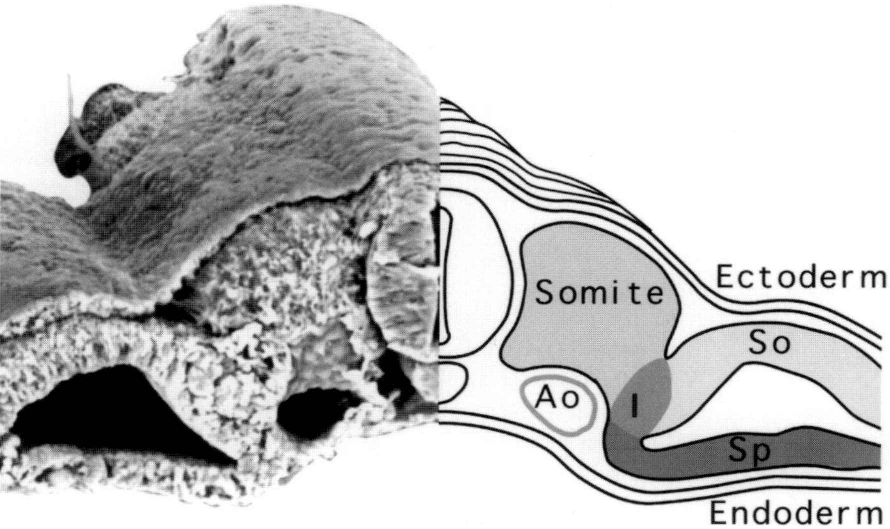

Fig. 2 Mesodermal subsets in a 36-hour avian embryo. On the left, scanning electron micrograph of the fractured trunk (courtesy of P. Coltey). On the right, scheme of the corresponding structures. Ao = aorta. I = intermediate cell mass. So = somatopleural mesoderm. Sp = splanchnopleural mesoderm.

ment proceeds, mesoderm partitions progressively into subdivisions with distinct fates, namely notochordal, somitic, nephrogenic, and lateral (Fig. 2). Somites will eventually divide into dermomyotome and sclerotome, then disintegrate by emigration of their cells. Lateral plate will split into splanchnopleural and somatopleural layers when the coelom forms. The deep, or ventral, splanchnopleural layer is tightly associated to the endoderm; it will give rise to the spleen and to the connective envelope and smooth muscle of visceral organs. The superficial, or dorsal, somatopleural mesoderm will provide the supporting tissues of the body wall and limbs. Cells immigrating from the somite into this somatopleural layer of the mesoderm will give rise to the skeletal muscle of the body wall and limbs, as established by means of the quail/chicken system (Chevallier et al., 1977; Jacob et al., 1979; Christ et al., 1983; Bagnall et al., 1988; Schramm and Solursh, 1990; review in Christ and Ordahl, 1995).

Protooncogene c-ets1 Expression, a Marker for Early EC

c-ets1, the cellular progenitor of one of the two oncogenes present in the genome of the avian erythroblastosis virus (Nunn et al., 1983; Leprince et al., 1983), encodes a DNA-binding protein and is actively transcribed in adult lymphoid organs. In situ hybridization carried out on the chicken embryo unexpectedly revealed an endothelial-specific expression (Vandenbunder et al., 1989; Quéva et al., 1993). The signal appears early in development, in the mesodermal layer while completely excluded from epithelia, then becomes restricted to endothelia, at the time the latter begin expressing QH1 affinity. c-ets1 mRNA is also abundant in the migrating cephalic neural crest. From the third embryonic day (E3) onward, c-ets1 becomes restricted to segmental arteries and smaller vessels, while disappearing from the aorta. Eventually the c-ets1 endothelial signal disappears completely around E10-12. These patterns of expression indicate that the c-ets1 is transcribed during the expansion phase of EC development, a conclusion confirmed by expression in EC of the mouse embryo (Maroulakou et al., 1994), and in EC of the human placenta (Luton et al., 1997).

The expression of another protooncogene, c-myb, was analyzed in parallel with that of c-ets1 in the chicken embryo (Vandenbunder et al., 1989). In the embryo, like in the adult, c-myb is expressed by expanding hematopoietic cells. In the E3-4 avian embryo, a bright c-myb signal is restricted to the ventrolateral aspects of the aorta, where it seems to overlap with the c-ets1 signal. A more precise study will be necessary to establish whether some cells, possibly hemangioblasts, do express both genes. In any case, the messengers of these two genes are excellent early markers for the commitment to these lineages.

When c-ets1 expression was analyzed at earlier stages of ontogeny, the signal was found to be stronger on the cells that are in contact with the endoderm than on cells near the ectoderm (Fig. 3, upper diagram). When the mesoderm splits to form the coelomic cavity, c-ets1 dramatically increases in the splanchnopleural layer, with an even more intense signal in all the cells of the extraembryonic blood islands, whether they have a hematopoietic or an endothelial fate (Fig. 3, lower diagram) (Pardanaud and Dieterlen-Lièvre, 1993b). Interestingly these early blood islands do not express detectable amounts of the c-myb gene (Vandenbunder et al., 1989). These findings in the chicken embryo are supported by the phenotype of c-myb knockout mice, which display normal yolk sac erythropoiesis, and die at E13-15 because definitive hematopoiesis does not proceed (Mucenski et al., 1991). In the somatopleural mesoderm the c-ets1 signal is restricted to small endothelial profiles. The intermediate cell mass, or presumptive kidney material, regularly displays a very intense hybridization signal (Fig. 3).

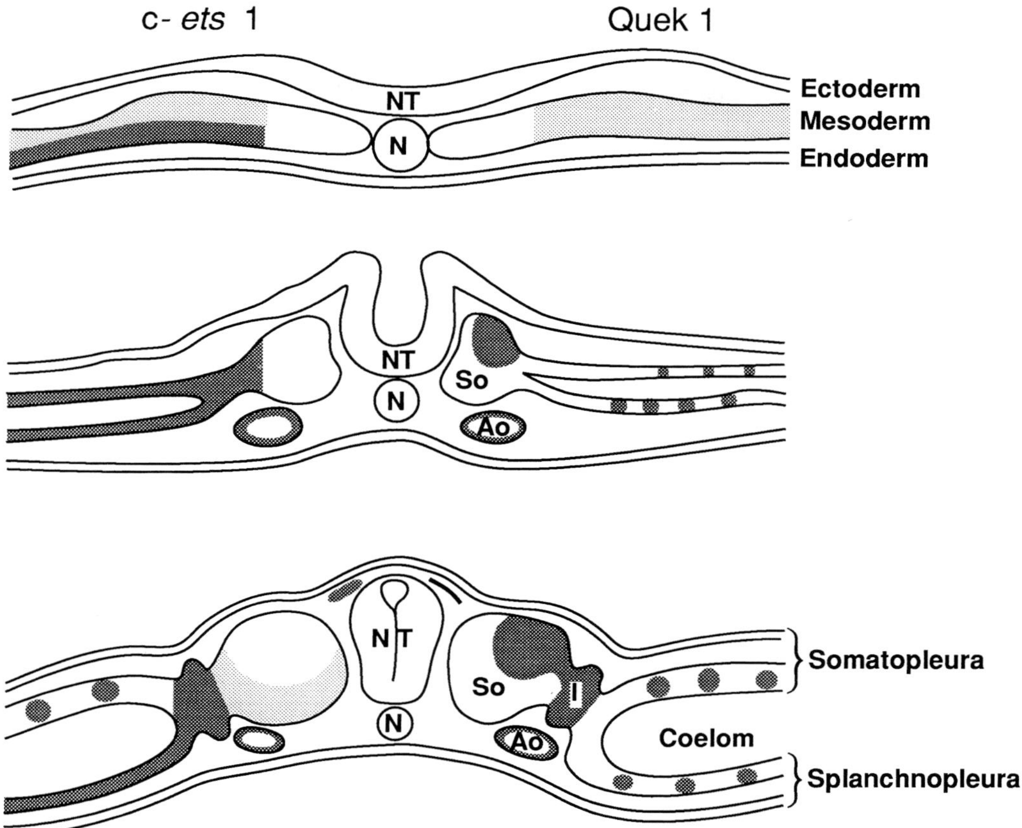

Fig. 3 Comparative scheme of the expression patterns of the c-ets1 protooncogene (left side) and the Quek1 tyrosine kinase receptor gene (right side) at different steps of mesoderm compartmentalization. The two genes are first expressed laterally in the mesoderm, then gradually become restricted to endothelial precursors and EC. The expression patterns overlap extensively except in the somites (see text). Ao = aorta. I = intermediate cell mass. N = notochord. NT = neural tube. So = somite.

Quek1, Alias VEGF-R2, Detects Quail and Chicken Angioblasts

Vascular Endothelial Growth Factor (VEGF) has been recognized as the only molecule whose growth-promoting activity acts exclusively on EC. In the last few years, a number of receptors for this molecule with tyrosine kinase activity have been cloned. In the quail, Quek1 was found to be homologous to mouse flk-1, also called VEGF-R2 (Eichmann et al., 1993, 1996). Quek1, expressed earlier than QH1, has a widespread distribution in the early mesoderm of the quail embryo, and secondarily becomes restricted to QH1^{+} EC. Because of this pattern this gene expression has been interpreted as a marker for the hemangioblast (Eichmann et al., 1993). Compared to c-ets1, Quek1 signals have a similar distribution in the lateral plate and in the intermediate cell mass. In the lateral plate, the two expressions patterns overlap, being present mainly in mesodermal cells associated with the endoderm. However, in the somites, Quek1 is expressed in the dorsolateral quadrant, opposite to the c-ets1 signal (Fig. 3, lower diagram). The gene for another tyrosine kinase, that was isolated in the same screen and designated as Quek2, becomes expressed slightly later than Quek1, namely at the same time and in the same cells as QH1 (Eichmann et al., 1993).

Two Vascularization Mechanisms at Work in the Embryo

We have described above the progressive appearance of single QH1$^+$ EC and their associations into a network. This emergence is ubiquitous rather than the result of cell migrations, as demonstrated by two groups of manipulations. As early as 1915, Reagan excised small portions of blastodisc and showed that both excised tissue and remaining tissue acquire an endothelial network. In the second group of manipulations, "yolk sac chimeras" were constructed by grafting the central region of the quail blastodisc, i.e., the presumptive body of the embryo, on the chicken extraembryonic area; this experimental model was extensively used to analyze the ontogeny of hematopoiesis (review in Dieterlen-Lièvre, 1994). In these chimeras, the quail/chick frontier in the endothelial network coincided exactly with the boundary between the two avian tissues (Beaupain et al., 1979).

Do endothelial precursors also become committed within all the mesodermal subsets? Our experiments first addressed this question at the organogenesis period by transplanting organ rudiments from quail to chicken or inversely (Pardanaud et al., 1989). Since QH1 mAb used to analyze these transplantations recognizes only quail, depending on the combination, QH1$^+$ cells are either intrinsic to the rudiment, in the case of a quail rudiment, or originate from the host when a chicken rudiment was transplanted. The results depended on the germ layer origin of the rudiments: limb bud rudiments were always colonized by endothelial precursors from the host, while visceral organ rudiments proved capable of producing their own endothelial network. In other words, limb buds exhibited a QH1$^+$ endothelial network, when chicken buds were grafted in quail, whereas transplanted quail spleen, pancreas, or lung, etc. developed a QH1$^+$ positive network. Thus, the mesoderm of visceral rudiments—splanchnopleural mesoderm—produced EC, while body wall—somatopleural—mesoderm did not. It is interesting that the hematopoietic cell population differentiating in the transplants was of host origin regardless of the explant source.

Our next question concerned the potentialities of earlier tissues, that is, potentialities of the somatopleural and splanchnopleural layers of mesoderm at the time of gastrulation. Regarding the production of EC, these potentialities were the same: only the endoderm-associated mesoderm produced EC, while ectoderm-associated mesoderm became colonized by extrinsic EC precursors. Regarding hematopoiesis, however, the whole splanchnopleural mesoderm appeared capable of producing HSC (Pardanaud and Dieterlen-Lièvre, 1993a). Thus, hematopoietic potential is initially distributed in the whole splanchnopleural layer and later becomes restricted to the paraaortic region.

Two Distinct Endothelial Lineages Contribute to Vascularization During Ontogeny

We then attempted to define the origin of EC precursors that colonize somatopleural mesoderm. A possible site of origin was the somites, wherefrom many groups had previously described the production of EC (Jacob et al. 1979; Solursh et al., 1987; Noden, 1989; Lance-Jones, 1988; Huang et al., 1994; Wilting et al., 1995). We first verified through "immunoectomy" (Solter and Knowles, 1975) that somites really do produce EC. We deemed this demonstration necessary because, as described earlier, during normal development, the emergence of QH1$^+$ stainable cells is synchronized with segmentation of the somites and these cells appear to be tightly associated with the ventral face of newly formed somites. Thus, after mechanical dissociation, somites are likely to carry adhering EC. Indeed, even after a further pancreatin treatment, in most cases one or two QH1$^+$ cells were still present (Fig. 4a). But when pancreatin-cleaned somites were treated with QH1 and then with complement, they appeared devoid of EC (Fig. 4b). Nevertheless, when such somites were grafted to the limb bud of chicken hosts, they were capable of producing EC (Pardanaud and Dieterlen-Lièvre, 1995). To establish definitively the contribution of somites to the endothelial network of the embryo, orthotopic transplantations were

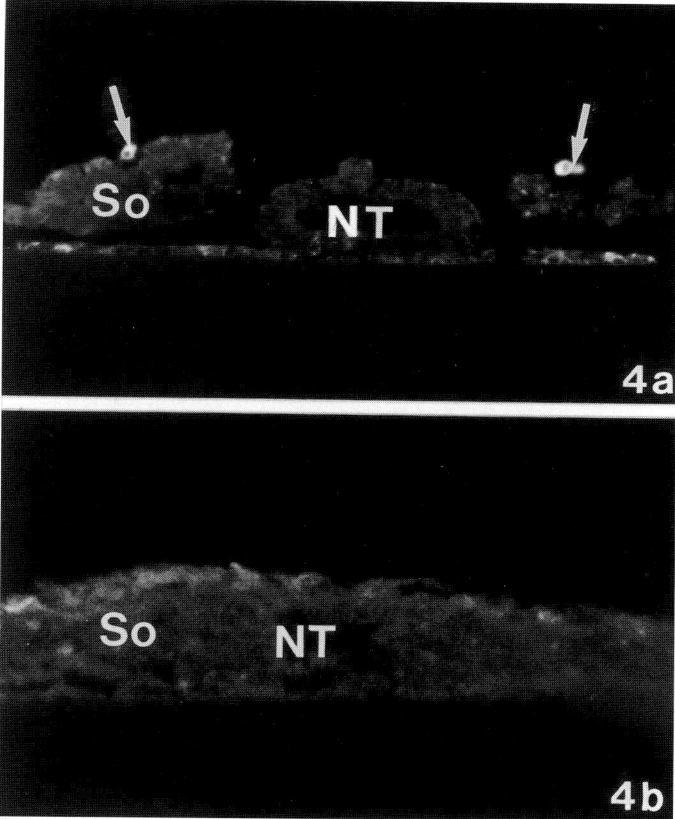

Fig. 4 Immunectomy experiment. (a) Trunk of a 36-hour quail embryo. The axial structures have been isolated by mechanical dissection and treated by pancreatin. QH1 immunofluorescence. Transverse section. Three strongly fluorescent QH1$^+$ EC (arrows) remain adhering to the ventral face of the somites, here seen at the top of the picture. (b) Treatment with QH1 followed by complement. Positive cells are no longer detected by QH1 immunostaining. The structures of the embryo have lost their definition due to the treatment. NT = neural tube. So = somites. × 250.

performed from the quail into chick hosts. Depending on cases, one to three somites were substituted in that way. The transfer was always carried out at the level of the caudalmost somites, that is, we used somites that had just become segmented, but were not yet dissociating prior to emigration. The results were clearcut: the endothelial networks became chimeric at the level of the graft, but only in "dorsal" structures, such as the body wall and the periphery of neural tube and kidney. With respect to vessels, $QH1^+$ EC integrated into the cardinal veins, the intersomitic arteries and in the roof and sides of the aorta. EC of somitic origin never entered the visceral organs and strikingly avoided the floor of the aorta (Pardanaud et al., 1996).

Quail somatopleural or splanchnopleural mesoderm was then transplanted in place of host somites or deposited onto the host splanchnopleura at E2. The somatopleural mesoderm (obtained after colonization by EC precursors) gave rise to a restricted EC pattern, that corresponded exactly to the pattern obtained from somitic transplants (Fig. 5). In contrast, splanchnopleural EC spread ubiquitously, settling in "dorsal" structures but also invading visceral domains and the floor of the aorta (Fig. 6). In the latter location, and there only, hematopoietic cells were seen budding into the vascular lumen.

Thus two distinct endothelial lineages (Fig. 7) are defined by these experiments; these lineages differ from each other by two important features. Firstly, they display distinct affinities, probably supported by distinct adhesion molecules. Secondly, only the lineage of the ventralmost origin is linked with hematopoiesis, while the dorsal lineage is purely endothelial. These results account for the fact that hematopoietic clusters are always located on the floor of the aorta and never on the roof.

In the region of the head and neck, as indicated above, there are no somites, and most supporting tissues are of neural crest or mesectodermal origin. There is however on each side of the neural tube a narrow strip of paraxial mesoderm. By orthotopic transplantations, Couly and coworkers (1995) could map the derivatives of this mesoderm, demonstrating that it is mostly devoted to an endothelial fate: apart from giving rise to some muscles and a few of the skull bones, this mesoderm provides the entire complement of EC necessary to vascularize the brain.

Conclusions and Perspectives

In order to analyze the ontogeny of the endothelial network, we have called upon a well codified experimental strategy, that consists of labeling a rudiment by transplantation and following the cells contributed to a defined lineage. Cells were followed throughout their homing and differentiation process. It is possible in this way to obtain an accurate representation of cell origins, cell migrations, and cell interactions. Avian embryos are particularly useful for this approach; furthermore, the information obtained serves as a basis that can be extended to other classes of vertebrates. The modalities of vascular development recognized in birds should prompt mammalian investigations, which will help to interpret the disturbances of endothelial or hematopoietic differentiation resulting from gene knockouts. Several genes, encoding VEGF or VEGF receptors (VEGF-R), have thus been found to be critical for the development of both the endothelial and hematopoietic systems. The endothelial tree does not form in embryos lacking VEGF, and is heavily impaired in $VEGF^{+/-}$ heterozygotes. Both types of embryos die at midgestation (Carmeliet et al., 1996). In contrast, only the homozygote state is lethal in VEGF-R knockouts. Knockout of flk-1 (or VEGF R2), which is the Quek1 homologue, is responsible for anomalies that affect both the endothelial network and the hematopoietic lineage (Shalaby et al., 1995). Flt-1 (VEGF R1) knockout affects only the endothelial lineage (Fong et al., 1995). Expression studies carried out in the quail embryo (Eichmann et al., 1993; Flamme et al., 1995), like those in

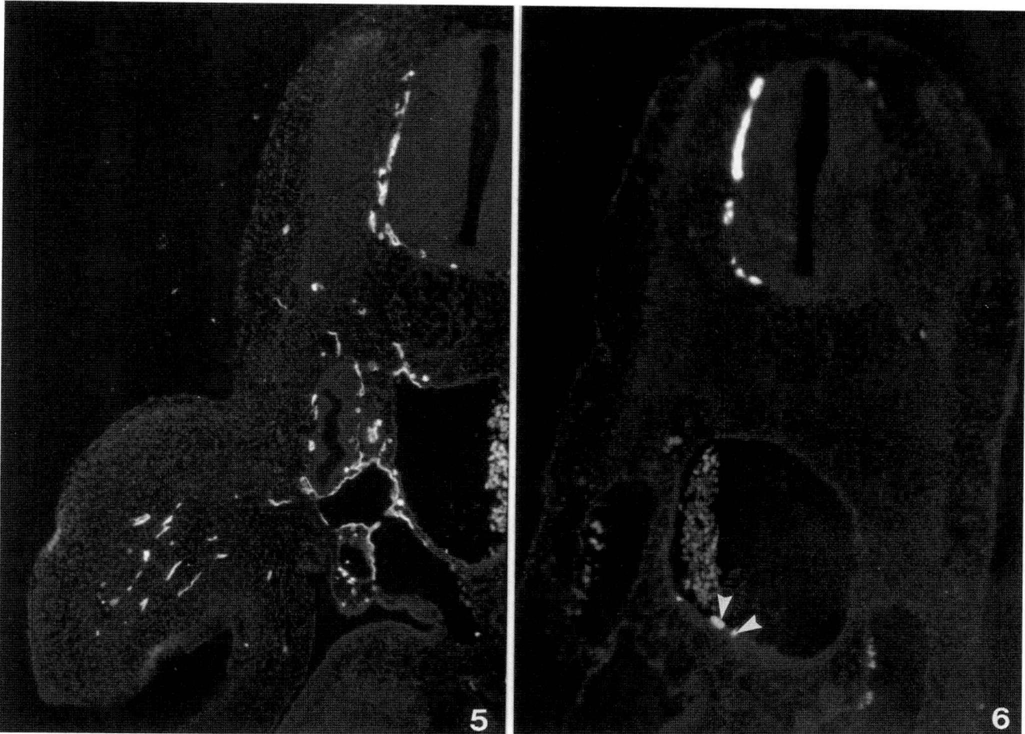

Fig. 5 4-day chick host engrafted with 2-day quail somatopleural mesoderm 2 days earlier. The graft was deposited onto the host splanchnopleural mesoderm. QH1 immunofluorescence. Transverse section at truncal level. $QH1^+$ EC have settled around the neural tube and invaded the body wall, wing, and kidney on the side of the graft. Some $QH1^+$ cells have integrated into the roof of the aorta, but none penetrated into the visceral organs. × 130.

Fig. 6 4-day chick embryo in which two somites were replaced by quail splanchnopleural mesoderm 2 days earlier. $QH1^+$ EC have settle around the neural tube, invaded the body wall and also integrated in the floor of the aorta (arrow heads). The level of the host embryo was selected to demonstrate cells of graft origin in the floor of the aorta. The body wall contains few $QH1^+$ cells at this particular level. × 130.

the mouse embryo, implicate VEGF and its receptors in the growth of the endothelial network, while FGF appears responsible primordial for induction of the hemangioblastic anlage (Flamme and Risau, 1992).

We have shown that the development of the endothelial network entails two very distinct modalities. One mechanism involves migration of precursors, while the other is defined by in situ commitment and differentiation. It will be interesting, therefore, to determine whether different growth factor combinations are expressed in the mesodermal layers of the embryo, as a function of which mechanism is operative.

Certainly the expression pattern of many genes implicated in EC emergence and organization is still to be detected. Two of them, that encode proteins with very different functions,

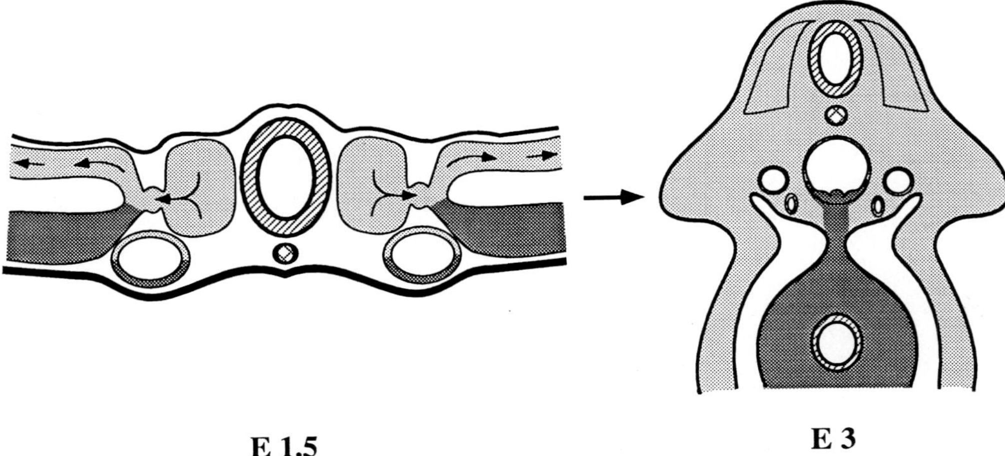

Fig. 7 Probable scheme of the origins of the two endothelial lineages responsible for vascularization of the embryo. The results do not formally exclude that cells of splanchnopleural origin do not also colonize the roof of the dorsal aorta. Light grey = regions invaded by angioblasts migrating from the somites. Dark grey = regions where endothelial precursors arise in situ.

Quek1 (or VEGF-R2) and c-ets1, have been carefully studied in the quail embryo. Functional information about c-ets1 indicates that this protooncogene plays a role in the command of migration, probably by activating metalloproteinases (Vandenbunder et al., 1996). In the present state of the art, these two genes provide convenient early markers because their expression becomes restricted to EC in the embryo. Prior to this restriction, the mesodermal expression of these two genes mostly overlaps. The only discrepancy is in somites where the nonoverlapping expression patterns are unexplained, especially if taking into account the demonstration by Wilting et al. (1995) that all quadrants of the somite are equivalent in their endothelial potential. Both c-ets1 and Quek1 are strongly expressed in the intermediate cell mass, the nephrogenic mesodermal subset. More work will be necessary to determine whether cells expressing c-ets1 and Quek1 there are migrating on their way from the somite to the somatopleural mesoderm, or become committed in situ.

In view of our findings, it will be interesting to determine whether precursors derived from the two different sources give rise to cells with different physiological properties. In this regard, it is noteworthy that the myogenic lineages deriving respectively from the somites and the splanchnopleura, that is, skeletal and smooth muscle, have vastly different properties and functions. Other problems to investigate in the future concern the developmental relationship between endothelial and hematopoietic lineages and the recognition mechanisms barring EC of somitic origin from entering territories of splanchnopleural origin.

References

Auerbach R., Kubai L., Knighton D., Folkman J. (1974): A simple procedure for the long-term cultivation of chicken embryos. Dev Biol 41:391–394.

Bagnall K.M. Higgins S., Sanders E.J. (1988): The contribution made by a single somite to the vertebral column: experimental evidence in support for resegmentation using the chick-quail chimaera model. Development 103:69–85.

Beaupain D., Martin C., Dieterlen-Lièvre F. (1979): Are developmental hemoglobin changes related to the origin of stem cells and site of erythropoiesis? Blood 53:212–225.

Brand-Saberi B., Seifert R., Grim M., Wilting J., Kühlewein M., Christ B. (1995): Blood vessel formation in the avian limb bud involves angioblastic and angiotrophic growth. Dev Dyn 202:181–194.

Carmeliet P., Ferreira V., Breier G., Pollefeyt S., Kieckens L., Gertsenstein M., Fahrig M., Vandenhoeck A., Harpal K., Ederhardt C., Declercq C., Pawling J., Moons L., Collen D., Risau W., Nagy A. (1996): Abnormal blood vessel development and lethality in embryos lacking a single VEGF allele. Nature 380:435–439.

Chevallier A., Kieny M., Mauger A. (1977): Limb-somite relationship: origin of the limb musculature. J Embryol Exp Morphol 41:245–258.

Christ B., Ordahl C.P. (1995): Early stages of chick somite development. Anat Embryol 191:381–396.

Clark E.R. (1918): Studies on the growth of blood vessels in the tail of the frog larva by observation and experiment on the living animal. Am J Anat 23:37–88.

Coffin J.D., Poole T.J (1988): Embryonic vascular development: immunohistochemical identification of the origin and subsequent morphogenesis of the major vessel primordia. Development 102:735–748.

Coffin J.D., Harrison J., Schwartz S., Heimark R. (1991): Angioblast differentiation and morphogenesis of the vascular endothelium in the mouse embryo. Dev Biol 148:51–62.

Coltey M., Jotereau F.V., Le Douarin N.M. (1987): Evidence for a cyclic renewal of lymphocyte precursor cells in the embryonic chick thymus. Cell Diff 22:71–82.

Couly G., Coltey P., Eichmann A., Le Douarin N.M. (1995): The angiogenic potentials of the cephalic mesoderm and the origin of brain and head blood vessels. Mech Dev 53:97–112.

Cuadros M.A., Moujahid A., Quesada A., Navascués J. (1994): Development of microglia in the quail optic tectum. J Comp Neurol 346:1–18.

Dieterlen-Lièvre F. (1994): Hematopoiesis during avian ontogeny. Poultry Sci Rev 5:273–305.

Dieterlen-Lièvre F., Martin C. (1981): Diffuse intraembryonic hematopoiesis in normal and chimeric avian development. Dev Biol 88:180–191.

Drake C.J., Davis L.A., Walters L., Little C.D. (1990): Avian vasculogenesis and the distribution of collagens I, IV, laminin and fibronectin in the heart primordia. J Exp Zool 255:309–322.

Drake C.J., Davis L.A., Little C.D. (1992): Antibodies to β1-integrins cause alterations of aortic vasculogenesis, in vivo. Dev Dyn 193:83–91.

Drake C.J., Cheresh D.A., Little C.D. (1995): An antagonist of integrin alpha (v) beta (3) prevents maturation of blood vessels during embryonic neovascularization. J Cell Sci 108:2655–2661.

Eichmann A., Marcelle C., Bréant C., Le Douarin N.M. (1993): Two molecules related to the VEGF receptor are expressed in early endothelial cells during avian embryonic development. Mech Dev 42:33–48.

Eichmann A., Marcelle C., Bréant C., Le Douarin N.M. (1996): Molecular cloning of Quek1 and 2, two quail vascular endothelial growth factor (VEGF) receptor-like molecules. Gene 174:3–8.

Emmel V.E. (1916): The cell clusters in the dorsal aorta of mammalian embryos. Am J Anat 19:401–421.

Evans H.M. (1909): On the development of the aorta, cardinal and umbilical veins, and other blood vessels of embryos from capillaries. Anat Rec 3:498–518.

Flamme I. (1989): Is extraembryonic angiogenesis in the chicken embryo controlled by the endoderm? Anat Embryol 180:259–272.

Flamme I., Risau W. (1992): Induction of vasculogenesis and hematopoiesis in vitro. Development 116:435–439.

Folkman J. (1974): Tumour angiogenesis. Adv Cancer Res 19:331–358.

Fong G.H., Rossant J., Gertsenstein M., Breitman M.L. (1995): Role of the Flt-1 receptor tyrosine kinase in regulating the assembly of vascular endothelium. Nature 376:66–70.

Garcia-Martinez V., Schoenwolf G.C. (1993): Primitive-streak origin of the cardiovascular system in avian embryos. Dev Biol 159:706–719.

Garcia-Porrero J.A., Godin I.E., Dieterlen-Lièvre F. (1995): Potential intraembryonic hemogenic sites at pre-liver stages in the mouse. Anat Embryol 192:425–435.

Hirakow R., Hiruma T. (1981): Scanning electron microscopy study on the development of primitive blood vessels in the chicken embryos at the early somite-stage. Anat Embryol 163:299–306.

Hirakow R., Hiruma T. (1983): TEM-studies on development and canalization of the dorsal aorta in the chicken embryo. Anat Embryol 166:307–315.

His W. (1886): Untersuchungen über die erste Anlage des Wirbeltierleibes. Leipzig, Germany: Vogel.

Houser J.W., Ackerman G.A., Knouff R.A. (1961): Vasculogenesis and erythropoiesis in the living yolk sac of the chicken embryo. Anat Rec 140:29–43.

Huang R., Zhi Q., Wilting J., Christ B. (1994): The fate of somitocoele cells in avian embryos. Anat Embryol 190:243–250.

Inagaki T., Garcia-Martinez V., Schoenwolf G.C. (1993): Regulative ability of the prospective cardiogenic and vasculogenic areas of the primitive streak during avian gastrulation. Dev Dyn 197:57–68.

Jacob M., Christ B., Jacob H.J. (1979): The migration of myogenic stem cells from the somites into the leg region of avian embryos. Anat Embryol 157:291–309.

Jordan H.E. (1916): The microscopic structure of the yolk-sac of the pig embryo with special reference to the origin of the erythrocytes. Am J Anat 19:277–299.

Jotereau F.V., Le Douarin N.M. (1978): The developmental relationship between osteocytes and osteoclasts: a study using the quail-chick nuclear marker in endochondral ossification. Dev Biol 63:253–265.

Labastie M.C. (1989): MB1, a quail leukocyte-endothelium antigen: further characterization of soluble and cell-associated forms. Cell Diff Dev 27:151–162.

Lance-Jones C. (1988): The somitic level of origin of embryonic chick hind limb muscles. Dev Biol 126:394–407.

Le Douarin N.M. (1973): A biological cell labelling technique and its use in experimental embryology. Dev Biol 30:217–222.

Le Douarin N.M., Jotereau F. (1973): Origin and renewal of lymphocytes in avian embryo thymuses. Nature New Biol 246:25–27.

Le Lièvre C., Le Douarin N.M. (1975): Mesenchymal derivatives of the neural crest: analysis of chimaeric quail and chicken embryos. J Embryol Exp Morphol 34:125–154.

Leprince D., Gegonne A., Coll J., De Taisne C., Schneeberger A., Lagrou C., Stéhelin D. (1983): A putative second cell-derived oncogene of the avian leukemia virus E26. Nature 306:395–397.

Linask K.K., Lash J.W. (1993): Early heart development: dynamics of endocardial cell sorting suggests a common origin with cardiomyocytes. Dev Dyn 196:62–69.

Luton D., Sibony O., Oury J.F., Blot P., Dieterlen-Lièvre F., Pardanaud L. (1996): The c-ets1 protooncogene is expressed in human trophoblast during the first trimester of pregnancy. Early Hum Dev 47:147–156.

Maroulakou I.G., Papas T.S., Green J.E. (1994): Differential expression of ets-1 and ets-2 protooncogenes during murine embryogenesis. Oncogene 9:1551–1565.

Metcalf D., Moore M.A.S. (1971): Embryonic aspects of haemopoiesis. In: Haemopoietic cells, Neuberger A, Tatum EL, eds. Amsterdam: North Holland Publ.

Mucenski M.L., McLain K., Kier A.B., Swerdlow S.H., Schreiner C.M., Miller T.A., Pietryga D.W., Scott W.J., Potter S. (1991): A functional c-myb gene is required for normal murine fetal hepatic hematopoiesis. Cell 65:677–689.

Murray P.D.F. (1932): The development in vitro of the blood of the early chicken embryo. Proc R Soc Lond Biol 111:497–521.

Noden D.M., (1989): Embryonic origins and assembly of embryonic blood vessels. Ann Rev Pulmon 140:1097–1103.

Noden D.M. (1991): Origins and patterning of avian outflow tract endocardium. Development 111:867–876.

Nunn M.F., Seeburg P.M., Moscovici C., Duesberg P. (1983): Tripartite structure of the avian erythroblastosis virus E26. J Virol 63:398–402.

Pardanaud L., Altmann C,. Kitos P., Dieterlen-Lièvre F., Buck C. (1987): Vasculogenesis in the early quail blastodisc as studied with a monoclonal antibody recognizing endothelial cells. Development 100:339–349.

Pardanaud L., Yassine F., Dieterlen-Lièvre F. (1989): Relationship between vasculogenesis, angiogenesis and haemopoiesis during avian ontogeny. Development 105:473–485.

Pardanaud L., Dieterlen-Lièvre F. (1993a): Emergence of endothelial and hematopoietic cells in the avian embryo. Anat Embryol 187:107–114.

Pardanaud L., Dieterlen-Lièvre F. (1993b): Expression of c-ets1 in early chicken embryo mesoderm: relationship to the hemangioblastic lineage. Cell Adh Comm 1:151–160.

Pardanaud L., Dieterlen-Lièvre F. (1995): Does the paraxial mesoderm of the avian embryo have hemangioblastic capacities? Anat Embryol 192:301–308.

Pardanaud L., Luton D., Prigent M., Bourcheix L.M., Catala M., Dieterlen-Lièvre F. (1996): Two distinct endothelial lineages in ontogeny, one of them related to hematopoiesis. Development 122:1363–1371.

Patan S., Haenni B., Burri P.H. (1993): Evidence for intussusceptive capillary growth in the chicken chorio-allantoic membrane (CAM). Anat Embryol 187:121–130.

Péault B., Thiery J.P., Le Douarin N.M. (1983): A surface marker for the hematopoietic and endothelial cell lineages in the quail species defined by a monoclonal antibody. Proc Natl Acad Sci USA 80:2976–2980.

Péault B., Coltey M., Le Douarin N.M. (1988): Ontogenic emergence of a quail leukocyte/endothelium cell surface antigen. Cell Diff 23:165–174.

Poole T.J., Coffin D. (1989): Vasculogenesis and angiogenesis: two distinct morphogenetic mechanisms establish embryonic vascular pattern. J Exp Zool 251:224–231.

Quéva C., Leprince D., Stéhelin D., Vandenbunder B. (1993): p54$^{c\text{-}ets1}$ and p68$^{c\text{-}ets1}$, the two transcription factors encoded by the c-ets1 locus, are differentially expressed during development of the chicken embryo. Oncogene 8:2511–2520.

Reagan F.P. (1915): Vascularization phenomena in fragments of embryonic bodies completely isolated from yolk sac blastoderm. Anat Rec 9:229–241.

Risau W., Flamme I. (1995): Vasculogenesis. Annu Rev Cell Dev Biol 11:73–91.

Risau W., Lemmon V. (1988): Changes in the vascular extracellular matrix during embryonic vasculogenesis and angiogenesis. Dev Biol 125:441–450.

Romanoff A. (1960): The Avian Embryo. New York: The MacMillan Company.

Roncali L., Virgintino D., Coltey P., Bertossi M., Errede M., Ribatti D., Nico B., Mancini L., Sorino S., Riva A. (1996): Morphological aspects of the vascularization in intraventricular neural transplants from embryo to embryo. Anat Embryol 193:191–203.

Sabin F. (1917): Origin and development of the primitive vessels of the chick and of the pig. Contrib Embryol 18:62–124.

Sabin F. (1920): Studies on the origin of the blood vessels and of chick during the second day of incubation. Control Embryol 9:215–262.

Schramm C., Solursh M. (1990): The formation of premuscle masses during chick wing bud development. Anat Embryol 182:235–247.

Shalaby F., Rossant J., Yamaguchi T., Gertsenstein M., Breitman M.L., Schuh A.C. (1995): Failure of blood island formation and vasculogenesis in Flk-1 deficient mice. Nature 376: 62–66.

Solursh M., Drake C., Meier S. (1987): The migration of myogenic cells from the somites at the wing level in avian embryos. Dev Biol 121:389–396.

Solter D., Knowles B.B.(1975): Immunosurgery of mouse blastocyst. Proc Natl Acad Sci USA 72:5099–5102.

Sugi Y., Markwald R.R. (1996): Formation and early morphogenesis of endocardial endothelial precursor cells and the role of endoderm. Dev Biol 175:66–83.

Tavian M., Coulombel L., Luton D., San Clémente H., Dieterlen-Lièvre F., Péault B. (1996): Aorta-associated CD34+ hematopoietic cells in the early human embryo. Blood 87: 67–72.

Vandenbunder B., Pardanaud L., Jaffredo T., Mirabel M.A., Stéhelin D. (1989): Complementary patterns of expression of c-ets1, c-myb and c-myc in the blood-forming system of the chicken embryo. Development 107:265–274.

Vandenbunder B., Quéva C., Desbiens X., Wernert N., Stéhelin D. (1996): Expression of the transcription factor c-ets1 correlates with the occurrence of invasive processes during normal and pathological development. Inv Metast 95:198–209.

Wilting J., Brand-Saberi B., Huang R., Zhi Q., Köntges G. (1995): Angiogenic potential of the avian somite. Dev Dyn 202:165–171.

1.4

Methodology for the Study of Vascular Morphogenesis In Vivo

M.C. DeRuiter, R.E. Poelmann, and A.C. Gittenberger-de Groot

Introduction

Approximately 1% of children are born with a congenital heart disease, often in combination with abnormal vascular patterning and vessel wall anomalies (Hoffman, 1995). This shows that the development of heart and blood vessels is a critical event during embryonic and fetal life. From transgenic mice with a lethal phenotype, about 50% die of cardiovascular maldevelopment in utero (Copp, 1995). Development of diagnostic techniques as echo Doppler ultrasound and nuclear magnetic resonance has greatly aided clinical examination of vascular anomalies. Interpretation of morphological data, however, is facilitated if possible variations and abnormalities of the vascular system are placed into the broader context of normal development. Moreover, knowledge on in vivo cardiovascular development is necessary to understand the pathogenesis of the cardiovascular system and the role of specific gene products in the numerous transgenic animals produced by the molecular biologist.

With this in mind we examined the origin of a pulmonary vascular system that is present in pulmonary atresia (Fig. 1A). If the outflow tract of the right ventricle is obstructed, systemic venous blood from the body cannot flow directly toward the lungs for gas exchange. Systemic blood has to flow to the left side of the heart by an open foramen ovale in the atrial septum and/or a ventricular septal defect between the left and right ventricle. In these cases, blood supply to the lungs can be incurred by abnormal vessels. These may consist of a patent ductus arteriosus or by systemic pulmonary collateral arteries (SPCAs). The SPCAs connect the descending aorta, subclavian artery or aortic arch on the one hand and the intrapulmonary vasculature on the other hand (Liao et al., 1985; Rossi et al., 1988; Sotomora and Edwards 1978; Thiene et al., 1977).

The patent ductus arteriosus is anomalous in that it fails to close after birth. The normal ductus, therefore, must be specialized. It connects the pulmonary trunk with the aortic arch, which shunts blood from the right ventricle to the systemic circulation. Although the ductus arteriosus arises from the system of pharyngeal arch arteries, its muscular histology and the presence of intimal thickenings (Gittenberger-de Groot et al., 1980, 1985) differ completely from that of the other derivatives such as the elastic aortic arch, pulmonary trunk, and carotid arteries (Bergwerff et al., 1996). The histology of the ductus arteriosus facilitates its functional closure after birth. Sponta-

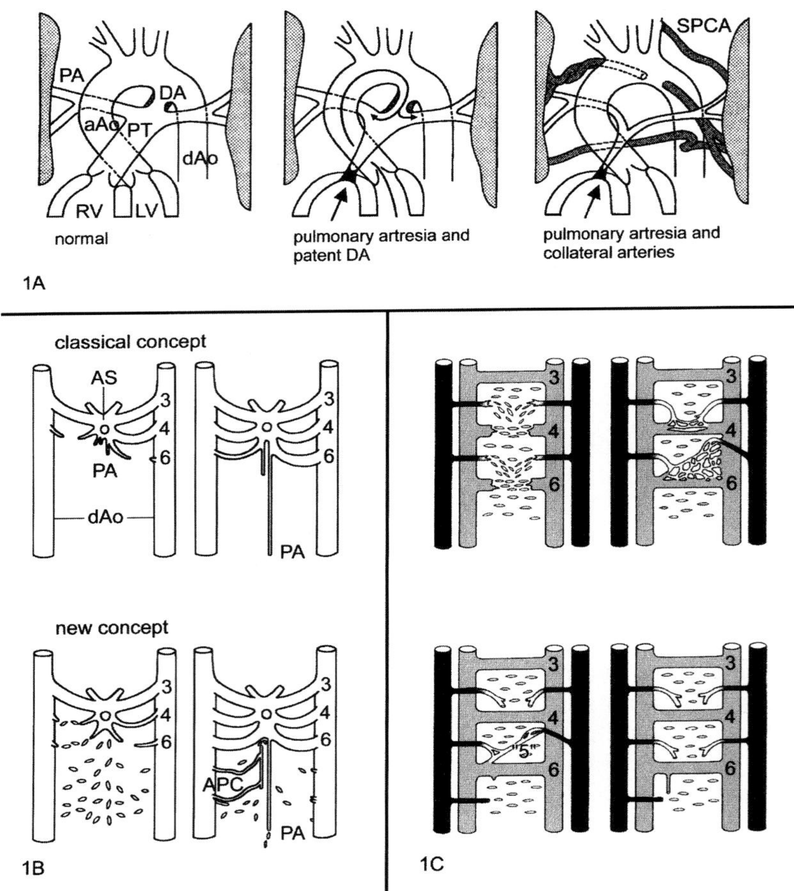

Fig. 1 (A)The lungs are supplied by the pulmonary trunk (PT) which divides into a left and right pulmonary artery (PA). Before birth the ductus arteriosus (DA) shunts the blood to the systemic circulation. After birth the ductus arteriosus closes. In case of pulmonary atresia (arrow) the lungs are supplied either by a patent ductus arteriosus or by a number of systemic-pulmonary collateral arteries (SPCAs). Ascending aorta (aAo); descending aorta (dAo). (B) The classical concept describes that the pulmonary artery (PA) grows as a lumenized sprout by angiogenesis from the aortic sac (AS) into the lung area. In the meantime pairs of pharyngeal arch arteries (3,4,6) develop as vessels around the foregut connecting the arterial pole of the heart with the two dorsal/descending aortae (dAo). New data, based on QH1 staining show that the foregut is covered with endothelial precursors, which lumenize and give rise to transient aortopulmonary connections (APC). Ventrally these vessels interconnect and give rise to the pulmonary artery, which retrogradely joints the aortic sac. (C) The splanchnic plexus around the foregut of the embryo gives rise to the pharyngeal arch arteries (3,4,6). During development each pharyngeal arch artery is connected to the surrounding veins (black vessels). After establishment of the artery the arterial-venous connections disappear. The plexus between the fourth and the sixth pharyngeal arch artery is often interpreted as the missing fifth pharyngeal arch artery ("5"). In none of the cases studied was found evidence for an additional mesenchymal arch.

neous closure of the ductus arteriosus is initialized by a cascade of molecular and cellular processes to ensure that all the right ventricular blood will flow to the lungs immediately after birth (de Reeder et al., 1988; Slomp et al., 1992).

In cases of pulmonary atresia, the ductus arteriosus can supply the lungs by a retrograde blood flow before birth; whereas normal closure of the ductus after birth will hamper the blood flow resulting in anoxia of the child. SPCAs are often present in patients with pulmonary atresia lacking a ductus arteriosus (DeRuiter et al., 1994; Rossi et al., 1988; Liao et al., 1985). Although the course of these SPCAs is highly variable, the histology of such vessels resembles the ductus arteriosus in that they have a muscular vessel wall or at least muscular segments; furthermore, the vessels usually develop intimal thickenings (Sullivan et al., 1988; Puga et al., 1989; DeRuiter et al., 1994).

We recently addressed the question whether SPCAs are enlarged bronchial arteries, which are remodeled into collaterals during embryonic development, or whether they result from persistence of normally transient aortopulmonary vessels. Classic literature (Bremer, 1912; Buell, 1922; Congdon, 1922; Huntington, 1920) cannot explain the origin of congenital SPCAs satisfactorily. We investigated the development of the system of pharyngeal arch arteries and pulmonary vessels using serial sections and India ink injections at successive developmental stages. Three-dimensional reconstructions were employed to demonstrate the changing relationships between these vascular systems. This method allows visualization of the luminal part of the developing vasculature only (Part 2).

The above methods, however, are not sufficient to describe early formation of the vasculature. Endothelial cells, or rather, endothelial precursors presage the formation of the vascular system. The first indication of newly developing vessels can be acquired by immunohistochemistry. The availability of an antibody (QH1) (Pardanaud et al., 1987) that recognizes quail endothelial cells and their precursors was a major advance in the study of vascular morphogenesis in vivo (Part 3).

Using QH1, we observed a remarkable phenomenon—during media formation differentiating smooth muscle cells still retain the QH1 antigen. This finding led to the hypothesis that endothelial cells can transdifferentiate into smooth muscle cells. Differentiation independent cell-lineage markers are required to study this hypothesis. The contribution of endothelial cells to the tunica media and the shift from an endothelial fate toward a smooth muscle specific expression pattern could be studied. Colloidal gold endocytosed by endothelial cells and a retroviral-based infection of the LacZ reportergene of endothelial cells, in combination with immunohistochemistry, were used in this part of the study (Part 4).

Established Techniques

The Origin of Collateral Arteries

One of the oldest, and most important methods for microscopical examination of the developing embryo is serially sectioning of embryos of successive stages. The nuclei and cytoplasm can be visualized by standard stainings like hematoxylin/eosin, vital red, or azan (smooth muscle cells), while certain extracellular matrix components can be studied with resorcin/fuchsin (elastin) and again azan (fibrillar collagens). The development of the vascular system in the embryo starts with the formation of simple tubes, consisting only of a monolayer of endothelial cells. At later stages surrounding mesenchyme will differentiate into condensed layers of smooth muscle cells. Using classical staining techniques the vessels without a media can only be recognized by the presence of a lumen they include. In combination with plastic embedding techniques to obtain thin sections, a very small lumen diameter of 2 μm can be detected in developing endothelial vessels.

We recently reinvestigated the normal development of the pulmonary and bronchial vasculature using rodent and avian embryos as models for human development to explain the variability in course, origin, and histology of the SPCAs (DeRuiter et al., 1989, 1993a). Rat and chick embryos' vascular beds were injected with a diluted solution of India ink. After injection the embryos were cultured for 20 minutes to achieve a maximal attachment of the India ink to the luminal surface of the endothelium. The advantage of this technique is that complicated three-dimensional aspects of the vascular system can be studied in toto, and even the smallest lumenized vessels can be traced in histological sections. Classical theories (Buell 1922; Congdon, 1922) describe that the pulmonary arteries arise by capillary outgrowth from the aortic sac that invests the lung (Fig. 1B). We found that, before the lumen of the intrapulmonary arteries is outlined, numerous lumenized vessels exist between the dorsal aortae (the precursors of the future descending aorta) and the lungs. From this plexus, the pulmonary arteries develop, which in turn retrogradely connect with the aortic sac. After establishment of a pulmonary circulation, the aortopulmonary connections disappear. Later, the aortic sac will be septated into the pulmonary trunk and the ascending aorta. From these data, it can be hypothesized that in case of pulmonary-outflow-tract obstruction the aortopulmonary connections can persist as SPCAs. The development of an animal model with induced pulmonary atresia will be of great help to study the histogenesis, and mechanism of progressive stenosis.

The Myth of the Fifth Pharyngeal Arch Artery

Recent work established that the development of the arterial system is matched to that of the veins and, therefore, these events cannot be considered as two separate processes (DeRuiter et al., 1993a). Each time an artery, e.g., pharyngeal arch artery or pulmonary artery, arises the developing sprout is connected to various neighboring venous vessels. These arterial-venous connections will disappear during further establishment of the artery (Fig. 1C). The existence of transient connections to the venous system seems to be a general phenomenon that is obligatory to the formation of organ-related arterial systems in the embryo. This observation implies that the initial endothelial vasculature exhibits a high degree of plasticity to remodel into more definitive vessels.

The plasticity can probably explain the enigma of the existence of the fifth pair of pharyngeal arch arteries in amniotes. Many investigators have speculated on the existence of this pair of vessels (Congdon, 1922; Gerlis et al., 1989; Macartney et al., 1974; Reagan, 1912). The fifth artery can be considered as an "outsider" in the pharyngeal arch system. The arches are widely considered to contain six pairs of arteries. The fifth pair, with its highly variable course, develops only after the establishment of the sixth artery. Each of the developing pharyngeal arch arteries (II to VI) initially contacts the venous system by a complex of small vessels (DeRuiter et al., 1993a). In case of the fifth arch, these additional vessels can give the impression of the presence of an additional arch artery. This confusion is understandable; but in our view, it is incorrect to interpret the plexus between the fourth and the sixth pharyngeal arch arteries as a transient fifth artery. We base this statement on the fact that such connections, although less distinct, are present in the mesenchyme of all the other arches. The function of these transient vessels is as yet unknown, but their presence suggests the possibility that anomalous definitive structures could arise from such vessels. During experiments in which we ligated a major yolk sac vein, reversal of blood flow through connecting capillaries was observed. Alteration in blood flow through preferential capillaries will cause a detour of blood via one of the other vessels in the plexus (Hogers et al., 1997). With this observation in mind we hypothesize that slight changes in normal vessel patterning might result in pharyngeal arch anomalies such as double-lumen left aortic arch or SPCAs from various derivatives of the pharyngeal arch arteries.

Scanning Electron Microscopy and Immunohistochemistry

Formation of the Initial Vascular Network

The formation of a new lumenized vessel can be described by two processes: vasculogenesis and angiogenesis. The first process concerns the in situ differentiation of endothelial cells in the splanchnic mesodermal compartment (Risau, 1991; Poole and Coffin, 1989, 1991). Experimental data indicate the necessity of an endoderm–mesoderm interface for vasculogenesis (Pardanaud et al., 1989). The initial population of endothelial cells arises in both the splanchnic mesoderm of the intra-embryonic and extra-embryonic yolk sac and extends toward the embryo (Poole and Coffin, 1991; Drake and Jacobson, 1988). The endothelial precursor cells proliferate and align to form the lining of the vitelline vessels, the first pair of pharyngeal arch arteries, the dorsal aortae, the endocardium, and the primary vascular network. It remains to be elucidated whether the endocardium arises from a special subpopulation within the heart-forming fields.

New vessels in the somatic mesoderm (mesoderm–ectoderm interface) may arise by outgrowth of sprouts from existing lumenized vessels after the establishment of the initial intra-embryonic vasculature (Pardanaud et al., 1989; Poole and Coffin 1991). At the tip of angiogenic sprouts, isolated endothelial precursors are present as shown by a combination of vascular corrosion casts and immunohistochemistry (DeRuiter et al., 1991). The corrosion casts were obtained by injecting a prepolymerized methacrylate, Mercox®, into the vascular system. Polymerization of the plastic was followed by digestion of the surrounding tissue. The casts were studied in the scanning electron microscope. Blind-ended sprouts were often observed. At the tip of similar sprouts, a number of endothelial precursors were identified by immunohistochemistry. The precursor cells appear to assemble and become incorporated in the developing sprout. This process is called angiogenesis and is thought to be the major mechanism of vessel formation during physiological and pathological vessel formation during life (Risau, 1991).

Vasculogenesis can be demonstrated elegantly by scanning electron microscopy (DeRuiter et al., 1993b). Endothelial precursors, both strands and isolated cells, appear as elongated cells. These are located in the splanchnic mesoderm where they differentiate and delaminate from the mesothelium. Careful dissection of mesothelium or endoderm exhibits the alignment of the endothelial precursor cells in the yolk sac and around the foregut of the embryo (Fig. 2A). Quail endothelial cells express a partially characterized cytoplasmic and cell surface epitope, which is recognized by the QH1 antibody (Pardanaud et al., 1987). The antigen is also specifically expressed by hemopoietic cells and the endothelial precursors and can be used to great advantage to study vascular morphogenesis in vivo (Fig. 2B). The development of the left and right splanchnic mesodermal compartments from laterocaudal to mediocranial is seen in successive embryonic stages (DeRuiter et al., 1992, 1993). Both compartments fuse cranially from the buccopharyngeal membrane, resulting in a singular horseshoe-shaped plexus. The next step in vasculogenesis is the formation of endothelial-lined tubes, arising in the yolk sac as the

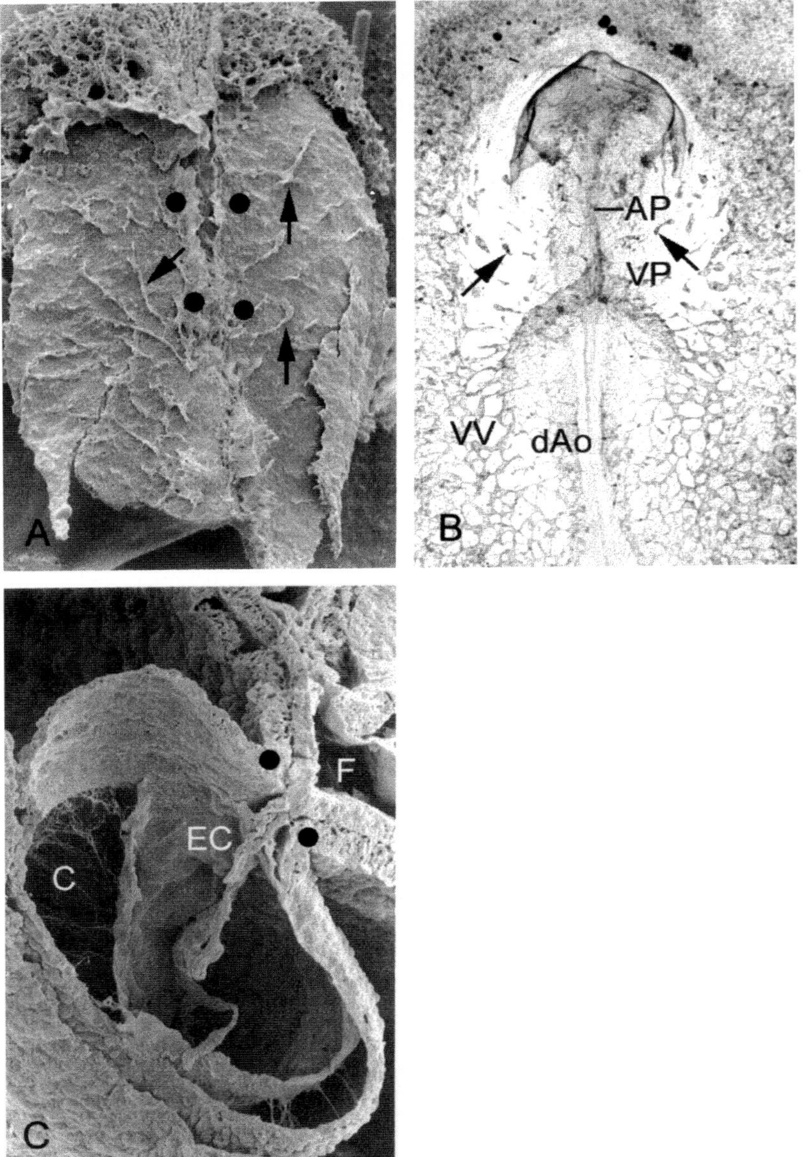

Fig. 2 (A) SEM picture of the splanchnic plexus consisting of isolated and strands of endothelial precursors (arrows) around embryonic foregut. The heart tube of this quail embryo (7-sometimes) has been removed. (B) Quail embryo (5-somites) incubated in toto with the QH1 antibody. Both the vitelline vessels (VV), the dorsal/descending aortae (dAo), and the venous pole of the heart (VP) are lumenized. The formation of the arterial pole (AP) is still under progress, but does not consist of two tubes. Endothelial precursors (arrows). (C) SEM picture of a transversely sectioned quail embryo, approximately 10-somites. In the area of the dorsal mesocardium (between the dots) (see also Fig. 2A), the endocardial cells (EC) are firmly attached to the endoderm of the forgut (F).

vitelline veins. The spatial remodeling of the developing foregut enforces the lumenizing plexus to change its spatial connections concomitantly. Superficial observation might give the erroneous impression of two fusing endocardial heart tubes in the midline of the embryo. The remaining parts of the vascular plexus (i.e., splanchnic plexus) is translocated around the foregut where it serves as a template for the pharyngeal arch arteries and the dorsal aortae. After formation of the dorsal aorta and the first pair of pharyngeal arch arteries, many endothelial precursors from the initial vascular plexus are still located around the foregut (DeRuiter et al., 1993a). It is possible that these precursors can be translocated and used as the elements for future vessels, such as the second to sixth pair of pharyngeal arch arteries and the pulmonary vessels. Outgrowth of endothelial sprouts from existing vessels by angiogenesis into an area lacking endothelial precursors has never been observed. It cannot be ruled out, however, that vasculogenesis plays an additional role in the formation of the splanchnic vessels.

Vessel formation by Deposition of Endothelial Cells and by Splitting up of Vessels

A clear example of vessel formation by deposition of endothelial cells has been observed during the disruption of the dorsal mesocardium of the heart (Fig. 4C). At first, the endocardial heart tube is connected over its full length to the endoderm of the foregut. Subsequently, the heart tube, between the arterial and venous pole, disconnects from the foregut. A strand of endocardial cells remains positioned ventral to the foregut after the endocardial tube is separated from the endoderm. This so-called midpharyngeal endothelial strand (MPES) stays connected to the sinus venosus and the aortic sac region and remains part of the earlier described splanchnic plexus. During further development, the MPES lumenizes and gives rise to the common pulmonary vein (DeRuiter et al., 1995). This method of vessel formation has received little attention up till now. However, this mechanism may be more generally used than previously assumed. This possibility is supported by the increase of capillary vessels in the lung (Burri and Tarek, 1990) and heart (Van Groningen et al., 1991) in newborn rat and in the chicken embryonic chorioallantoic membrane (Pattan et al., 1993). During fast postnatal growth the number of capillaries increases by splitting up a large lumenized capillaries into multiple smaller ones. This process starts with small pits in the endothelial wall and ends up with holes. Thereafter the vessels are pushed apart by the myocytes, resulting in a doubling of the vessel number, while angiogenesis and vasculogenesis do not play a role.

Identifying Endothelial Cells

Endothelial Cell Mobility

Although many approaches are available to determine the fate of embryonic cellular compartments, or even individual cells, only a few methods can be used for the study of endothelial cell lineage in the embryo. The use of quail-chicken chimeras has shown to be a powerful tool to study the potency of vasculogenesis and angiogenesis in various embryonic tissues (see Chapter 1.3) and the mobility of endothelial cells within the vascular system. This last process is beautifully demonstrated by Noden (1989) who transplanted quail head mesenchyme into a chick host. Endothelial precursors were present in the head mesenchyme at the moment of transplantation. Grafted endothelial cells were observed to take part in the endothelial lining of the pharyngeal arch arteries and also in the endocardium of the heart. This implies that the endothelial cells can move upstream along the arterial vessel wall.

Similar results were obtained with wing bud transplantation experiments. Quail endothelial cells of the isotopically transplanted quail wing bud were traced in the descending aorta of the chicken host. No grafted endothelial cells were found in the adjacent cardinal veins. Injection of a peptide containing the integrin-blocking RGD sequence proximal to the site of transplantation blocks the adhesion of endothelial cells to the extracellular matrix. In those cases the grafted endothelial cells were not incorporated into the aorta (Christ et al., 1990). A possible explanation is that the arteries increase in size by recruitment of endothelial cells and their precursors from the peripheral pool of endothelial cells in contrast to the veins. This is in line with the low proliferation rate of the endothelial cells in the descending aorta (33%). The mitotic activity of endothelial cells both in peripheral capillary vessels and the veins is much higher (69% and 52%, respectively) (Gittenberger-de Groot et al., 1995). The latter portions of the vascular bed appear to be the most active sites in vascular remodeling as described in the previous section.

Tunica Media Formation Receives Contributions from Three Cell Lineages

The descending aorta is the first intra-embryonic vessel that acquires a tunica media, consisting of smooth muscle cells (Hughes, 1942). Media formation starts at morphogenetically quiet sites that are no longer involved in angiogenetic and vascular remodeling processes (DeRuiter et al., 1990). At first, the endothelial tube is wrapped by mesenchymal cells (Thayer et al., 1995). Most of these cells are believed to derive from the splanchnic mesoderm ventrally to the aorta (Hungerford et al., 1996). During migration, these cells start to express proteins as (α-smooth) muscle actins (DeRuiter et al., 1990), desmin (Sumida, 1988), and the 1E12 antigen (Hungerford et al., 1996), which are also present in adult smooth muscle cells. After reaching their endothelial target the mesenchymal cells form compact cellular layers. During subsequent maturation other proteins are expressed to complete the smooth muscle cell phenotype (Duband et al., 1993; Owens, 1995; Slomp et al., 1997; Glukhova and Koteliansky 1995). At the onset of media formation we observed a remarkable phenomenon in the descending aorta of a stage 14 quail embryo (DeRuiter et al., 1997). Next to the endothelium, a population of subendothelial mesenchymal cells expressed a thin line of endothelium-specific QH1. Ultrastructural analysis of this population showed that the expression of the antigen was cell-membrane bound, while cytoplasmic antigen expression, characteristic or endothelial cells, was not detectable. Subsequently, (stage 15–16), the subendothelial cells started to coexpress α-smooth muscle actin (Fig. 3A). The subendothelial QH1 expression was lost after condensation of the mesenchyme into a media. These data led to the hypothesis that during the earliest stages of media formation, endothelial cells undergo a mesenchymal transformation. The transforming cells delaminate from the endothelium and migrate into the subendothelial space, where they loose endothelial specific expression and begin to express contractile proteins. We experimentally labeled the endothelial cells with wheatgerm agglutinin conjugated to gold particles (10–12 nm) to follow endothelial–mesenchymal transformation in the descending aorta. The endothelial cells endocytose the colloidal-gold complexes that become aggregated into large intracellular vesicles. Such vesicles do not participate in exocytosis. Leakage of colloidal gold through fenestrae between the endothelial cells was not observed. These observations imply that the fate of the previous endothelial cells could be traced after they had lost the endothelial-specific phenotype. Labeled mesenchymal cells expressing (α-smooth) muscle actin were traced to a subendothelial position (Fig. 3B) within 24 hours after injection of the colloidal-gold into the vascular system of the quail embryo. This indicates that endothelial cells contribute to the formation of the mesenchymal vessel wall and most probably

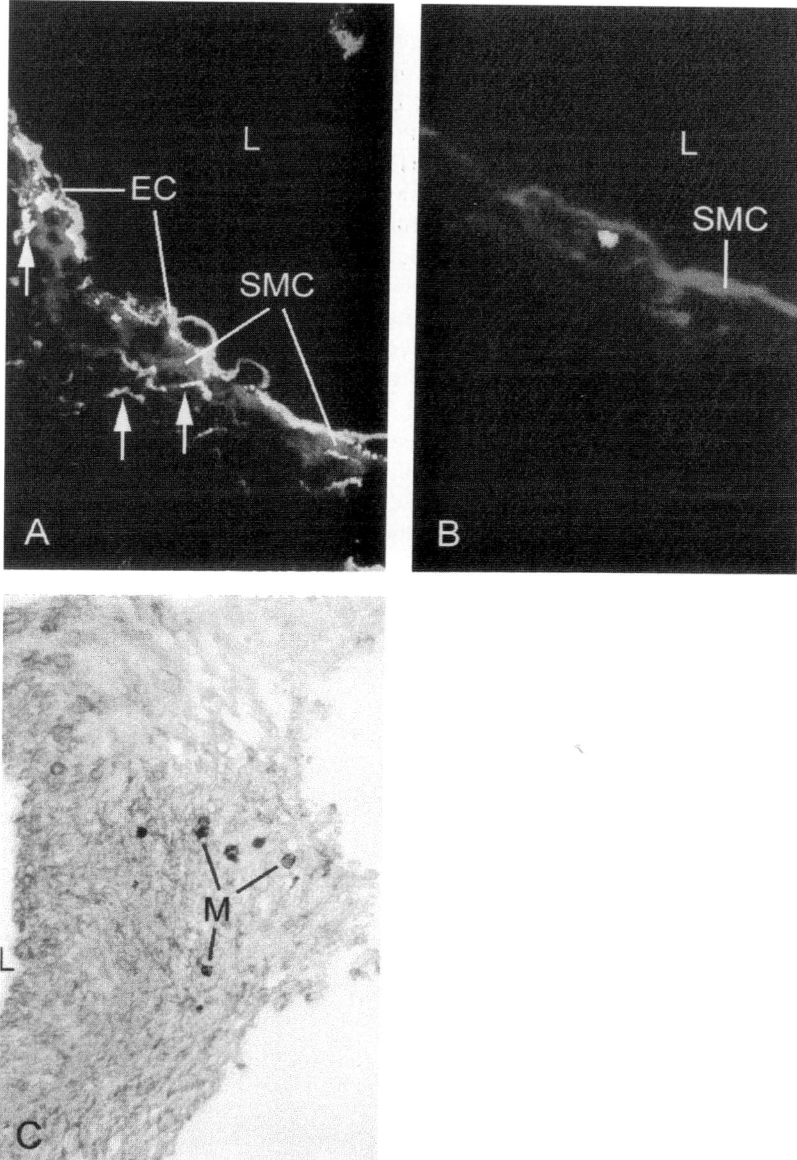

Fig. 3 (A) Descending aorta of a quail embryo incubated with the endothelial specific QH1 antibody (white) and antibody against alpha smooth muscle actin (grey). QH1 is not only expressed by the endothelial cells (EC), but also in the cellular membranes (arrows) of the differentiating smooth muscle cells (SMC). (B) The endothelial cells were labeled with colloidal gold by an injection into the lumen (L) of the vasculature. After 24 hours, labeled mesenchymal cells containing a colloidal gold vesicle (white spot) are traced subendothelially. These cells also express muscle specific actins (SMC). (C) If endothelial cells are infected with a replication incompetent retrovirus carrying LacZ labelled medial cells (M), expressing alpha mooth muscle actin, they can be traced 10 days after injection.

differentiate into smooth muscle cells. However, with the colloidal-gold technique, the fate of the endothelial-derived mesenchymal cells could only be studied up to 24 hours after injection. In this limited time period, the mesenchymal cells did not express SM22, h-caldesmon, calponin (Duband et al., 1993), or smootheline (van der Loop et al., 1996), molecules that would define them conclusively as fully differentiated smooth muscle cells. Moreover, during this short time span, no definitive outer and inner border of the tunica media can be determined. In a subsequent study, we tagged chicken endothelial cells with a replication incompetent retrovirus containing a LacZ reporter gene (Mikawa et al., 1991). With the LacZ tag, the contribution of the endothelium to the vessel wall could be studied now over a much longer period. For example, neural crest cells were labeled to trace the fate of the pharyngeal arch smooth muscle cells of the region and the mesenchyme of the aortopulmonary septum (Gittenberger-de Groot et al., 1995).

Retroviral labeling of endothelial cells confirmed the data found with colloidal gold. In the descending aorta at stage 36, labeled cells were traced in the entire tunica media (Fig. 3C) where they appeared to be randomly distributed. In addition, labeled cells express to (α-smooth) muscle actins, vinculin and desmin. In those areas with a neural crest smooth muscle cell contribution (Le Lièvre and Le Douarin, 1975; Philips et al., 1987; Rosenquist and Beall, 1990), endothelial-derived smooth muscle cells were not distributed at random, but were mainly situated in the tunica intima and the outermost layer of the tunica media.

Epilogue

Further investigations are necessary to detect the full impact of the multifaceted origin of smooth muscle cells. What is the significance of three separate smooth muscle lineages (i.e., splanchnic mesoderm, cardiac neural crest, and endothelium) for maturation of the vessels, pathogenesis, and congenital malformations? Is the topological prevalence or degree of atherosclerosis related to smooth muscle origin? Whatever the case, these data show that the role of embryonic endothelial cells has to be extended. The potency of endothelial cells to transform into mesenchymal cells suggest a new role in vascular processes.

References

Bergwerff M., DeRuiter M.C., Poelmann R.E., Gittenberger-de Groot A.C. (1996): Onset of elastogenesis and downregulation of smooth muscle actin as distinguishing phenomena in artery differentiation in the chick embryo. Anat Embryol 194:545–557.

Bremer J.L. (1912): Aorta and aortic arches in rabbits. Am J Anat 13:111–128.

Buell C.E. (1922): Origin of the pulmonary vessels in the chick. Carnegie Inst Contrib Embryol 14:13–26.

Burri P.H., Tarek M.R. (1990): A novel mechanism of capillary growth in the rat pulmonary microcirculation. Anat Rec 228:35–45.

Christ B., Poelmann R.E., Mentink M.M.T., Gittenberger-de Groot A.C. (1990): Vascular endothelial cells migrate centripetally within embryonic arteries. Anat Embryol 181: 333–339.

Congdon E.D. (1922): Transformation of the aortic arch system during the development of the human embryo. Carnegie Contrib Embryol 14:47–110.

Copp A.J. (1995): Death before birth: clues from gene knockouts and mutations. Trends Genet 11:87–93.

De Reeder E.G., Girard N., Poelmann R.E., van Munsteren J.C., Patterson D.F., Gittenberger-de Groot A.C. (1988): Hyaluronic acid accumulation and endothelial cell detachment in intimal thickening of the vessel wall. The normal and genetically defective ductus arteriosus. Am J Pathol 132:574–585.

DeRuiter M.C., Gittenberger-de Groot A.C., Rammos S., Poelmann R.E. (1989): The special status of the pulmonary arch artery in the branchial arch system of the rat. Anat Embryol 179:319–325.

DeRuiter M.C., Poelmann R.E., van Iperen L., Gittenberger-de Groot A.C. (1990): The early development of the tunica media in the vascular system of rat embryos. Anat Embryol 181:341–349.

DeRuiter M.C., Hogers B., Poelmann R.E., VanIperen L., Gittenberger-de Groot A.C. (1991): The development of the vascular system in quail embryos: A combination of microvascular corrosion casts and immunohistochemical identification. Scanning Microsc 4:1081–1090.

DeRuiter M.C., Poelmann R.E., VanderPlas-de Vries, I., Mentink M.M.T., Gittenberger-de Groot A.C. (1992): The development of the myocardium and endocardium in mouse embryos. Fusion of two heart tubes? Anat Embryol 185:461–473.

DeRuiter M.C., Gittenberger-de Groot A.C., Poelmann R.E., van Iperen L., Mentink M.M.T. (1993a): Development of the pharyngeal arch system related to the pulmonary and bronchial vessels in the avian embryo. Circulation 87:1306–1319.

DeRuiter M.C., Poelmann R.E., Mentink M.M.T., van Iperen L., Gittenberger-de Groot A.C. (1993b): Early formation of the vascular system in quail embryos. Anat Rec 235:261–274.

DeRuiter M.C., Gittenberger-de Groot A.C., Bogers A.J.J.C. (1994): The restricted surgical relevance of morphologic criteria to classify systemic-pulmonary collateral arteries in pulmonary atresia with ventricular septal defect. Thorac Cardiov Surg 108:692–699.

DeRuiter M.C., Gittenberger-de Groot A.C., Wenink A.C.G., Poelmann R.E., Mentink M.M.T. (1995): In normal development pulmonary veins are connected to the sinus venosus segment in the left atrium. Anat Rec 243:84–92.

DeRuiter M.C., Poelmann R.E., VanMunsteren J.C., Mironov V., Markwald R.R., Gittenberger-de Groot A.C. (1997): Embryonic endothelial cells transdifferentiate into mesenchymal cells expressing smooth muscle actins in vivo and in vitro. Circ Res 80:444–451.

Drake C.J., Jacobson A.G. (1988): A survey by scanning electron microscopy of the extracellular matrix and endothelial components of the primordial chick heart. Anat Rec 222:391–400.

Duband J.L., Gimona M., Sartore S., Small J.V. (1993): Calponin and SM22 as differentiation markers of smooth muscle: spatiotemporal distribution during avian embryonic development. Differentiation 55:1–11.

Gerlis L.M., Ho S.Y., Anderson R.H., Da costa P. (1989): Persistent 5th aortic arch—a great pretender: three new covert cases. Int J Cardiol 23:239–247.

Gittenberger-de Groot A.C., van Ertbrugger I., Moulaert A.J.M.G. (1980): The ductus arteriosus in the preterm infant: histological and clinical observation. J Pediatr 96:88–93.

Gittenberger-de Groot A.C., Strengers J.L.M., Mentink M.M.T., Poelmann R.E., Patterson D.F. (1985): Light and electronmicroscopic studies on normal an persistent ductus arteriosus in the dog. J Am Coll Cardiol 6:394–404.

Gittenberger-de Groot A.C., DeRuiter M.C., Poelmann R.E. (1995): Vasculogenesis and vessel wall differentiation in the embryo. BAM 6:5–12.

Glukhova M.A., Koteliansky V.E. (1995): Integrins, cytoskeletal and extracellular matrix proteins in developing smooth muscle cells of human aorta. In: The vascular smooth muscle cell, Schwartz S.M., Mecham R., eds. San Diego: Academic Press.

Hoffman J.I.E. (1995): Incidence of congenital heart disease: I. Postnatal incidence. Pediatr Cardiol 16:103–113.

Hogers B., DeRuiter M.C., Gittenberger-de Groot A.C., Poelmann R.E. (1997): Unilateral vitelline vein ligation alters intracardiac blood flow patterns and morphogenesis in the chick embryo. Circ Res 80:473–481.

Hughes A.F.W. (1942): The histogenesis of the arteries of the chick embryo. J Anat 77:266–287.

Hungerford J.E., Owens G.K., Argraves W.S., Little C.D. (1996): Development of the aortic vessel wall as defined by vascular smooth muscle and extracellular matrix markers. Dev Biol 178:375–392.

Huntington G.S. (1920): The morphology of the pulmonary artery in the mammalia. Anat Rec 17:165–201.

Le Lièvre C.S., Le Douarin N.M. (1975): Mesenchymal derivatives of the neural crest: analysis of chimaeric quail and chick embryos. J Embryol Exp Morphol 34:125–154.

Liao P., Edwards W.D., Julsrud P.R., Puga F.J., Danielson G.D., Feldt R.H. (1985): Pulmonary blood supply in patients with pulmonary atresia and ventricular septal defect. J Am Coll Cardiol 6:1343–1350.

Macartney F.J., Scott O., Deverall P. (1974): Haemodynamic and anatomical characteristics of pulmonary blood supply in pulmonary atresia with ventricular septal defect—including a case of persistent fifth aortic arch. Br Heart J 36:1049–1060.

Mikawa T., Fischman D.A., Dougherty J.P., Brown A.M.C. (1991): In vivo analysis of a new lacZ retrovirus vector suitable for cell lineage marking in avian and other species. Exp Cell Res 195:516–523.

Noden D.M. (1989): Embryonic origins and assembly of blood vessels. Am Rev Respir Dis 140:1097–1103.

Owens G.K. (1995): Regulation of differentiation of vascular smooth muscle cells. Phys Rev 75:487–517.

Pardanaud L., Altmann C., Kitos P., Dieterlen-Lièvre F., Buck C.A. (1987): Vasculogenesis in the early quail blastodisc as studied with a monoclonal antibody recognizing endothelial cells. Development 100:339–349.

Pardanaud L., Yassine F., Dieterlen-Lièvre F. (1989): Relationship between vasculogenesis angiogenesis and haemopoiesis during avian ontogeny. Development 105:473–485.

Pattan S., Haenni B., Burri P.H. (1993): Evidence for intussusceptive capillary growth in the chicken chorio-allantoic membrane (CAM). Anat Embryol 187:121–130.

Philips M.T., Kirby M.L., Forbes G. (1987): Analysis of cranial neural crest distribution in the developing heart using quail-chick chimeras. Circ Res 60:27–30.

Poole T.J., Coffin J.D. (1989): Vasculogenesis and angiogenesis: two distinct morphogenetic mechanisms establish embryonic vascular pattern. J Exp Zool 251:224–231.

Poole T.J., Coffin J.D. (1991): Morphogenetic mechanisms in avian vascular development. In: The development of the vascular system, Feinberg R.N., Auerbach R., eds. Issues in biomedicine, Vol 14 Stolte H., Kinne R.K.H., Bach P.H., eds. Basel: Karger.

Puga F.J., Leoni F.E., Julsrud P.R., Mair D.D. (1989): Complete repair of pulmonary atresia, ventricular septal defect, and severe peripheral arborization abnormalities of the central pulmonary arteries. J Thorac Cardiov Surg 98:1018–1029.

Reagan F. (1912): The fifth aortic arch of mammalian embryos; the nature of the last pharyngeal evagination. Am J Anat 12:493–514.

Risau W. (1991): Vasculogenesis, angiogenesis and endothelial cell differentiation during embryogenesis development. In: The development of the vascular system, Feinberg R.N., Sherer G.K., Auerbach R., eds. Issues in biomedicine, Vol 14. Stolte H., Kinne R.K.H., Bach P.H., eds. Basel: Karger.

Rosenquist T.H., Beall A.C. (1990): Elastogenic cells in the developing cardiovascular system. Smooth muscle, nonmuscle, and cardiac neural crest. Ann NY Acad Sci 558:106–119.

Rossi M., Filho R.R., Yen Ho S. (1988): Solitary arterial trunk with pulmonary atresia and arteries with supply to the left lung from both an aterial duct and systemic-pulmonary collateral arteries. Int J Cardiol 20:145–148.

Slomp J., van Munsteren J.C., Poelmann R.E., de Reeder E.G., Bogers A.J.J.C., Gittenberger-de Groot A.C. (1992): Formation of intimal cushions in the ductus arteriosus as a model for vascular intimal thickening. An immunohistochemical study of changes in extracellular matrix components. Atherosclerosis 93:25–39.

Slomp J., Gittenberger-de Groot A.C., Glukhova M.A., VanMunsteren J.C., Kockx M.M., Schwartz S.M., Koteliansky V.E. (1997): Differentiation, dedifferentiation and apoptosis of smooth muscle cells during the development of the human ductus arteriosus. Atherscler Thromb Vasc Biol (in press).

Sotomora R.F., Edwards J.E. (1978): Anatomic identification of so-called absent pulmonary artery. Circulation 57:624–633.

Sullivan I.D., Wren C., de Leval M.R., Macartney F.J., Deanfield J.E. (1988): Surgical unifocalization in pulmonary atresia and ventricular septal defect. A realistic goal? Circulation 78 (Suppl):III5–13.

Sumida H. (1988): Study of abnormal formation of the aortic arch in rats: by methylacrylate casts method and by immunohistochemistry for appearance and distribution of desmin, myosin and fibronectin in the tunica media. Hiroshima J Med Sci 37:19–36.

Thayer J.M., Meyers K., Giachelli C.M., Schwartz S.M. (1995): Formation of the arterial media during vascular development. Cell Mol Biol Res 41:251–262.

Thiene G., Bortolotti U., Gallucci V., Valente M.L., Dalla Volta S. (1977): Pulmonary atresia with venticular septal defect. Further anatomical observations. Br Heart J 39:1223–1233.

Van der Loop F.T.L., Schaart G., Timmer E.D.J., Ramaekers F.C.S., van Eys G.J.J.M. (1996): Smoothelin, a novel cytoskeletal protein specific for smooth muscle cells. J Cell Biol 134:401–411.

Van Groningen J.P., Wenink A.C.G., Testers L.H.M. (1991): Myocardial capillaries: increase in number by splitting of existing vessels. Anat Embryol 184:65–70.

1.5

Growth Factors in Vascular Morphogenesis: Insights from Gene Knockout Studies in Mice

Chitra Suri and George D. Yancopoulos

The basic processes leading to the generation of a mature vascular network are conserved between avians and mammals, however, more is known about the actual players in these processes in mammals due to the relative ease of genetic manipulations in mice.

Growth Factors Required in the Early Stages of Vascular Development

The process by which angioblasts differentiate from mesodermal precursors, transform into endothelial cells, proliferate and organize themselves into tubes either in situ or upon migration to different sites within the organism, is known as vasculogenesis. Vascular endothelial growth factor A (VEGF-A) and its receptors (Flk-1/VEGF-R2/KDR and Flt-1/VEGF-R1), that are found only on endothelial cells and some blood cells, have been shown to be crucial for these early events to occur. In mice lacking Flk-1, there is no functional endothelium, leading to a complete absence of blood vessels. The embryo dies between E8.5 and E9.5 due to this and other abnormalities in the hematopoietic system (Shalaby et al., 1995). On the other hand, in the absence of Flt-1, some endothelial cells do manage to form vascular channels but these are abnormally large and disorganized. The mutant embryo dies halfway through gestation, between E9.0 and E11.5, probably due to inefficient yolk sac circulation (Fong et al., 1995). The different results obtained in mice lacking either of the two receptors is perhaps not surprising since even though VEGF-A binds with high affinity to both receptors, it initiates different downstream signaling events in each case (Waltenberger et al., 1994). In this study, when Flk-1 was introduced into porcine aortic endothelial cells, which normally do not contain VEGF receptors, it led to impressive changes in their cellular morphology, chemotaxis, and cell division upon the addition of VEGF. In contrast, Flt-1 had no effect in the same assays, though it could be activated by VEGF. In the absence of any obvious effect on endothelial cells (Waltenberger et al., 1994; Keyt et al., 1996), the exact role of Flt-1 activation on endothelial cells was not known until the gene was inactivated in mice. Analyses of mice lacking Flt-1 clearly reveal that, while not apparently required for endothelial cell proliferation, Flt-1 instead plays a pivotal role in their maturation and structural organization. However, VEGF-A may not be the only ligand to activate the two receptors in vivo. Placenta growth factor (PlGF) is a recently identified molecule that bears significant homology to VEGF-A and binds with high affinity to Flt-1 (Park et al., 1994). Interestingly, although PlGF does not demonstrate any direct effects on endothelial cells in vitro, in conjunction with VEGF-A, it appears to enhance the permeability of endothelial cells. Gene knockout studies also suggest that VEGF-A is not the sole activator of Flt-1 and Flk-1 in a developing organism (Carmeliet et al., 1996a; Ferrara et al., 1996). Mice that lack this gene die a few days later than those that lack either of the two receptors, albeit from abnormal endothelial cell development. Unexpectedly, the dosage of the VEGF-A gene is a critical factor in this process since mice that lack only one copy of the gene also exhibit gross vascular abnormalities, demonstrating the tight control of the vasculogenic process. In addition, since these heterozygous mice live a few days longer than their homozygous littermates, they show deficits in the later stages of vascular development or angiogenesis (next section) as well, suggesting that VEGF-A is required at various stages in the development of blood vessels.

Growth Factors Required in the Late Stages of Vascular Development

Vasculogenesis results in the formation of an immature vascular network wherein the endothelial cell tubes interconnect to form a "primary capillary plexus" containing similarly sized vessels. The process by which this network is modified by sprouting/branching/regression mechanisms and stabilized by interactions with the extracellular matrix to finally create a mature vascular plexus is known as angiogenesis. Numerous growth factors have been shown to participate in this process. Most of these such as transforming growth factor β1 (TGFβ1), neuregulins, and the platelet derived growth factors (PDGFs) have receptors that are widely distributed in various cell types and are therefore also involved in processes other than angiogenesis. Besides the VEGF receptor family, there is only one known growth factor receptor family, namely the Tie receptor family, that is restricted to the endothelial/hematopoietic lineage. In fact, the Tie family has been found to be crucial to the later stages of vascular development.

The Tie receptor family consists of two receptors, Tie-1 and Tie-2 (or Tek) that are 85% homologous in their tyrosine kinase domains. This receptor family belongs to the superfamily of tyrosine kinase receptors which also contains Flk-1 and Flt-1. The genes for both the Tie receptors were recently inactivated in mice (Dumont et al., 1994; Puri et al., 1995; Sato et al., 1995), leading to clear vascular deficits with the death of the affected animals either at E10.5 (Tie-2) or E13.5-P1 (Tie-1). In either case, vascular development is arrested at a later stage compared to the Flk-1 or Flt-1 null mutations, consistent with the later onset of the Tie receptors during embryogenesis (Dumont et al., 1995). In mice lacking Tie-1, the endothelial cells are generated and they form the large and small vessels typical of an E13.5 embryo, however, the small vessels, typically the capillaries, do not maintain their integrity beyond a certain age. At that age, which could be E13.5 (Puri et al., 1995) or P1 (Sato et al., 1995) depending on the strain of the mouse, the embryo shows widespread haemorrhages and dies shortly thereafter. The first sign of abnormality actually appears to be a subcutaneous edema most likely due to leaky vessels, which then progresses into a full-blown haemorrhage (Sato et al., 1995). When mice are created with the absence of the Tie-2 receptor, they exhibit severe abnormalities in their vasculature with the embryos dying at E9.5-E10.5 (Dumont et al., 1994; Sato et al., 1995). The most obvious defect occurs in the heart. Both the endothelial layer (endocardium) and the muscle layer (myocardium) appear as simplified structures. The endocardium does not develop a close association with the myocardium, and the myocardium itself does not develop its typical trabeculations. Since Tie2 receptors are only expressed by the endothelial cells, the myocardial effects are most likely secondary to the effects on the endocardium. Vascular beds at other sites such as in the head and yolk sac are also morphologically abnormal since they do not exhibit the typical distribution of large and small vessels but instead portray a network of rather similarly sized vessels. The closely related Tie-1 and Tie-2 receptors are therefore essential for the development of a functioning circulatory system, but are required during distinct phases of angiogenesis. Tie-2, in the earlier phases, appears to be involved in establishing the correct vascular patterning. Tie-1 apparently has a later role in maintaining vessel integrity and preventing vascular haemorrhages.

The story, however, gets more complex! Recently, a new family of ligands for the Tie-2 receptor were identified and manipulations of these have afforded us new insights into their roles in angiogenesis and provided us with a possible model of blood vessel formation. Two of these ligands, Angiopoietin-1 (Ang1) and Angiopoietin-2 (Ang2) have been studied in detail (Davis et al., 1996; Suri et al., 1996; Maisonpierre et al., 1997). Transgenic mice that lack Ang1 do not survive beyond E12.5 and exhibit vascular deficits similar to those seen in mice lacking the Tie-2 receptor, indicating that Ang1 is a major bonafide ligand for the receptor (Suri et al., 1996). For instance, the hearts of these transgenic mice show the same developmental retardation manifested by similar endocardial and myocardial defects; the forebrain and yolk sac show the abnormal large dilated vessels. In fact, the vasculature in general is aberrant; for example, vessels that are normally large (such as those supplying the head and those feeding into the umbilicus) are thinner and fewer in number and they generate fewer and thinner branches. The branches follow rather straight courses, unlike normal vessels that always seem to meander. In addition, in some cases, vessels (such as those between somites) even seem to regress. The observed defects may result from a failure of the endothelial cells to interact appropriately with the neighboring cells and the surrounding extracellular matrix. The discovery of Ang1 combined with the subsequent generation of mice lacking this ligand has suggested that Ang1, acting on the Tie-2 receptor, is involved with the maturation and stabilization of the overall vascular pattern.

The role of Ang2, on the other hand, is not quite so straightforward (Maisonpierre et al., 1997). Ang2 binds to the Tie-2 receptor with the same affinity as Ang1 but it serves, at least on some endothelial cells, as an inhibitor of the Tie-2 receptor in vitro. Consistent with Ang2 acting as an antagonist of Ang1, when transgenic mice are created that overexpress Ang2 at sites of Ang1 expression, both the heart and blood vessel formation are greatly disrupted. In fact, the mice display abnormalities similar to those seen in the absence of Ang1. This would suggest that Ang2 has the potential to antagonize Ang1 in vivo. During development, Ang2 displays a more restricted expression pattern compared to Ang1 but there are sites (such as the aorta) where both ligands colocalize. If Ang1 is seen as a maturation and stabilizing protein, then Ang2 (since it antagonizes Ang1) must act as a destabilizing protein. One hypothesis then is that Ang2 expression is involved in initiating vascular reorganizations. This hypothesis is supported by data that shows that in the adult, Ang2 can be detected only in tissues that undergo extensive vascular remodeling, such as the ovary, uterus, and placenta. In the ovary, a tissue that has been analyzed in detail, Ang1, Ang2, and VEGF expression follow a cyclical pattern correlating with distinct stages of follicular vascularization. The chronology of events suggests that Ang2 is indeed a protein that destabilizes blood vessels but it is not simply an inhibitor of vascular growth! The outcome of Ang2 inhibition of Ang1 depends on the presence of other angiogenic molecules such as VEGF. It seems that the destability that is created by Ang2 can lead to vessel regression if there is no VEGF present; however, it can lead to sprouting or frank vessel outgrowth if there is VEGF in the vicinity, perhaps because an unstable vasculature is more responsive to surrounding angiogenic signals. Ang2 thus maybe a unique molecule in angiogenesis that possesses a dual role of either promoting or restraining vascular growth depending on the context.

In addition to the endothelial-specific growth factors, the VEGF and angiopoietin family members, there exist a diverse assortment of players that affect the overall vascular pattern in the developing embryo. None of these additional growth factors is specific for the heart or the blood vessels and their role in the development of these structures has only been brought to light by the generation of mice that lacked them. Platelet derived growth factor B (PDGF B) and its two receptors (PDGFα and PDGFβ), transforming growth factor β1 (TGFβ1) and neuregulin are particularly noteworthy. PDGF B appears to play an autocrine role for endothelial cell growth in capillaries and a paracrine role for mesangial cells, the support cells for capillaries in kidney glomeruli, as well as a paracrine role for the smooth muscle cells of the arteries and heart (Leveen et al., 1994; Soriano, 1994; Morrison-Graham et al., 1992). TGFβ1 seems to be of vital importance to the development of the yolk sac vasculature as well as the hemopoietic system (Dickson et al., 1995). In the null mutant mice, the yolk sac is comprised of endothelial cells that are only weakly associated with each other which might explain why the vessels appear disorganized or are completely absent. Mice lacking neuregulin or its receptor (Meyer and Birchmeir, 1995; Gassmann et al., 1995; Lee et al., 1995; Kramer et al., 1996) exhibit a cardiac phenotype that is very reminiscent of that seen in mice lacking Ang1 or its receptor. Notably, the ventricular myocardium shows profoundly reduced trabeculae. Interestingly, neuregulin is expressed by the endocardium and its receptors, erbB2 and erbB4, are found in the myocardium while Ang1 is expressed by the myocardium and its Tie-2 receptor is found on the endocardial cells. This reciprocal expression pattern suggests an intriguing possibility—it may be that the mechanism of cardiac morphogenesis involves the sequential activation of these two disparate ligands, with one acting directly on the endocardial cells and the other on the myocardial cells.

It is worth mentioning that there are a few other recently identified genes that do not code for growth factors but which upon inactivation compromise angiogenesis by either disrupting intracellular signaling in endothelial cells that may occur downstream of the growth factors (gene for GTP-ase activating protein (Henkemeyer et al., 1995)) or by affecting vascular smooth muscle cell development (genes for tissue factor (TF; Carmeliet et al., 1996b) and arylhydrocarbon-receptor nuclear translocator (ARNT; Maltepe et al., 1997)). In the absence of tissue factor, a multifunctional receptor for one of the coagulation factors, smooth muscle cells or pericytes are not sufficiently accumulated around the endothelial cells leading to fragile vessels. On the other hand, ARNT is a transcription factor which perhaps normally exerts its effect by activating genes such as VEGF-A and TF in response to hypoxic conditions during tissue growth.

So far, only the development of the vascular system has been considered. However, there exists another system of vessels, namely the lymphatic system, that is thought to originate from the venous system at specific sites during embryogenesis (Sabin, 1909). Not much is known about lymphangiogenesis or angiogenesis of the lymphatic system (see Chapter by Wilting et al. for a current update) except that VEGF-C (homologous to VEGF-A in its N-terminal domain), acting via VEGF-R3 (Flt-4) or VEGF-R2/VEGF-R3 heterodimeric receptors, appears to be involved (Kukk et al., 1996; Jeltsch et al., 1997); Flt-4 is rather specifically expressed in lymphatic endothelial cells. Mice that overexpress this ligand in the skin develop dilated lymphatic vessels in the dermis that have been attributed to the proliferation of endothelial cells lining these vessels.

The numerous genetic manipulations in mice have thus significantly enhanced our understanding of vascular development. It is possible now to construct a working model of blood vessel development in vivo (see Fig. 1). VEGF activates its receptors and acts in concert with other growth factors such as TGFβ1 to lay the foundation early in development by generating simple vascular channels. These primitive vessels are then manipulated by the Tie-2 receptor pathway, which may be activated or inhibited by the Angiopoietins, as well as by other proteins namely TF,

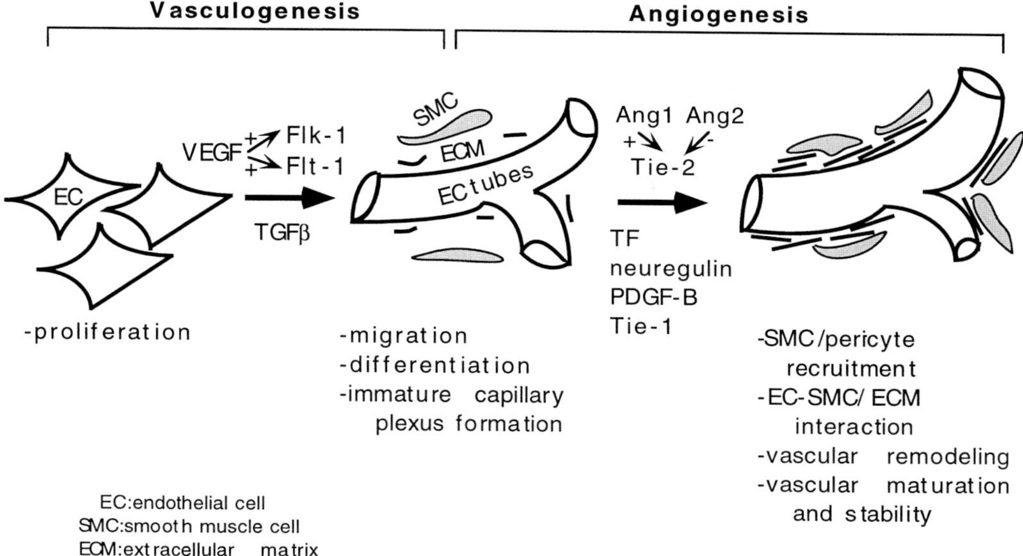

Fig. 1 Schematic showing the major processes controlling the development of a mature vasculature and the genes that regulate these processes.

neuregulin, PDGF-B and Tie-1, to create a mature vasculature that is comprised of a stably interacting endothelium, mesenchyme and extracellular matrix.

References

Carmeliet, P., Ferreira, V., Breier, G., Pollefeyt, S., Kieckens, L., Gertsenstein, M., Fahrig, M., Vandenhoeck, A., Harpal, K., Eberhardt, C., Declercq, C., Pawling, J., Moons, L., Collen, D., Risau, W., and Nagy, A. (1996a). Abnormal blood vessel development and lethality in embryos lacking a single VEGF allele. Nature 380, 435–439.

Carmeliet, P., Mackman, N., Moons, L., Luther, T., Gressens, P., Van Vlaenderen, I., Demunck, H., Kasper, M., Breier, G., Evrard, P., Muller, M., Risau, W., Edginton, T., and Collen, D. (1996b). Role of tissue factor in embryonic blood vessel development. Nature 383, 73–75.

Davis, S., Aldrich, T.H., Jones, P.F., Acheson, A., Compton, D.L., Jain, V., Ryan, T.E., Bruno, J., Radjiewski, C., Maisonpierre, P.C., and Yancoppoulos, G.D. (1996). Isolation of Angiopoietin-1, a ligand for the angiogenic TIE2 receptor, by secretion-trap expression cloning. Cell 87, 1161–1169.

Dickson, M.C., Martin, J.S., Cousins, F.M., Kulkarni, A.B., Karlsson, S., and Akhurst, R.J. (1995). Defective haematopoiesis and vasculogenenis in transforming growth-factor-β1 knock out mice. Development 12, 1845–1854.

Dumont, D.J., Gradwohl, G., Fong, G.-H., Puri, M.C., Gerstenstein, M., Auerbach, A., and Breitman, M.L. (1994). Dominant-negative and targeted null mutations in the endothelial receptor tyrosine kinase, tek, reveal a critical role in vasculogenesis of the embryo. Genes Dev. 8, 1897–1909.

Dumont, D.J., Fong, G.-H., Puri, M.C., Gradwohl, G., Alitalo, K., and Breitman, M.L. (1995). Vascularization of the mouse embryo: a study of flk-1, tek, tie and vascular endothelial growth factor expression during development. Dev. Dyn. 203, 80–92.

Ferrara, N., Carver-Moore, K., Chen, H., Dowd, M., Lu, L., O'Shea, K.S., Powell-Braxton, L., Hillan, K.J., and Moore, M.W. (1996). Heterozygous embryonic lethality induced by targeted inactivation of the VEGF gene. Nature 380, 439–442.

Fong, G.H., Rossant, J., Gertenstein, M., and Breitman, M.L. (1995). Role of the Flt-1 receptor tyrosine kinase in regulating assembly of vascular endothelium. Nature 376, 66–70.

Gassmann, M., Casagranda, F., Orioli, D., Simon, H., Lai, C., Klein, R., and Lemke, G. (1995). Aberrant neural and cardiac development in mice lacking the ErbB4 neuregulin receptor. Nature 378, 390–394.

Henkemeyer, M., Rossi, D.J., Holmyard, D.P., Puro, M.C., Mbamalu, G., Harpal, K., Shih, T.S., Jacks, T., and Pawson, T. (1995). Vascular system defects and neuronal apoptosis in mice lacking Ras GTPase-activating protein. Nature 377, 695–701.

Lee, K.-F., Simon, H., Chen, H., Bates, B., Hung, M.-C., and Hauser, C. (1995). Requirement for neuregulin receptor erbB2 in neural and cardiac development. Nature 378, 394–398.

Jeltsch, M., Kaipainen, A., Joukov, V., Meng, X., Lasko, M., Rauvala, H., Swartz, M., Fukumura, D., Jain, R.K., and Alitalo, K. (1997). Hyperplasia of lymphatic vessels in VEGF-C transgenic mice. Science 276, 1423–1425.

Keyt, B.A., Hung, N.V., Berleau, L.T., Duarte, C.M., Park, J., Chen, H., and Ferrara, N. (1996). Identification of vascular endothelial growth factor determinants for binding KDR and FLT-1 receptors. J. Biol. Chem. 271, 5638–5646.

Kramer, R., Bucay, N., Kane, D.J., Martin, L.E., Tarpley, J.E., and Theill, L.E. (1996). Neuregulins with an Ig-like domain are essential for mouse myocardial and neuronal development. Proc. Natl. Acad. Sci. USA 93, 4833–4838.

Kukk, E., Lymboussaki, A., Taira, S., Kaipainen, A., Jeltsch, M., Joukov, V., and Alitalo, K. (1996). VEGF-C receptor binding and pattern of expression with VEGFR-3 suggests a role in lymphatic vascular development. Development 122, 3829–3837.

Leveen, P., Pekny, M., Gebre-Medhin, S., Swolin, B., Larsson, E., and Betsholtz, C. (1994). Mice deficient for PDGF B show renal, cardiovascular, and hematological abnormalities. Genes Dev. 8, 1875–1887.

Maisonpierre, P.C., Suri, C., Jones, P.F., Bartunkova, S., Wiegand, S.J., Radziejewski, C., Compton, D., McClain, J., Aldrich, T.H., Papadopoulos, N., Daly, T.J., Sato, T.N., and Yancopoulos, G.D. (1997). Angiopoietin-2, a natural antagonist for Tie2 that disrupts in vivo angiogenesis. Science 277, 55–60.

Maltepe, E., Schmidt, J.V., Baunoch, D., Bradfield, C.A., and Simon, M.C. (1997). Abnormal angiogenenis and responses to glucose and oxygen deprivation in mice lacking the protein ARNT. Nature 386, 403–406.

Meyer, D., and Birchmeier, C. (1995). Multiple essential functions of neuregulin in development. Nature 378, 386–398.

Morrison-Graham, K., Schattenab, G.C., Bork, T., Bowen-Pope, D.F., and Weston, J.A. (1992). A PDGF receptor mutation in the mouse (Patch) perturbs the development of a non-neuronal subset of neural crest-derived cells. Development 115, 133–142.

Park, J.E., Chen, H.H., Winer, J., Houck, K.A., and Ferrara, N. (1994). Placenta growth factor. J. Biol. Chem. 269, 25646–25654.

Puri, M.C., Rossant, J., Alitalo, K., Bernstein, A., and Partanen, J. (1995). The receptor tryosine kinase TIE is required for integrity and survival of vascular endothelial cells. EMBO J. 14, 5884–5891.

Sabin, F.R. (1909). The lymphatic system in human embryos, with a consideration of the morphology of the system as a whole. Am. J. Anat. 9, 43–91.

Sato, T.N., Tozawa, Y., Deutsch, U., Wolburg-Buchholz, K., Fujiwara, Y., Gendron-Maguire, M., Gridley, T., Wolburg, H., Risau, W., and Qin, Y. (1995). Distinct roles of the receptor tryosine kinases Tie-1 and Tie-2 in blood vessel formation. Nature 376, 70–74.

Shalaby, F., Rossant, J., Yamaguchi, T.P., Gertsenstein, M., Wu, X.F., Breitman, M.L., and Schuh, A.C. (1995). Failure of blood-island formation and vasculogenesis in Flk-1-deficient mice. Nature 376, 62–66.

Soriano, P. (1994). Abnormal kidney development and hematological disorders in PDGF β-receptor mutant mice. Genes Dev. 8, 1888–1896.

Suri, C., Jones, P.F., Patan, S., Bartunkova, S., Maisonpierre, P.C., Davis, S., Sato, T.N., and Yancopoulos, G.D. (1996). Requisite role of Angiopoietin-1, a ligand for the TIE2 receptor, during enbryonic angiogenesis. Cell 87, 1171–1180.

Waltenberger, J., Claesson-Welsh, L., Siegbahn, A., Shibuya, M., and Heldin, C.-H. (1994). Different signal transduction properties of KDR and Flt1, two receptors for vascular endothelial growth factor. J. Biol. Chem. 269, 26988–26995.

Part II

Vascular Morphogenesis In Vitro

Introduction: Regulation of Vascular Morphogenesis

E. Helene Sage, Ph.D.

Blood vessels are thought to develop according to two fundamental processes. Vasculogenesis refers chiefly to the embryonic formation of the larger vessels, which arise in situ from primordial endothelial cells, and angiogenesis, to the growth of vessels from extant vasculature that occurs during development and during remodeling in the adult. It is likely that fundamental mechanisms directing endothelial cell adhesion, migration, and proliferation apply to both processes (1).

Although there is much to learn about how blood vessels grow, we now appreciate that angiogenesis is regulated at different levels as a function of injury, repair, or cyclic renewal. Capillary beds undergo extensive remodeling throughout life. Several classes of regulatory molecules translate information for the assembly of a three-dimensional structure (a microvessel with branches and a patent lumen) from individual endothelial cells that initiate a different repertoire of gene expression and assume an angiogenic phenotype. Figure 1 is a generalized scheme for the initiation of a capillary sprout from otherwise noncycling endothelium comprising the parent vessel (a general definition of angiogenesis). Comprehensive reviews have been written on most of the stages illustrated in Fig. 1 (2–7); however, the temporal sequence of events controlling vascular morphogenesis is not clear and likely reflects a considerable degree of redundancy.

Positive endogenous regulators of endothelial proliferation appear to be numerous (2,3). Prominent among this group are basic fibroblast growth factor (bFGF) and vascular endothelial growth factor (VEGF), the latter essentially specific for the endothelium (8). In addition to the direct signaling provided to endothelial cells by receptors for growth factors, indirect regulation of the cell cycle by extracellular macromolecules is also an important feature of growth control (9). Certain proteins and proteoglycans of the extracellular matrix, cell surface-associated glycoproteins, and plasma components have affinities for growth factors that modulate both their availability and activation. In contrast, several interesting negative regulators of angiogenesis have been characterized, e.g., thrombospondin 1 & 2, SPARC, and tissue inhibitors of metalloproteases, although our understanding of their intracellular targets remains limited (1,2,10).

A recurring theme underlying cellular responses to injury and developmental signals is the necessity for a change in the morphology of adherent cells. It has long been appreciated that cell-surface adhesion macromolecules are critical components of anchorage-dependent growth and cell-cycle traverse. Adhesion proteins that mediate angiogenesis include integrins, selectins,

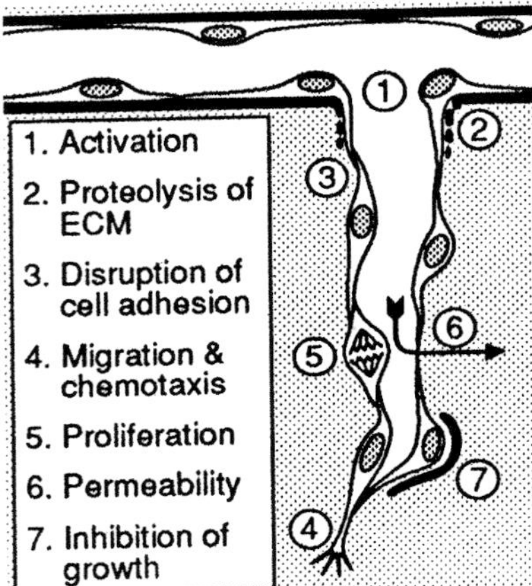

Fig. 1 The process of angiogenesis. An angiogenic sprout growing from a patent microvessel is shown. Numbers refer to various stages of the process but are not necessarily sequential. Basement membrane is shown as a thick black line; interstitium is stippled. Not shown are macrophages, mast cells and pericytes within the interstitium that contribute to endothelial activation and vascular morphogenesis. Reproduced from Ref. 1, with permission.

cadherins, and a class of cell adhesion molecules termed CAMs (11–14). The biological activities that regulate cell shape (and the consequences of its alteration) have been attributed to yet another group of macromolecules termed matricellular (15). Matricellular proteins are defined as secreted components that interact with cell-surface receptors, extracellular matrix, and/or growth factors and proteases, but do not in themselves subserve structural roles. Lastly, the spectrum of collagens produced by endothelial cells in vitro and the relevance of particular collagen types to endothelial cell origin and behavior have been recognized for almost 20 years (1). What is the significance of collagen production to angiogenesis in vivo? Basal lamina proteins, including collagen IV, are expressed sequentially during wound angiogenesis (16). The prevalent interstitial collagens I & III, and the interstitial collagenase, matrix metalloprotease (MMP)-1, are typical of fetal angiogenesis (17,18), and collagen I is prominent in microvessels of restenotic lesions (19). Mov 13 mice bearing a retroviral insertion in the COL1A1 gene, which renders it transcriptionally inactive in most tissues, die of vessel rupture at midgestation (20).

A fundamental component of angiogenesis is the directed migration of endothelial cells that occurs with invasion (Fig.1). Mitogens, adhesion molecules, extracellular matrix, proteases, and their inhibitors have been accorded regulatory roles in this process. Growth factor activity is controlled in part by the extracellular matrix, which sequesters factors in a latent form for concerted release and/or activation at concentrations higher than those present under steady-state conditions (9). Consistent with the directed timing and localization of growth factor delivery is the concept of controlled proteolysis at the cell surface, which provides for localized turnover of extracellular matrix as well as the activation of specific signaling pathways (21). Although numerous proteases participate in matrix remodeling, angiogenesis has generally been associated with two classes: serine proteinases (especially plasminogen activators) and MMPs (6,22). The ratio between a specific protease and its inhibitor can thus determine the extent of vascular invasion as a function of endothelial cell adhesion, proliferation, receptor availability, and permissiveness of the extracellular matrix (23).

There are several tissue culture models that simulate vascular morphogenesis, and each implicates the extracellular matrix as requisite for the formation of endothelial cords and tubes from single cells or cell monolayers (3,11,24,25). One important model involves the invasion of collagen gels by activated microvascular endothelial cells, which degrade the matrix, invade, and form vessel-like structures with lumens (24). This system recapitulates many of the stages shown in Fig. 1 and appears to be an accurate representation of angiogenesis, especially for invasion and lumen formation. The design of the invasion model is highly conducive to the analysis of potential angiogenesis regulators.

Another approach exploits collagen gels as a medium for (a) invasion by endothelial cells from "organ cultures" (26), or (b) assembly of tubelike structures from cultured microvascular cells plated within the gel (27). In (a), intimal cells from rat aortic rings sprout into the collagen and exhibit differential responses to angiogenesis regulators that reflect the heterogeneity of the cell population. The noninvasive model (b) exploits the uniformity of a well-characterized cell strain and the induction of a differentiated phenotype concomitant with cessation of proliferation. Alternatively, the morphogenesis of certain capillary beds is thought to proceed via cellular migration onto "matrical tracks" that consist partially of collagen I secreted by the cells themselves (25; see Chapter 3.2 in this volume by Murray et al.). Although this type of planar vascular morphogenesis differs from that characterized by cellular invasion, it is apparent that adhesion between endothelial cells and their extracellular ligand(s) can modulate vessel-like growth in vitro.

In the following three chapters are presented models for the assay of vascular morphogenesis and function in vitro. Drs. Montesano and Pepper (Chapter 2.1) provide an in-depth discussion of their three-dimensional system, which is characterized by the invasion of native collagen gels by cultured endothelial cells, with the subsequent formation of tubular structures that resemble capillaries. In contrast, Dr. Nicosia (Chapter 2.2) describes an aortic ring assay of angiogenesis, in which explants of rat aorta are cultured in matrices of different biological components, including plasma clots. The final chapter in this section (Chapter 2.3), by Drs. Hirschi and D'Amore, presents a novel, two-dimensional model, in which vascular endothelial cells are established in coculture with either 10T1/2 cells (mesenchymal cell precursors), mural cells (pericytes), or astrocytes.

Whereas each of these models represents a viable (and often preferable) alternative to the study of angiogenesis in vivo, it is clear that no system in vitro recapitulates the process of vascular morphogenesis in its entirety. Several of the stages through which endothelial cells must progress for the completion of a vascular network (e.g., as illustrated in Fig.1) appear to be represented accurately by one or more of the angiogenesis models described in the succeeding chapters. Ultimate demonstration of the efficacy of any of these assays depends on the rate of success by which agents with regulatory properties on endothelium in vitro fulfill their predicted function on vascular growth or inhibition in vivo. In several instances, this rate has been gratifyingly high (1).

References

Sage E.H. (1996) Angiogenesis inhibition in the context of endothelial cell biology. Adv. Oncol. 12(2):17–29.

Folkman J. (1995) Clinical applications of research on angiogenesis. New Engl. J. Med. 333:1757–1763.

3. D'Amore P.A. (1995) Mechanisms of retinal and choroidal neovascularization. Invest. Ophthalmol. Vis. Sci. 35:3974–3979.

4. Lauffenburger D.A., Horwitz A.F. (1996) Cell migration: a physically integrated molecular process. Cell 84:359–369.

5. Powell W.C., Matrisian L.M. (1996) Complex roles of matrix metalloproteinases in tumor progression. In Attempts to Understand Metastasis Formation I, U. Günthert, W. Birchmeier, Eds., Springer, Berlin, pp. 1–21.

6. Mignatti P., Rifkin D.B. (1996) Plasminogen activators and angiogenesis. In Attempts to Understand Metastasis Formation I. U. Günthert, W. Birchmeier, Eds., Springer, Berlin, pp 33–50.

7. Simmons D.L. (1993) Dissecting the modes of interactions amongst cell adhesion molecules. Development 1993:193–203.

8. Senger D.R. (1996) Molecular framework for angiogenesis. Am. J. Pathol. 149:1–7.

9. Flaumenhaft R., Rifkin D.B. (1992) The extracellular regulation of growth factor action. Mol. Biol. Cell. 3:1057–1065.

10. Volpert O.V., Tolsma S.S., Pellerin S., Feige J-J., Chen H., Mosher D.F., Bouck N. (1995) Inhibition of angiogenesis by thrombospondin-2. Biochem. Biophys. Res. Commun. 217:326–332.

11. Bischoff J. (1995) Approaches to studying cell adhesion molecules in angiogenesis. Trends Cell Biol. 5:69–74.

12. Dejana E., Corada M., Lampugnani M.G. (1995) Endothelial cell-to-cell junctions. FASEB J. 9:910–918.

13. Polverini P.J. (1996) Cellular adhesion molecules. Am. J. Pathol. 148:1023–1029.

14. Schwartz M.A., Schaller M.D., Ginsberg M.H. (1995) Integrins: Emerging paradigms of signal transduction. Annu. Rev. Cell Dev. Biol. 11:549–599.

15. Bornstein P. (1995) Diversity of function is inherent in matricellular proteins: An appraisal of thrombospondin 1. J. Cell Biol. 130:503–506.

16. Sephel G.C., Kennedy R., Kudravi S. (1996) Expression of capillary basement membrane components during sequential phases of wound angiogenesis. Matrix Biol. 15:263–279.

17. Kitaoka M., Iyama K-I., Hoshioka H., Monda M., Usuku G. (1994) Immunohistochemical localization of procollagen types I and III during placentation in pregnant rats by type-specific procollagen antibodies. J. Histochem. Cytochem. 42:1453–1461.

18. Karelina T.V., Goldberg G.I., Eisen A.Z. (1995) Matrix metalloproteinases in blood vessel development in human fetal skin and in cutaneous tumors. J. Invest. Dermatol. 105:411–417.

19. Rekhter M.D., O'Brien E., Shah N., Schwartz S.M., Simpson J.B., Gordon D. (1996) The importance of thrombus organization and stellate cell phenotype in collagen I gene expression in human, coronary atherosclerotis and restenotic lesions. Cardiovasc. Res. 32:496–502.

20. Löhler J., Timpl R., Jaenisch R. (1984) Embryonic lethal mutation in mouse collagen I gene causes rupture of blood vessels and is associated with erythropoietic and mesenchymal cell death. Cell 38:597–607.

21. Basbaum C.B., Werb Z. (1996). Focalized proteolysis: spatial and temporal regulation of extracellular matrix degradation at the cell surface. Curr. Opin. Cell Biol. 8:731–738.

22. Stetler-Stevenson W.G. (1996) Dynamics of matrix turnover during pathologic remodeling of the extracellular matrix. Am. J. Pathol. 148:1345–1350.

23. Pepper M.S., Montesano R., Mandriota S.J., Orci L., Vassalli J.D. (1996) Angiogenesis: A paradigm for balanced extracellular proteolysis during cell migration and morphogenesis. Enz. Prot. 49:138–162.

24. Pepper M.S., Mandriota S.J., Vassalli J-D., Orci L., Montesano R. (1996) Angiogenesis-regulating cytokines: Activities and Interactions. In Attempts to Understand Metastasis Formation II, U. Günthert, W. Birchmeier, Eds., Springer, Berlin, pp 31–67.

25. Vernon R.B., Sage E.H. (1995) Between molecules and morphology: Extracellular matrix and the creation of vascular form. Am. J. Pathol. 147:873–883.

26. Nicosia R.F., Nicosia S.V., Smith M. (1994) Vascular endothelial growth factor, platelet-derived growth factor, and insulin-like growth factor-1 promote rat aortic angiogenesis in vitro. Am. J. Pathol. 145:1023–1029.

27. Marx M., Permutter R.A., Madri J.A. (1994) Modulation of platelet-derived growth factor receptor expression in microvascular endothelial cells during in vitro angiogenesis. J. Clin. Invest. 93:131–139.

2.1

Three-Dimensional In Vitro Assay of Endothelial Cell Invasion and Capillary Tube Morphogenesis

Roberto Montesano and Michael S. Pepper

Introduction

The establishment and maintenance of a vascular supply is an absolute requirement for the growth of normal and neoplastic tissues and, as might be expected, the cardiovascular system is the first organ system to develop and to become functional during embryogenesis. Both during development and in postnatal life, all blood vessel begin as simple endothelial-lined capillaries. Although some remain as capillaries, many of these newly-formed vessels develop into larger vessels through the concentric addition of smooth muscle cells and fibroblasts. Capillary blood vessels are formed by two processes: (a) vasculogenesis, in which a primary capillary plexus is formed from endothelial cells which differentiate in situ from mesodermal precursors, and (b) angiogenesis, the formation of new capillary blood vessels by a process of sprouting from preexisting vessels (Risau et al., 1988; Pardenaud et al., 1989). While both processes are required for formation of the vascular system during embryonic development, neovascularization which occurs in postnatal life is attributed to angiogenesis. In adult tissues, capillary proliferation is tightly controlled, and occurs in female reproductive organs (e.g., in the corpus luteum and regenerating endometrium), in the placenta and mammary gland during pregnancy, during exercise-induced muscle hypertrophy, in the wound healing process and in response to tissue hypoxia associated with vessel occlusion. Angiogenesis may however be detrimental to the organism. This occurs in pathological conditions such as proliferative retinopathy and juvenile hemangioma. Angiogenesis is also necessary for the continued growth of solid tumors, and allows for the hematogenous dissemination of tumor cells and the formation of metastases (reviewed by Folkman and Shing, 1992; Rak et al., 1993; Fidler and Ellis, 1994; Folkman, 1995a).

The series of morphogenetic events which result in the formation of new capillary blood vessels has been well described (Ausprunk and Folkman, 1977; Folkman and Klagsburn, 1987; Folkman and Shing, 1992; Paku and Paweletz, 1991). Angiogenesis begins with localized breakdown of the basement membrane of the parent vessel (usually a postcapillary venule). Endothelial cells then migrate into the surrounding extracellular matrix (ECM) within which they form a capillary sprout. As the sprouts elongate by migration and proliferation of endothelial cells, a lumen is gradually formed proximal to the migrating front. Contiguous tubular sprouts subsequently anastomose to form functional capillary loops, and vessel maturation is accomplished by reconstitution of the basement membrane (Fig. 1A). Angiogenesis is thus characterized by alterations in at least three endothelial cell functions: (1) modulation of interactions with the ECM, which requires alterations of cell-matrix contacts and the production of matrix-degrading proteolytic enzymes; (2) an initial increase and subsequent decrease in locomotion (migration), which allows the cells to translocate toward the angiogenic stimulus and to stop once they reach their destination; and (3) an increase in proliferation, which provides new cells for the growing and elongating vessel, and a subsequent return to the quiescent state once the vessel is formed. Together, these cellular functions contribute to the process of capillary morphogenesis, i.e., the formation of patent tubelike structures.

In view of the physiological and pathological importance of angiogenesis, much work has been dedicated over the last decade to the identification of factors capable of regulating this process. A number of polypeptide growth factors produced both by normal and tumor cells have been shown to stimulate formation of new capillary blood vessels in vivo. The most thoroughly characterized of these angiogenic cytokines are acidic and basic fibroblast growth factors (aFGF, bFGF), vascular endothelial growth factor (VEGF) (also known as vascular permeability factor), transforming growth factor-β1 (TGF-β1), and tumor necrosis factor-alpha (TNF-α) (reviewed by Klagsburn and D'Amore, 1991; Folkman and Shing, 1992; Ferrara et al., 1992a; Neufeld et al., 1994; Dvorak et al., 1995; Thomas, 1996; Pepper et al., 1996a). However, while bFGF and VEGF are mitogenic for endothelial cells, TGF-β and TNF-α inhibit their growth in vitro. The angiogenic effect of the latter two cytokines is therefore believed to be mediated in part by direct-acting positive regulators released from stromal and inflammatory cells. In this context, TGF-β and TNF-α are considered to be indirect angiogenic factors. A variety of other cytokine and noncytokine factors have been reported to regulate angiogenesis in vivo (reviewed by Folkman and Klagsbrun, 1987; Klagsbrun and D'Amore, 1991; Folkman and Shing, 1992).

Despite our detailed knowledge of the sequential steps of the neovascularization process from descriptive in vivo studies and the identification of a host of angiogenesis-modulating cytokines, the molecular mechanisms of angiogenesis are incompletely understood. The two most widely used assays for studying angiogenesis in vivo are the chick chorioallantoic membrane (Klagsburn et al., 1976) and the rabbit corneal micropocket (Gimbrone et al., 1974). More recently described quantitative in vivo assays involve subcutaneous implantation of various three-dimensional substrates to which angiogenesis-regulating factors can be added. These include polyester sponges (Andrade et al., 1987), polyvinyl-alcohol foam discs covered on both sides with Millipore filters (the disc angiogenesis system; Fajardo et al., 1988) and Matrigel, a basement membrane-rich ECM (Passaniti et al., 1992). These assays have been used for many years to describe the morphologically identifiable events which occur during angiogenesis, and have been important in the identification of positive and negative regulators. The in vivo assays described above are essential to establish whether a given molecule stimulates blood vessel formation in the intact organism. However, their interpretation is frequently complicated by the fact that the experimental conditions may inadvertently favor inflammation, and that under these conditions the angiogenic response is elicited indirectly, at least in part, through the activation of inflammatory or other nonendothelial cells. Although this may be relevant to some settings in which angiogenesis occurs in vivo, it does not allow one to study the consequences of the direct interaction of angiogenesis regulators with endothelial cells. To circumvent this drawback, in vitro assays using populations of cultured endothelial cells have been developed for several of the cellular components of the angiogenic process, and based on the geometry of the assay, these can be classified as either two-dimensional or three-dimensional. Conventional two-dimensional assays include measurement of endothelial cell proliferation, migration, and production of pro-

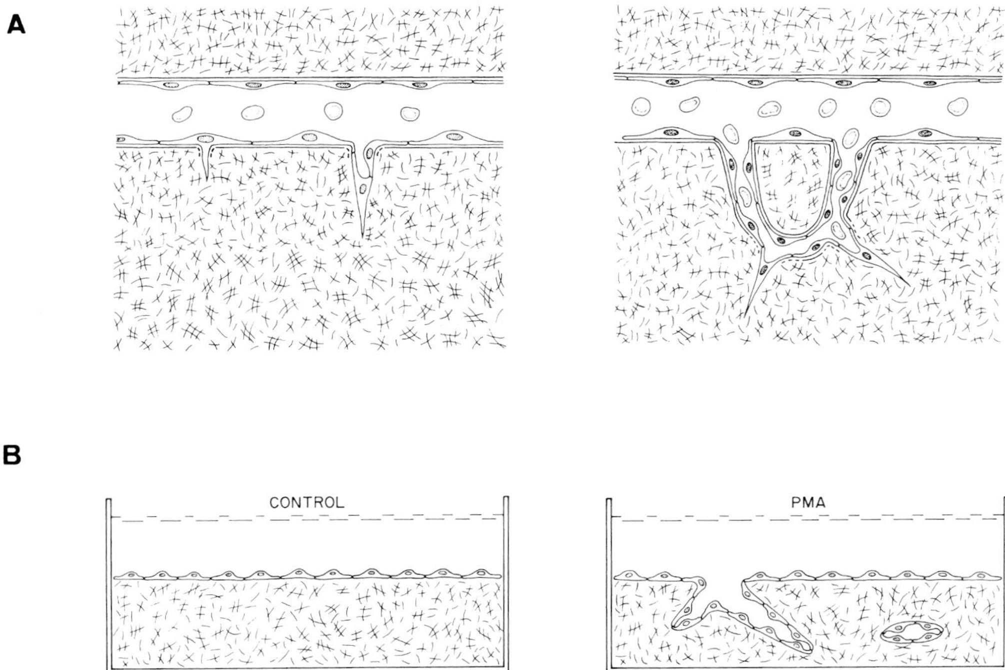

Fig. 1 Schematic representation of angiogenesis in vivo (A) and of the in vitro collagen gel invasion assay (B). (A) In response to an angiogenic stimulus, endothelial cells lining an existing capillary (or postcapillary venule) focally degrade the adjacent basement membrane and extend cytoplasmic processes in the direction of the stimulus (left panel). The cells then migrate into the surrounding ECM, within which they form a capillary sprout. Lumen formation commences in the proximal part of the sprout. The hollow sprouts thus generated elongate by migration of endothelial cells at their tip and proliferation of endothelial cells below the tip (not illustrated). Contiguous tubular sprouts subsequently anastomose with each other (right panel) to form capillary loops through which blood begins to flow. New sprouts then arise from each loop and eventually give rise to a new capillary network. Sprout maturation is completed by reconstitution of the basement membrane. (B) Endothelial cells grown on the surface of a three-dimensional collagen gel form a confluent monolayer without invading the underlying matrix (Control, left panel). Addition of the tumor promoter PMA (right panel) induces the cells to invade the underlying gel and to form capillary-like tubular structures. Invasion and tube formation are also induced by sodium orthovanadate, a potent inhibitor of phosphotyrosine phosphatases, and by the physiological angiogenic factors bFGF and VEGF (see text). (A and B are from Montesano et al., 1994, with copyright permission from Ares-Serono-Symposia.)

teolytic enzymes, such as matrix metalloproteinases (MMPs) and plasminogen activators (PAs) (Moscatelli et al., 1986). Three-dimensional assays have as their end point the formation of capillary-like cords or tubes by endothelial cells cultured either on the surface of (planar models) or within simplified extracellular matrices. These assays include: (i) long-term culture of endothelial cells in dishes coated with a thin layer of ECM proteins (Folkman and Haudenschild, 1980; Maciag et al., 1982; Madri and Williams, 1983; Ingber and Folkman, 1989; Iruela-Arispe et al., 1991; Iruela-Arispe and Sage, 1993; Battegay et al., 1994); (ii) short-term culture of endothelial cells on a thick gel of basement membrane-like matrix (Kubota et al., 1988; Grant et al., 1989; Vernon and Sage, 1995; Vernon et al., 1995); (iii) suspension of endothelial cells within a three-dimensional collagen gel (Madri et al., 1988); (iv) radial growth of branching microvessels from rings of rat aorta embedded in collagen or fibrin gels (Nicosia and Ottinetti, 1992; see the chapter by Nicosia et al. in this book), or (v) radial growth of tubular sprouts from endothelial cells grown on microcarrier beads and embedded in a fibrin gel (Nehls and Drenckhahn, 1995).

Since in the living organism angiogenesis occurs in a three-dimensional ECM microenvironment, we have designed culture systems which allow the re-establishment of three-dimensional interactions between microvascular endothelial cells and the surrounding ECM. In this review, we summarize work from our laboratory on the utilization of these in vitro assays which accurately recapitulate the invasive nature of the angiogenic process and are also permissive for histotypic morphogenesis, i.e., for the formation of patent capillary-like tubes whose abluminal surface is in direct contact with the ECM.

Collagen Matrix Promotes the Organization of Endothelial Cells into Capillary-Like Tubules

In vivo, microvascular endothelial cells experience a different ECM environment, depending on whether they are in a resting state or are undergoing sprouting and migration during angiogenesis. In the normal quiescent state, endothelial cells rest on a specialized ECM, the basement membrane, which contains predominantly type IV collagen and laminin. During angiogenesis, however, these cells focally degrade their investing basement membrane, and subsequently migrate into the interstitial matrix of the surrounding connective tissue, which consists mainly of type I collagen (Ausprunk and Folkman, 1977; Madri and Pratt, 1988). In an attempt to understand the role of the ECM in the process of angiogenesis, we have studied the interactions of endothelial cells with three-dimensional gels of reconstituted type I collagen fibrils. When a monolayer of microvascular endothelial cells on the surface of a collagen gel is covered with a second layer of collagen, it reorganizes within a few days into a network of branching and anastomosing tubules, demonstrating that a three-dimensional interaction with collagen fibrils plays an important role in driving capillary morphogenesis (Montesano et al., 1983).

Induction of the Invasive Phenotype

In the studies described above, formation of capillary-like tubules was experimentally induced by embedding endothelial cells within a three-dimensional collagen matrix. During angiogenesis in vivo, however, the morphogenetic events that culminate in the formation of new capillary blood vessels are intimately associated with the activation of an invasive process, i.e., the local breakdown of the microvascular basement membrane and the penetration of endothelial sprouts into the interstitial ECM. Central to the understanding of the mechanisms of angiogenesis is therefore the question of how normally quiescent endothelial cells acquire invasive properties that endow them with the ability to breach the mechanical barriers of the ECM. Cell invasiveness in angiogenesis and in other biological processes is believed to require the elaboration of extracellularly acting proteolytic enzymes, which include essentially MMPs (e.g., interstitial collagenases, gelatinases, and stromelysins) and serine proteases, in particular the PA/plasmin system (Alexander and Werb, 1991; Liotta et al., 1991; Mignatti and Rifkin, 1996; Pepper et al., 1996b). The central component of the PA/plasmin system is plasmin, which is capable of degrading directly, or indirectly through the activation of latent MMPs, most ECM proteins. Plasmin is generated from its inactive precursor plasminogen by the activity of two PAs, urokinase-type PA (uPA) and tissue-type PA (tPA). uPA activity can be localized to the cell surface through binding to a specific high affinity receptor (uPAR), and uPA and tPA are subject to inhibition by specific physiological PA inhibitors (PAI-I and PAI-2) (Vassalli et al., 1991).

Our earlier studies described above (Montesano et al., 1983) had shown that endothelial cells grown on a collagen gel form a monolayer on the surface of the gel and do not invade the underlying matrix. We next asked whether induction of matrix-degrading protease synthesis might allow endothelial cells to penetrate into the collagen matrix, as they do during angiogenesis in vivo. To investigate this possibility, confluent monolayers of microvascular endothelial cells on collagen gels were treated with phorbol myristate acetate (PMA), a tumor promoter that markedly stimulates the production of collagenase and PAs (Gross et al., 1982). Whereas control endothelial cells were confined to the surface of the gels, PMA-treated endothelial cells invaded the underlying collagen matrix, within which they formed capillary-like tubular structures (Fig. 1B; Table I). The finding that invasion and tube formation are associated with degradation of collagen fibrils (Montesano and Orci, 1985) and are prevented either by the metal chelator 1,10-phenanthroline (Montesano and Orci, 1985), by synthetic low molecular weight MMP inhibitors (Fischer et al., 1994; R. Montesano and M.S. Pepper, unpublished data), or by recombinant tissue inhibitors of metalloproteinases (TIMPs) (Fischer et al., 1994; Anand-Apte et al., 1997) (Table II), suggests an important role for interstitial collagenases and/or other metalloproteinases in angiogenesis.

During angiogenesis, new capillary blood vessels arise exclusively from existing capillaries or postcapillary venules, not from arteries or veins. The reason for this selective origin are

Table I Inducers of angiogenesis in collagen gel model

Cytokines:	
bFGF	Montesano et al., 1996
aFGF + heparin	unpublished
$VEGF_{165}$	Pepper et al., 1992a
$VEGF_{121}$	unpublished
Activators of protein kinase C:	
PMA	Montesano & Orci, 1985
Bryostatin I (200 ng/ml)	unpublished
Mezerein (100 ng/ml)	unpublished
Others:	
Sodium orthovanadate	Montesano et al., 1988
Hyaluronan oligosaccharides	Montesano et al., 1996

Table II Inhibitors of angiogenesis in collagen gel model

Cytokines:	
TGF-β*	Pepper et al., 1990, 1993b
TNF-α*	unpublished
IFN-α2a	Pepper et al., 1994c
LIF	Pepper et al., 1995
Steroids and metabolites:	
2-methoxyestradiol	Fotsis et al., 1994
Synthetic angiostatic steroids (U-24067, U-42129)	Pepper et al., 1994c
Dexamethasone (1 nM–1 μM)	unpublished
Hydrocortisone (100 ng/ml–10 μg/ml)	unpublished
Protease inhibitors:	
TIMP-2, TIMP-3	Anand-Apte et al., 1997
Synthetic MMP inhibitors	unpublished
α2-antiplasmin	Mandriota and Pepper, 1997
Others:	
Genistein and isoflavonoid analogues	Fotsis et al., 1993, 1997
Retinoic acid	Pepper et al., 1994c
Heparin	Pepper et al., 1994c
Suramin*	Pepper et al., 1994c
Protamine (100 μg/ml)	unpublished
dbcAMP (1 mM)	unpublished
IBMX (1.5 mM); Theophylline (1.5 mM)	unpublished
LiCl (10–30 mM)	unpublished

*Biphasic effect.

not known. Are the endothelial cells of large vessels less responsive to angiogenic stimuli than capillary endothelial cells, or are there other factors, such as the presence of a smooth muscle cell layer, that prevent angiogenesis from occurring in these vessels? To establish whether the potential to invade a collagen matrix as capillary-like sprouts is specifically restricted to microvascular endothelium, endothelial cells isolated from large vessels (human umbilical vein and calf pulmonary artery) were grown to confluence on collagen gels and subsequently treated with PMA. Both types of large vessel endothelial cells were able to invade the underlying collagen matrix as vessel-like tubular structures following treatment with PMA, indicating that endothelial cells not normally involved in the neovascularization process can be induced to express an angiogenic phenotype in response to appropriate signals (Montesano and Orci, 1987).

The collagen gel invasion assay described above has been utilized by several groups to investigate the effect of various cytokines or tumor cell conditioned media, as well as the role of proteolytic enzymes in angiogenesis in vitro (see for example Leibovich et al., 1987; Yasunaga et al., 1989; Mawatari et al., 1991; Gajdusek et al., 1993; Sato et al., 1993; Laaroubi et al, 1994; Wang et al., 1994; Fischer et al., 1994; Deroanne et al., 1996). In addition, a modified version of our collagen gel invasion assay using Millipore chambers has been described by Okamura et al. (1992) and by Sakuda et al. (1992).

Basic Fibroblast Growth Factor Induces Angiogenesis In Vitro

The experiments described above indicated that endothelial cells, even after repeated passage in culture, retain the potential to express a latent "angiogenic program," that may be switched on by appropriate signals. To establish whether physiological messengers could elicit an angiogenic response similar to that induced by PMA, microvascular endothelial cells grown on collagen gels were treated with bFGF, one of the best characterized angiogenic polypeptides (reviewed by Baird and Klagsbrun, 1991; Basilico and Moscatelli, 1992; Pepper et al., 1996a). As previously observed in response to PMA, bFGF induced endothelial cells to invade the underlying collagen matrix and to form capillary-like tubules (Fig. 2; Table I) Concomitantly, bFGF stimulated the endothelial cells to produce PAs (Montesano et al., 1986). These results showed that, in vitro, bFGF can induce two essential components of angiogenesis, namely invasion of a three-dimensional ECM and morphogenesis of endothelial tubules. In addition, these studies highlighted the importance of three-dimensional cell-ECM interactions in the response of endothelial cells to angiogenic factors.

Since various growth factors, including bFGF, bind to and activate tyrosine kinase receptors, we subsequently investigated the potential role of protein tyrosine phosphorylation in the angiogenic response. We found that sodium orthovanadate, an agent which inhibits phosphotyrosine phosphatases (enzymes that reverse the action of tyrosine kinases), mimicked the effect of bFGF by inducing endothelial cells to form capillary-like tubes within collagen gels (Table I) and to produce increased amounts of PAs. These results suggested that tyrosine phosphorylation may represent a signal transduction pathway involved in the regulation of angiogenesis (Montesano et al., 1988). This hypothesis is further supported by the finding that genistein, an inhibitor of tyrosine kinases, inhibits bFGF-induced angiogenesis in the collagen gel invasion assay (Table II) (Fotsis et al., 1993).

Protease Inhibitors Play an Important Permissive Role in Angiogenesis

The coordinate modulation of invasive behavior and PA production by PMA, bFGF, and sodium orthovanadate (see above), together with our demonstration of an increase in uPA and uPAR expression (Pepper et al., 1987, 1993a) in endothelial cells migrating from the edge of an experimental wound in vitro, supported the proposed role for the PA/plasmin system (Gross et al., 1982) in angiogenesis. The expression of proteolytic activity by endothelial cells must, however, be tightly controlled in order to prevent inappropriate matrix degradation. That the balance between proteases and protease inhibitors might be important for normal capillary morphogenesis was first demonstrated in experiments in which fibrin gels were substituted for collagen gels, the reasoning being that angiogenesis often occurs in a fibrin-rich matrix, for example during wound healing or as a consequence of vascular hyperpermeability in inflammation and tumors. In striking contrast to what we had observed with collagen gels, we noted that upon addition of the angiogenic stimulus, endothelial cells progressively lysed the underlying fibrin gel. The absence of a three-dimensional substrate therefore precluded invasion and the formation of capillary-like tubules. However, inhibition of excessive fibrinolysis by addition of serine protease inhibitors allowed the preservation of a three-dimensional matrix into which endothelial cells migrated to form tubelike structures (Fig. 3). This study highlights the notion that although

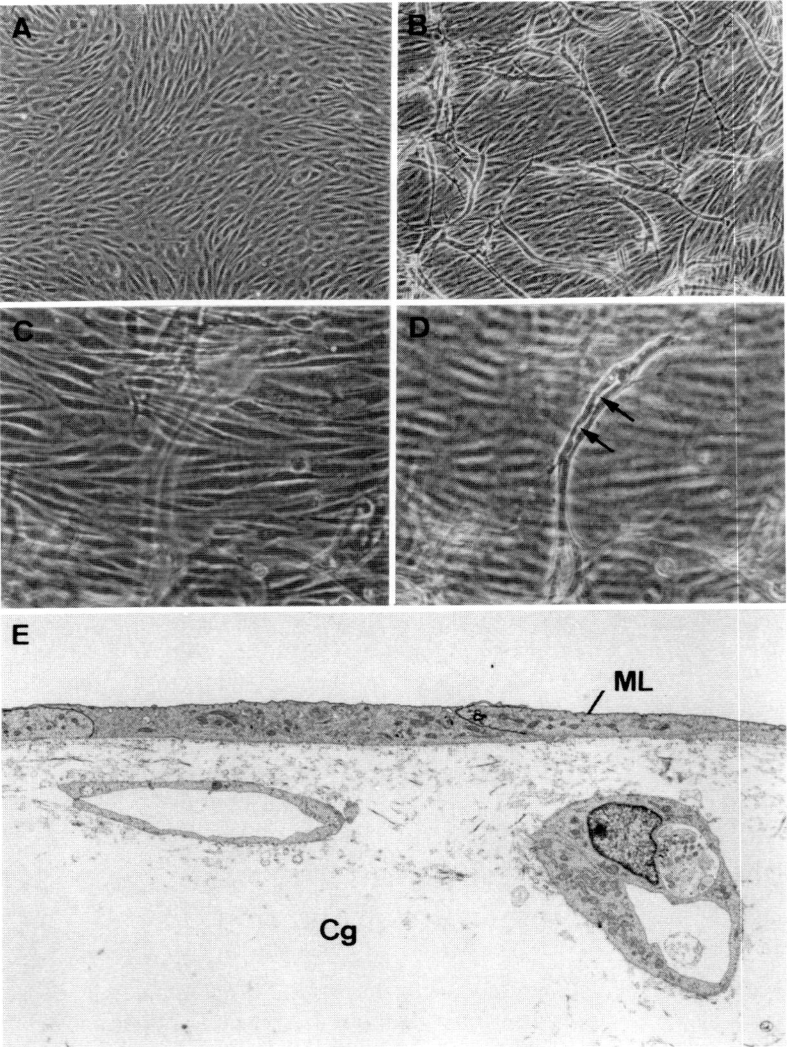

Fig. 2 bFGF induces in vitro angiogenesis in the collagen gel invasion assay. (A) Microvascular endothelial cells grown on a collagen gel form a monolayer of closely apposed cells. (B) Three days after addition of bFGF, the cells have formed a network of branching cords. (C,D) Higher magnification of a bFGF-treated culture. The same field is shown at two different planes of focus. In (C), the focus is on the gel surface. By focusing beneath the surface monolayer (D), an endothelial cell cord containing a lumen-like translucent space (arrows) can be seen. The cord appears blurred in (C). When viewed in cross-section by electron microscopy (E), the tubular nature of the invading cell cords, which are morphologically similar to capillaries seen in vivo, can be appreciated. ML, endothelial monolayer; Cg, collagen gel. A,B = 70x; C,D = 175x; E = 3000x. (A–D are reproduced from Montesano et al., 1994, with copyright permission from Ares-Serono Symposia. E is reproduced from Pepper et al., 1996a, with copyright permission from Springer-Verlag.)

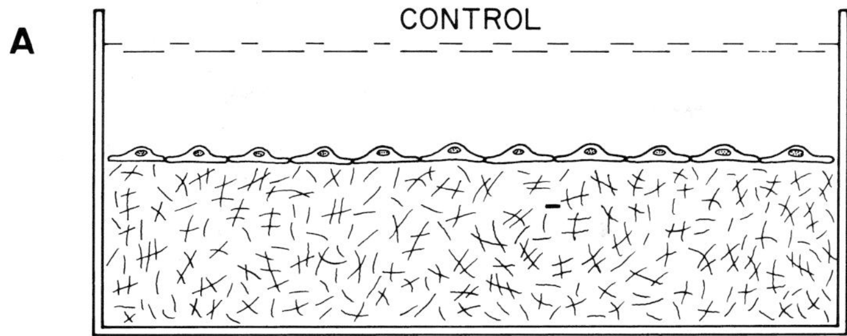

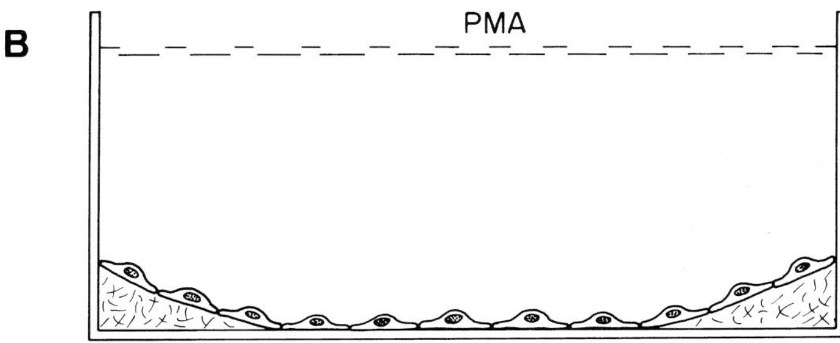

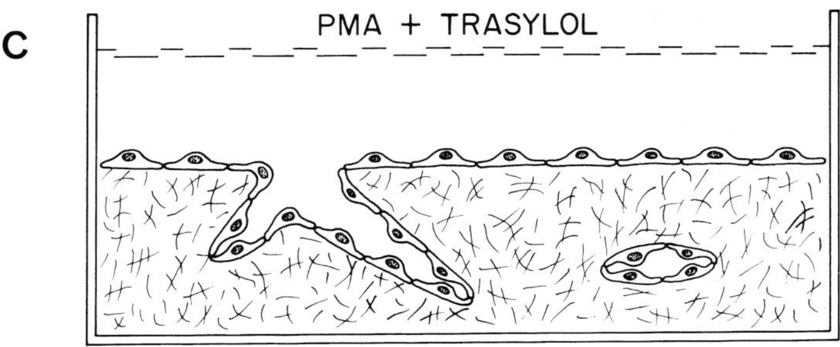

Fig. 3 Three-dimensional fibrin gel invasion assay. When a confluent monolayer of endothelial cells on a three-dimensional fibrin gel (A) is treated with an in vitro angiogenic stimulus, such as the tumor promoter PMA, the underlying matrix is progressively lysed (B). When a serine protease inhibitor such as Trasylol is added at the same time as the angiogenic stimulus, excessive matrix lysis is prevented, thereby ensuring the presence of an intact three-dimensional matrix scaffold into which stimulated endothelial cells can migrate to form capillary-like tubes (C). These observations suggest that protease inhibitors play an important permissive role during angiogenesis by preventing excessive ECM degradation and thereby preserving matrix integrity.

increased protease activity is clearly associated with the invasive phenotype, protease inhibitors play an equally important albeit permissive role in angiogenesis, by preventing excessive and unnecessary matrix destruction and thereby ensuring the integrity of the ECM scaffold (Montesano et al., 1987).

Additional support for the role of protease inhibitors in normal capillary morphogenesis has come from our observations on the behavior of endothelial cells expressing the polyoma virus middle-T (mT) oncogene. It had previously been shown that mT induces cyst-like endothelial tumors (endotheliomas) when expressed in chimeric or transgenic mice (Bautch et al., 1987; Williams et al., 1989). We have developed an in vitro correlate of endothelioma formation by embedding mT-expressing endothelial cells into three-dimensional fibrin gels. In contrast to normal endothelial cells, which formed a network of capillary-like tubules, mT-expressing endothelial cells formed large cystlike structures which bear striking resemblance to endotheliomas observed in vivo. When studying the proteolytic properties of mT-expressing endothelial cells, we found that these cells displayed increased PA activity when compared to non-mT-expressing endothelial cells, and that this could be accounted for by an increase in uPA and a decrease in PAI-1 activity. With these observations in mind, we asked what would happen if we attempted to reduce the excess proteolysis by adding protease inhibitors to the culture system. We found that when serine protease inhibitors were added to the cultures, the mT-expressing endothelial cells, instead of forming cysts, now formed branching capillary-like tubules (Montesano et al., 1990). These results demonstrated that excessive proteolytic activity is not compatible with normal capillary morphogenesis, but that by reducing this activity by the addition of protease inhibitors, normal morphogenetic properties can be restored to the cells. For a more detailed discussion of the properties of mT-expressing endothelial cells and their relevance to vascular tumors, see Pepper et al. (1997).

Proteolytic Balance and Capillary Lumen Formation

A protective role for protease inhibitors was also suggested by the finding that bFGF, PMA, and vanadate, three agents which induce invasion and tube formation in the three-dimensional in vitro model described above, increase not only uPA in microvascular endothelial cells (Gross et al., 1982; Montesano et al., 1986, 1988; Moscatelli et al., 1986), but also PAI-1 (Saksela et al., 1987; Pepper et al., 1990). Similarly, wound-induced two-dimensional migration is characterized by a concomitant increase in uPA and PAI-1 (Pepper et al., 1987, 1992b, 1993a). In an attempt to understand the respective roles of proteases and protease inhibitors in angiogenesis, we studied the modulation of uPA and PAI-1 by bFGF and TGF-β1, a cytokine which greatly increases PAI-1 production by endothelial cells (Saksela et al., 1987; Pepper et al., 1990). In addition to increasing PAI-1, we also observed that TGF-β1 increased uPA expression in microvascular endothelial cells. However, when using the uPA/PAI-1 mRNA ratio as a reflection of the potential proteolytic activity of the cells, the net response to TGF-β1 was always antiproteolytic, in marked contrast to the large increase in potential proteolysis observed in response to bFGF. When cells were exposed to both bFGF and TGF-β1, levels of proteolysis as represented by the uPA/PAI-1 mRNA ratio, mimicked those seen in controls (Pepper et al., 1990).

Knowing that TGF-β1 was capable of modulating bFGF-induced proteolysis, we next assessed its effect on bFGF-induced capillary-like tube formation in vitro. Experiments aimed at addressing this problem were performed in fibrin rather than collagen gels in order to assay more specifically for the PA system. In response to bFGF, the cells invaded the underlying gel, resulting in the formation of branching tubelike structures with large, ectatic lumina. When added alone, TGF-β1 had no effect. However, when co-added with bFGF, lumen diameter of capillary-like tubes was markedly reduced. Although this was true both at 500 pg/ml and 5 ng/ml TGF-β1, doses which increased and decreased bFGF-induced invasion respectively (see next section), the presence of a lumen was less frequently observed at 5 ng/ml than at 500 pg/ml. Furthermore, lumen size at 500 pg/ml TGF-β1 was reduced to a size which was physiologically more relevant (Pepper et al., 1990, 1993b). Since the creation of a hollow space (i.e., the lumen) within the fibrin gel is dependent on fibrinolysis, these findings suggest that the antiproteolytic effect of TGF-β1, which results from a large increase in PAI-1, is likely to be responsible, at least in part, for the reduction in lumen size.

In addition to the model described above, in which tubule formation is induced by exogenous stimuli, a three-dimensional model of spontaneous angiogenesis has been developed for the purpose of identifying potential physiological inhibitors. Endothelial cells are seeded onto a non-adhesive agarose substrate, which results in the formation of solid cell aggregates floating in the culture medium. These aggregates are then embedded into three-dimensional collagen or fibrin gels. After a few hours, endothelial sprouts begin to grow out spontaneously from the original aggregate, resulting after a few days in the formation of radially-disposed hollow endothelial tubes. The observation that the capillary lumen is devoid of ECM suggests that lumen formation is proteolysis-dependent. When the serine protease inhibitor Trasylol was added to fibrin gel cultures, the cells migrated out as solid endothelial cords, i.e., lumen formation was completely inhibited. These findings directly demonstrate that lumen formation in fibrin gels requires extracellular serine protease activity (Pepper et al., 1991a). Since cartilage is one of the few avascular tissues in the body, we also wished to determine whether chondrocytes might produce a factor which inhibits endothelial sprout formation in vitro using the model just described. When chondrocytes were co-incorporated in collagen or fibrin gels with endothelial cell aggregates, or added to the culture medium above the gel, sprout formation normally seen in controls was markedly inhibited. On the basis of antibody inhibition studies, we demonstrated that a chondrocyte-derived TGF-β1 was at least in part responsible for inhibition of sprout formation in this experimental system (Pepper et al., 1991a).

Synergism Between VEGF and bFGF in the Induction of Angiogenesis In Vitro

Very little is known about interactions between angiogenesis-modulating cytokines. It is highly likely however that endothelial cells are rarely (if ever) exposed to a single cytokine during physiological and pathological processes. In order to explore potential interactions between angiogenesis-modulating cytokines, we have used the collagen gel invasion assay described above, in which the extent of endothelial tube formation in response to angiogenic factors can be quantitated by measuring the total additive length of all cellular structures which have penetrated from the surface monolayer into the underlying gel. Using this three-dimensional model, we wished to assess the effect of simultaneous addition of bFGF and VEGF on the in vitro angiogenic response.

As mentioned earlier, a variety of cytokines have been implicated in the control of angiogenesis (Folkman and Shing, 1992). However, VEGF is unique among these angiogenic factors in that it is the only secreted cytokine discovered so far that is mitogenic specifically for endothelial cells (Ferrara et al., 1992a). Four isoforms, which vary in their relative proportions in different tissues, are generated through alternative splicing of a single mRNA. The most abundant and most extensively studied isoform contains 165 amino acids ($VEGF_{165}$). VEGF is produced by a variety of both normal and tumor cells, and its expression is upregulated by hypoxia (Shweiki et al., 1992). In addition, the expression of VEGF and its tyrosine kinase receptors correlates strongly with periods of vasculogenesis and angiogenesis during embryonic development (Breier et al., 1992; Millauer et al., 1993; Peters et al., 1993) and with programmed neovascularization that occurs during the female reproductive cycle (Shweiki et al., 1993). Finally, neutralization of VEGF (Kim et al., 1993) or inhibition of VEGF receptor function (Millauer et al., 1994) suppresses tumor growth. VEGF is therefore considered to be a major positive regulator of physiological and pathological angiogenesis (reviewed by Ferrera et al., 1992a; Klagsbrun and Soker, 1993; Neufeld et al., 1994; Dvorak et al., 1995; Thomas, 1996; Pepper et al., 1996a). We found that, like bFGF (Montesano et al., 1996), VEGF induces microvascular endothelial cells grown on collagen gels to invade the underlying matrix, within which they form capillary-like tubules (Table I) (Pepper et al., 1992a). The most striking effect of VEGF, however, was observed in combination with bFGF: when added simultaneously, VEGF and bFGF induced an in vitro angiogenic response which was far greater than additive and which occurred with greater rapidity than the response to either cytokine alone (Fig. 4A). These results demonstrate that, by acting in concert, these two cytokines have a potent synergistic effect on the induction of angiogenesis in vitro (Pepper et al., 1992a).

In attempting to understand the mechanisms responsible for this synergistic effect, we initially assessed the effect of co-addition of bFGF and VEGF in conventional two-dimensional in vitro assays of endothelial cell proliferation, migration and PA-mediated extracellular proteolysis. In none of these situations was the effect of simultaneous addition of bFGF and VEGF greater than additive (Pepper et al., 1994a; Mandriota et al., 1995; see also Yoshita et al., 1996), although others have been able to detect a synergistic effect of bFGF and VEGF on endothelial proliferation when the cells are grown in three dimensions (Goto et al., 1993). Our subsequent approach has been to determine whether bFGF and VEGF might modulate expression of receptors for FGF and VEGF in monolayer culture. We have found that while neither cytokine, either alone or in combination, is capable of modulating expression of FGF receptor-1 (FGFR-1), bFGF increases expression of VEGF receptor-2 (VEGFR-2 or Flk-1) in bovine endothelial cells (Mandriota and Pepper, unpublished observation).

What are the implications of these findings? Modulation of new capillary blood vessel formation may serve as an alternative/adjunct to current therapeutic modalities in several angiogenesis-associated diseases. At first sight, the redundancy of angiogenesis-regulating cytokines might suggest that therapeutic strategies based on neutralization of single angiogenic factors might be unrealistic. If however the synergism which we have observed in vitro is relevant to the endogenous regulation of angiogenesis in vivo, angiogenesis would be more prominent in tumors or other pathologic settings in which more than one angiogenic factor is produced. This may justify anti-angiogenesis strategies based on the neutralization of a single angiogenic factor, since this would reduce the synergistic effect. On the other hand, recent work has demonstrated that administration of angiogenic factors can enhance the growth of collateral vessels in animal models of coronary, peripheral, and cerebral arterial occlusion (reviewed by Höckel et al., 1993; Symes and Sniderman, 1994; Folkman, 1995b). We have suggested that the effect of co-addition of two cytokines whose interaction is synergistic would be greater than that derived from the addition of one of these cytokines alone. Support for this hypothesis has recently been provided by an in vivo study in which co-administered bFGF and VEGF synergized in the induction of collateral blood vessel formation in a rabbit model of hind limb ischemia (Asahara et al., 1995). In summary, our findings on the synergism between bFGF and VEGF may have relevance both to understanding the biology of angiogenesis (for a detailed discussion of this point, see Pepper et al.,

1996a) as well as to positive and negative therapeutic modulation of this process. Our observations also highlight the importance of a three-dimensional environment for the study of angiogenesis in vitro: had we relied exclusively on traditional two-dimensional assays of proliferation, migration, or proteolysis, the synergism between bFGF and VEGF would not have been detected.

Are cytokines which induce neovascularization in the blood vascular system also operative in the lymphatic system? This issue was addressed by assessing the in vitro angiogenic and proteolytic properties of endothelial cells isolated from bovine lymphatic vessels. We found that lymphatic endothelial cells respond to bFGF and VEGF in a manner similar to what has previously been observed with endothelial cells derived from the blood vascular system, i.e., they are induced to form capillary-like tubules in the collagen gel invasion assay and to increase their expression of uPA, uPAR, and PAI-1. These results suggest that bFGF and VEGF, cytokines with well-recognized angiogenic properties in the blood vascular system, may also be operative in the lymphatic system (Pepper et al., 1994b).

Biphasic Effect of TGF-β1 on In Vitro Angiogenesis

TGF-β is an angiogenesis-modulating cytokine that has variously been described as angiogenic or anti-angiogenic. In vivo, TGF-β is a potent inducer of angiogenesis (Roberts et al., 1985; Yang and Moses, 1990), whose effect is believed to be mediated by secretory products of TGF-β-recruited connective tissue and inflammatory cells (Wahl et al., 1987; Wisemann et al., 1988; Yang and Moses, 1990; Phillips et al., 1992). In vitro, however, TGF-β inhibits a number of essential components of the angiogenic process. These include endothelial cell proliferation (Baird and Durkin, 1986; Fràter-Schröder et al., 1986; Müller et al., 1987), migration (Heimark et al., 1986; Müller et al., 1987; Sato and Rifkin, 1989), and extracellular proteolytic activity (Saksela et al., 1987; Pepper et al., 1990, 1991b). Results from three-dimensional in vitro assays demonstrate that the response to TGF-β varies depending on the assay used. Thus, TGF-β inhibits endothelial cell invasion of three-dimensional collagen gels (Müller et al., 1987), as well as the invasion of the explanted amnion (Mignatti et al., 1989). In addition, as described in a previous section, TGF-β (5 ng/ml) inhibits bFGF-induced capillary lumen formation within three-dimensional fibrin gels (Pepper et al., 1990). These results support the notion that TGF-β is a direct-acting inhibitor of extracellular matrix invasion and tube formation. However, it has also been reported that TGF-β promotes organization of endothelial cells into tubelike structures (Madri et al., 1988; Merwin et al., 1990). These apparently conflicting results may be reconciled by considering that TGF-β might have different functions on vessel formation at different stages of the angiogenic process. Thus when acting directly on endothelial cells, it may inhibit invasion and vessel formation, and once sprout formation has occurred, TGF-β may be necessary for the inhibition of further endothelial cell replication and migration, and induce vessel organization and functional maturation. An additional possibility is that the direct effect of TGF-β on endothelial cell function is concentration dependent, particularly since this cytokine has been described as a bifunctional regulator in a variety of other biological processes (Nathan and Sporn, 1991).

The effect of a wide range of concentrations of TGF-β1 on bFGF- or VEGF-induced angiogenesis was assessed in our three-dimensional in vitro model. We found that in the presence of TGF-β1, bFGF- or VEGF-induced invasion was increased at 200–500 pg/ml TGF-β1 and decreased at 5–10 ng/ml TGF-β1 (Fig. 4B, Fig. 5, and Table II). The inhibitory effect at relatively high concentrations is in accord with previous studies in which endothelial cell invasion of three-dimensional collagen gels (Müller et al., 1987) or the explanted amnion (Mignatti et al., 1989) were inhibited by TGF-β1 at 1–10 ng/ml. These results clearly demonstrate that the effect of

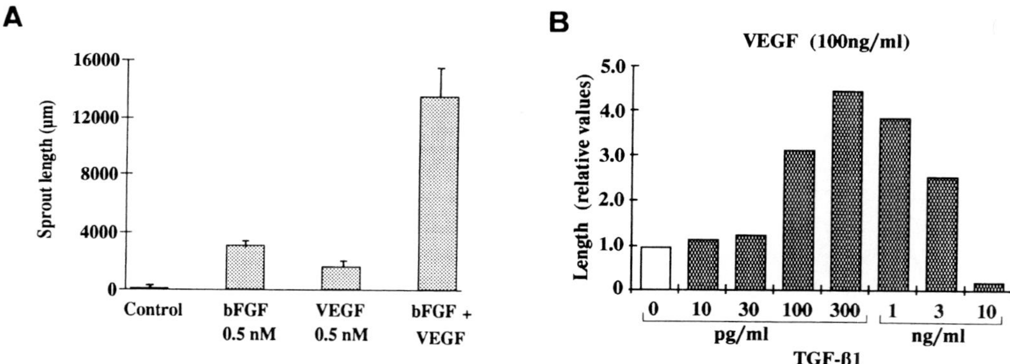

Fig. 4 Cytokine interactions and angiogenesis in vitro. (A) Synergistic effect of bFGF and VEGF on in vitro angiogenesis. Endothelial cell invasion was quantitated by measuring the length of all cell cords which had penetrated beneath the surface monolayer. At equimolar concentrations (0.5 nM), bFGF was about twice as potent as VEGF. Co-addition of the two cytokines induced an invasive response which was greater that additive. (From Pepper et al., 1992, with copyright permission from Academic Press, Inc.) (B) Quantitative analysis of the effect of TGF-β1 on VEGF-induced collagen gel invasion. Confluent monolayers of microvascular endothelial cells were treated for 4 days with both VEGF (100 ng/ml) and TGF-β1, or with VEGF alone, and the length of invading cell cords was determined. (From Pepper et al., 1993b, with copyright permission from Academic Press, Inc.)

TGF-β1 on bFGF- or VEGF-induced in vitro angiogenesis is concentration-dependent (Pepper et al., 1993b).

The mechanisms responsible for the biphasic effect of TGF-β1 are not known. One hypothesis is based on alterations in the net balance of extracellular proteolysis (Pepper and Montesano, 1990). Thus, at the dose of TGF-β1 which potentiates bFGF- or VEGF-induced invasion, an optimal balance between proteases and protease inhibitors may be achieved at the cell surface, which allows for focal pericellular matrix degradation, while at the same time protecting the matrix against excessive degradation and inappropriate destruction (Pepper et al., 1994b). However, we also have evidence to suggest that integrin expression is differentially affected at these different concentrations of TGF-β1 (G. Collo and M.S. Pepper, manuscript in preparation). The relative contribution of these parameters, namely proteases and integrins, is currently under investigation. Finally, concentrations of TGF-β1 which are inhibitory in our assay, also decrease expression of VEGF receptor-2 (Flk-1) in vascular endothelial cells (Mandriota et al., 1996). To summarize the effects of TGF-β on the angiogenic response, it could be stated that the direct effect of TGF-β on endothelial cells not only varies at different stages of the angiogenic process, but is also concentration-dependent. Thus, in addition to its indirect angiogenic effect, TGF-β could either promote or inhibit angiogenesis when acting directly on endothelial cells.

Leukemia Inhibitory Factor

Leukemia inhibitory factor (LIF), oncostatin M (OSM), interleukin-6 (IL-6), interleukin-11 (IL-11), and ciliary neurotrophic factor (CNTF) are cytokines which share a common signal

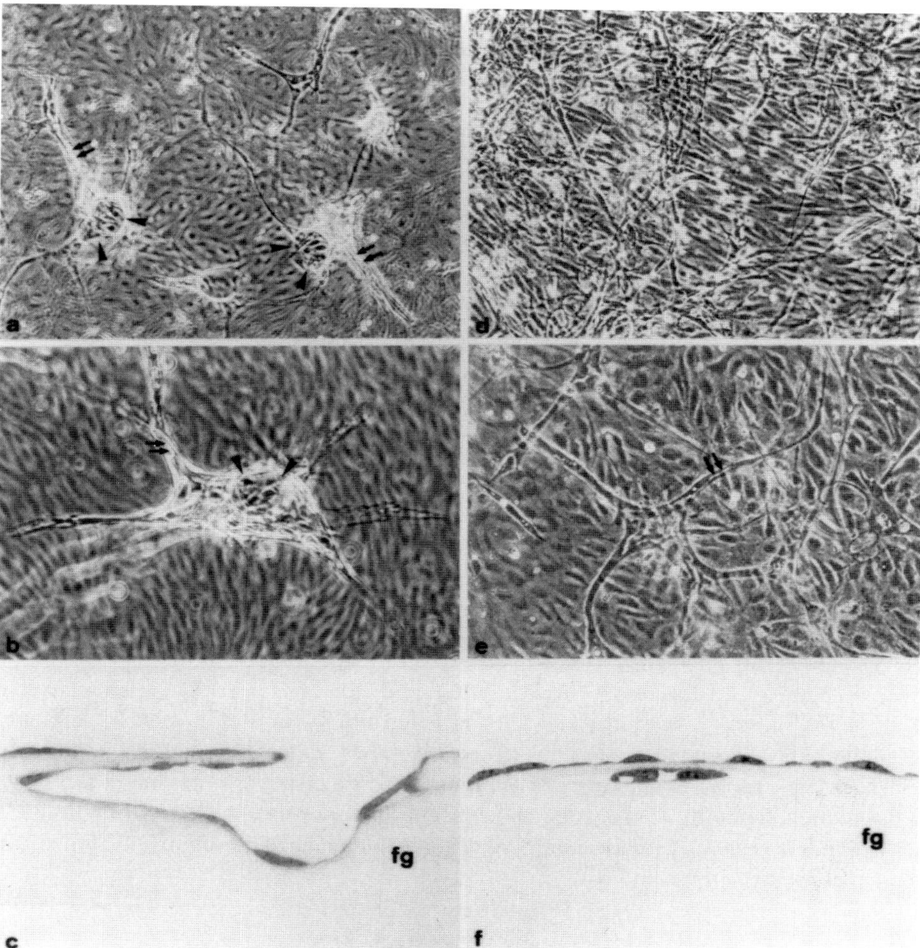

Fig. 5 Morphological analysis of invading cell cords induced by bFGF or co-addition of bFGF and TGF-β1. bFGF (30 ng/ml) was added without (a–c) or with 500 pg/ml TGF-β1 (d–f) to confluent monolayers of microvascular endothelial cells on fibrin gels. The resulting capillary-like tubular structures were viewed by phase-contrast microscopy (a, b, d, and e) and were further assessed by examination of semi-thin sections (c and f). bFGF induced endothelial cells to invade from a circular opening in the surface monolayer (arrowheads in a and b), to form well-organized cell cords with a clearly visible refringent lumen (arrows in a and b), which tapered down progressively in the distal part of the cords. These observations were confirmed by semi-thin sectioning, in which the proximal part of the cords was often seen to be cavernous (c). When 500 pg/ml TGF-β1 was co-added with bFGF, the total additive length of the invading cell cords was increased (compare a and d). Clearly distinguishable lumina were present beneath the surface monolayer (white refringent line indicated by the arrows in e), although lumen size was decreased to a more physiological size when compared to cultures treated with bFGF alone (compare c and f). Bars: a, b, d, and e, 100 μm; c and f, 20 μm. (From Pepper et al., 1993b, with copyright permission from Academic Press, Inc.)

transduction pathway that is mediated via the gp130 signal converter (reviewed by Stahl and Yancopoulos, 1993; Taga and Kishimoto, 1993). LIF is a highly glycosylated 180 amino acid single chain polypeptide varying in molecular weight from 38–67 kDa. Although LIF was initially purified and cloned using a bioassay based on its ability to induce monocyte differentiation, it has a multitude of effects on both hemopoietic and nonhemopoietic cells, and like OSM, IL-6 and IL-11, is one of a growing number of cytokines which are characterized by pleiotropy and functional redundancy (reviewed by Hilton, 1992; Alexander et al., 1994).

Based on the observation that LIF inhibits aortic endothelial cell proliferation in vitro (Ferrara et al., 1992b), we have assessed LIF's potential as an angiogenesis regulator using our three-dimensional in vitro model. LIF was found to inhibit invasion with an IC_{50} of less than 1 ng/ml, and to lack the concentration-dependent stimulatory effect characteristic of TGF-β1 (Table II). The inhibitory effect was observed on both microvascular and aortic endothelial cells and occurred irrespective of the angiogenic stimulus, which included bFGF, VEGF, or the synergistic effect of the two factors in combination (Pepper et al., 1995). Inhibition of invasion could be correlated with inhibition of proliferation of both microvascular and aortic endothelial cells, although this was more marked with the latter, confirming previous observations (Ferrara et al., 1992b). In addition, LIF decreased the proteolytic potential of both microvascular and aortic endothelial cells by increasing their expression of PAI-1 (Pepper et al., 1995). Since LIF shares a common signal transduction pathway with OSM, IL-6, IL-11, and CNTF, the effects of these cytokines in our in vitro invasion model are currently being assessed.

To summarize, using a three-dimensional model of in vitro angiogenesis, we have demonstrated that important interactions exist between different cytokines in the in vitro angiogenic response. Synergism was observed between bFGF and VEGF, TGF-β1 had a biphasic effect on bFGF- or VEGF-induced invasion, and LIF inhibited bFGF- or VEGF-induced angiogenesis. These interactions appear to be limited to the cytokines indicated, since a number of other cytokines we have tested in our collagen gel model had no effect (Montesano et al., 1993). We suggest that the temporally-coordinated and concentration-dependent activity of a limited number of cytokines is necessary for the control of different elements of the angiogenic process in specific and appropriate settings in vivo.

Modulation of Endothelial Cell Invasive and Proteolytic Properties by Inhibitors of Angiogenesis

Since, as discussed earlier, a tightly controlled increase in extracellular proteolysis is an important component of the angiogenic process, we investigated whether previously described inhibitors of in vivo angiogenesis alter the invasive and proteolytic properties of microvascular endothelial cells. We found that although synthetic angiostatic steroids (U-24067 and U-42129), heparin, suramin, interferon alpha-2a and retinoic acid all inhibit bFGF-induced invasion and tubule formation in collagen gels (Table II), each of these agents has distinct effects on the PA-dependent proteolytic system (Pepper et al., 1994c). Thus, although it has been shown that exogenously added protease inhibitors suppress angiogenesis both in vivo and in vitro (Mignatti et al., 1989; Moses et al., 1990), it does not necessarily follow that agents which inhibit angiogenesis will reduce extracellular proteolysis. A reduction in extracellular proteolysis would be expected to decrease the ability of endothelial cells to overcome the mechanical barriers imposed by the surrounding ECM. However, since proteases also modulate cytokine activity, an increase in proteolysis could activate latent inhibitory cytokines, which in turn would inhibit endothelial cell proliferation and migration and indirectly reduce extracellular proteolytic activity (Flaumenhaft et al., 1992). The spectrum of effects on different elements of the PA system observed in response to the agents assessed suggests that the role of modulations in extracellular proteolytic activity in anti-angiogenesis is likely to be varied and complex. On the other hand, while assessing the effect of angiogenesis inhibitors on endothelial cell proliferation and migration in conventional two-dimensional assays, we observed an excellent correlation between inhibition of bFGF-induced invasion of collagen gels and inhibition of migration, whereas a less consistent correlation was observed with proliferation (Pepper et al., 1994c). This suggests that although three-dimensional collagen gels are required for histotypic capillary-like tube formation, the two-dimensional migration assay might serve as an appropriate substitute in the screening of potential pharmacological inhibitors of angiogenesis.

Synergistic Effect of Hyaluronan Oligosaccharides and VEGF on Angiogenesis In Vitro

In addition to diffusible cytokines, ECM components including collagens (Madri and Williams, 1983; Montesano et al., 1983), laminin (Grant et al., 1989; Nicosia et al., 1994), and other glycoproteins (Ingber and Folkman, 1989; Lane et al., 1994; Canfield and Schor, 1995) have been shown to be important in angiogenesis. Hyaluronan (hyaluronic acid, HA), a glycosaminoglycan composed of repeating disaccharide units of D-glucuronate and N-acetylglucosamine, is one of the most abundant constituents of the ECM. Originally considered primarily as a structural moiety, HA has now emerged as an important signaling molecule (reviewed by Toole, 1990; Laurent and Fraser, 1992; Knudson and Knudson, 1993). Thus, HA is involved in a number of developmental processes (reviewed by Toole, 1991; Fenderson et al., 1993), and has been shown to promote cell proliferation (Yoneda et al., 1988), differentiation (Kujawa et al., 1986a, 1986b), and motility (West and Kumar, 1988; Boudreau et al., 1991; Turley et al., 1991; Savani et al., 1995). The diverse biological activities of HA are believed to be mediated, at least in part, through interaction with specific cell surface receptors such as CD44 and RHAMM (Aruffo et al., 1990; Culty et al., 1990; Miyake et al., 1990; Stamenkovic et al., 1991; Banerjee and Toole, 1991; Hardwick et al., 1992; Sherman et al., 1994) resulting in activation of intracellular signalling events (Hall et al., 1994; Slevin et al., 1996). HA obtained from different tissue sources exhibits considerable variation in size, and its biological activity has been shown to be critically dependent on molecular mass in a number of experimental systems, including angiogenesis (reviewed by Rooney et al., 1995, 1996). Thus, native high-molecular-weight HA is anti-angiogenic (Feinberg and Beebe, 1983), whereas HA degradation products of specific size (3–10 disaccharide units) stimulate endothelial cell proliferation (West and Kumar, 1989; Sattar et al., 1992; Slevin et al., 1996) and migration (West and Kumar, 1988; Sattar et al., 1994; Slevin et al., 1996), and induce angiogenesis in the chick chorioallantoic membrane assay (West et al., 1985), in rat skin (Sattar et al., 1994), and in a cryoinjured skin graft model (Lees et al., 1995). The involvement of HA in the regulation of angiogenesis is also supported by the observation that blocking the interaction of endogenous HA with cell surface HA-binding proteins inhibits both endothelial cell migration from the edge of a wounded monolayer and formation of tubular structures by endothelial cell clumps suspended in collagen/basement membrane gels (Banerjee and Toole, 1992). Finally, histochemical studies have demonstrated that newly-formed blood vessels in the rabbit cornea and chick chorioallantoic membrane are transiently surrounded by a HA-rich pericellular matrix (Ausprunk et al., 1981; Ausprunk, 1986).

We wished to determine whether HA and/or its degradation products influence an essential component of the angiogenic process, i.e., endothelial cell invasion of a three-dimensional ECM. Using our collagen gel assay, we found that like bFGF and VEGF (see above) oligosaccharides of HA (OHA) induce endothelial cells to invade the underlying gel within which they form capillary-like tubes, with an optimal effect at approximately 0.5–2 μg/ml OHA (Table I). Strikingly, co-addition of OHA (0.5–2 μg/ml) and VEGF (30 ng/ml), but not co-addition of OHA and bFGF (10 ng/ml), induced an in vitro angiogenic response which was greater than the sum of the effects elicited by either agent separately. In contrast to OHA, native high-molecular-weight HA (nHA) was consistently inactive, either whether added alone or in combination with VEGF or bFGF (Montesano et al., 1996). Since, as discussed earlier, endothelial cell invasion is believed to require extracellular proteolytic activity, we also investigated the effect of OHA on the PA/plasmin system. OHA (0.01–1 μg/ml), but not nHA, induced a dose-dependent increase in mRNA levels of uPA, uPAR and PAI-1, and a parallel increase in the functional activity of uPA and PAI-1, as determined by zymography and reverse zymography, respectively. The effects of OHA on proteolytic activity were additive with those of VEGF, but not with those of bFGF (Montesano et al., 1996).

The mechanisms by which OHA stimulate endothelial cell invasion of collagen gels and modulate their PA-mediated extracellular proteolytic activity are not known. However, in bovine endothelial cells, OHA have recently been found to induce phosphorylation and activation of MAP kinase (Slevin et al., 1996), as well as up-regulation of early response genes such as c-*fos*, c-*jun*, and *jun*-B (Rooney et al., 1995), which are known to control the expression of a number of other genes including those of matrix-degrading proteases. Since HA receptors, including a CD44-like transmembrane protein, have been identified in bovine endothelial cells (Sattar et al., 1992; Bourguignon et al., 1992; Banerjee and Toole, 1992), it is conceivable that OHA promotes endothelial cell invasion of collagen gels and tube formation by activating intracellular signaling pathways that ultimately result in modulation of pericellular proteolysis. This hypothesis is supported by the ability of OHA to increase uPA and uPAR mRNA at very low concentrations (10 ng/ml), as well as by our finding that OHA-induced endothelial tube formation in collagen gels is almost completely suppressed by pertussis toxin, an inhibitor of G-protein-mediated intracellular signalling (R. Montesano, unpublished observation). The molecular mechanisms responsible for the specific synergistic interaction between OHA and VEGF in the induction of angiogenesis in vitro are also unknown. OHA and VEGF might activate independent but converging intracellular signalling pathways, resulting in a synergistic effect, or OHA might upregulate expression of high affinity VEGF receptors such as Flk-1. Alternatively, as has been shown for heparin-like glycosaminoglycans (Tessler et al., 1994; Schlessinger et al., 1995; Faham et al., 1996; Ornitz et al., 1995), OHA may complex with VEGF molecules, thereby increasing ligand half-life or facilitating multivalent VEGF binding and receptor oligomerization. Although we cannot dismiss the possibility of a physical interaction between OHA and VEGF, we have found that OHA do not protect VEGF against proteolytic degradation.

While exogenously-added OHA promotes angiogenesis in in vitro and in vivo assays, it has not yet been clearly established whether endogenous OHA can act as a physiological regulator of angiogenesis. Several observations nonetheless suggest the potential involvement of OHA in angiogenesis associated with reparative and pathological processes. In a number of clinical settings, including wound healing, rheumatoid arthritis, vasoproliferative retinopathy, and cancer, angiogenesis occurs in close proximity to HA-rich tissues or fluids (Toole et al., 1979; Turley and Tretiak, 1984; Iozzo, 1985; Weigel et al., 1986; Knudson et al., 1989; Bertrand et al., 1992;

Ponting et al., 1993). HA catabolism has been shown to be very rapid: in skin for instance, up to 25% of injected HA is degraded locally in 24 hours (Laurent and Frazer, 1992). Although most vertebrate hyaluronidases so far characterized are lysosomal, HA-degrading activities with near neutral pH optima have recently been shown to be expressed by tumor cells and to induce angiogenesis in vivo (Lokeshwar et al., 1996; Liu et al., 1996). It is therefore conceivable that breakdown of high-molecular-weight HA occurs in the extracellular space during pathological processes. This would result in the production of HA oligosaccharides, which in addition to being angiogenic on their own, could synergize with VEGF, which has been shown to be overexpressed in all the clinical settings mentioned above (reviewed by Dvorak et al., 1995; Ferrara, 1995). Based on our in vitro studies, we therefore propose that the potential therapeutic effect of coadministration of VEGF and OHA deserves to be investigated in situations which would benefit from stimulation of angiogenesis, particularly in animal models of coronary or peripheral arterial insufficiency.

The Collagen Gel Invasion System as a Bioassay for the Identification of Additional Regulators of Angiogenesis

In addition to its use as an experimental system for investigating the mechanisms of angiogenesis, the collagen gel invasion model provides a convenient alternative to widely used endothelial cell proliferation assays for the identification of positive and negative regulators of angiogenesis (Tables I and II). Indeed, endothelial cell proliferation represents only one component of the angiogenic process. In contrast, collagen gel invasion assays rely on the detection of biological activities that induce both endothelial cell migration into a three-dimensional ECM substratum and morphogenesis of capillary tubes, which are two essential components of the angiogenic process. The use of the collagen gel invasion system as a bioassay is exemplified by the coculture experiments described below.

In an attempt to identify additional regulators of angiogenesis, we have developed a coculture system allowing the study of paracrine interactions between microvascular endothelial cells and other types of normal or tumoral cells. In this assay, cells that may potentially produce angiogenic factors are suspended within a collagen gel and overlaid with an additional collagen gel devoid of cells, onto which endothelial cells are subsequently seeded and grown to confluence (Fig. 6a). Among a number of cell types which we have cocultured with endothelial cells in this experimental system, Swiss mouse embryo 3T3 fibroblasts induced a robust in vitro angiogenic response. When microvascular or large vessel endothelial cells grown on the upper cell-free collagen gel layer attained confluence, numerous cell cords began to extend from the surface monolayer into the underlying collagen matrix. Over the next few days of coculture, these cords elongated, branched progressively, and developed patent lumina (Fig. 6b,d). Extensive radial outgrowth of endothelial sprouts was also observed in a second coculture model in which a collagen disc containing a suspension of endothelial cells was surrounded by an annular collagen gel containing Swiss 3T3 cells (Montesano et al., 1993). Conditioned medium from Swiss 3T3 cells mimicked the effect of coculture by inducing endothelial cell invasion of collagen gels and tube formation, and also increased endothelial cell uPA activity (Montesano et al., 1993). The nature of the factor(s) produced by Swiss 3T3 cells which induce(s) angiogenesis in vitro has not yet been determined, but available evidence suggest that it (they) is (are) likely to be different from a number of well-characterized angiogenic cytokines.

In addition to representing a convenient assay for the detection of positive regulators of angiogenesis in conditioned media, the collagen gel invasion model can also be exploited in the

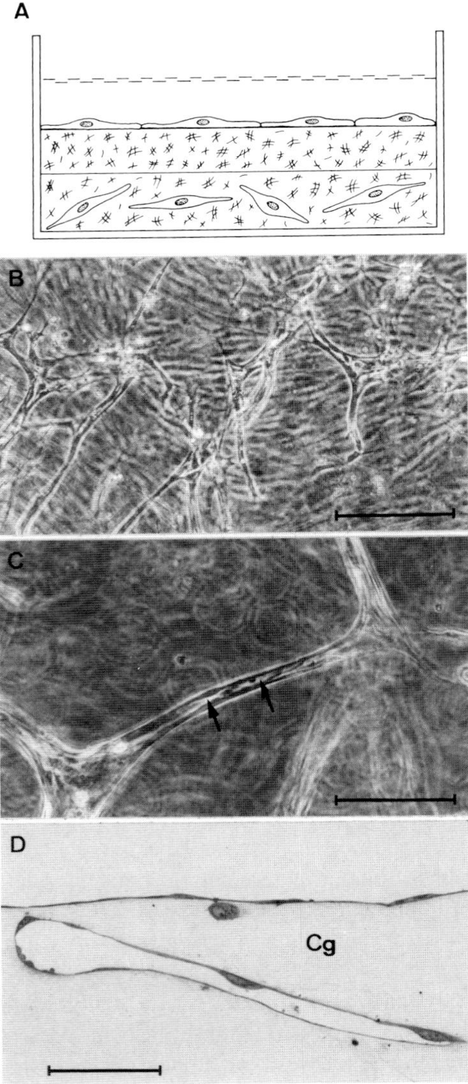

Fig. 6 Paracrine induction of angiogenesis in vitro by Swiss 3T3 fibroblasts. (A) Schematic diagram of the "collagen bilayer" coculture system. Fibroblasts or other cell types are suspended within an underlying collagen gel layer and overlaid with a cell-free gel, onto which endothelial cells are subsequently seeded and grown as a monolayer. (B) endothelial cells grown on a cell-free collagen gel cast on top of a gel layer containing Swiss 3T3 fibroblasts have invaded the gel, forming a network of interconnecting tubes (the plane of focus is beneath the surface monolayer). (C) detail of a capillary-like tubule with a patent lumen (arrows) formed by endothelial cells cocultured with Swiss 3T3 cells (phase contrast microscopy). (D) semithin section of collagen gel cocultures of endothelial cells and Swiss 3T3 cells. The endothelial cells have invaded the underlying collagen gel (Cg) and have formed tubular structures with wide lumina. (B) Bar = 200 μm; (C) bar = 100 μm; (D) bar = 50 μm. (From Montesano et al., 1993, with copyright permission from The Company of Biologists, Ltd.)

search of new agents, either pharmacologic or physiologic, which inhibit angiogenesis. Thus, Fotsis et al. (1993) have demonstrated that genistein, an isoflavonoid present in high concentration in the urine of subjects consuming a diet rich in soya, is a potent inhibitor of bFGF-induced angiogenesis in our three-dimensional in vitro model (Table II). The same investigators have subsequently identified an endogenous estrogen metabolite in human urine, namely 2-methoxyestradiol, which is a potent inhibitor of bFGF-induced tubule formation in collagen gels (Table II) and of tumor neovascularization in vivo (Fotsis et al., 1994).

Potential Clinical Implications of In Vitro Studies of Angiogenesis

Since angiogenesis plays an important role in a wide range of physiological and pathological processes, its modulation provides a useful alternative/adjunct to current therapeutic modalities in several diseases characterized by local hyper- or hypovascularity (Folkman, 1995b).

Inhibition of angiogenesis has long been recognized as a potential strategy for cancer treatment (Folkman, 1971). Inhibition of angiogenesis is also of potential benefit in the treatment of ocular neovascularization (e.g., diabetic proliferative retinopathy) and of life-threatening hemangiomas. The redundancy of angiogenesis-stimulating cytokines may however hinder therapeutic strategies based on neutralization of angiogenic factors. Since sprouting endothelial cells must invade and translocate across the extracellular matrix regardless of the nature of the angiogenic stimulus, targeting cellular processes such as extracellular proteolysis may overcome the problems of growth factor redundancy. We therefore believe that a better understanding of the factors which regulate the invasive and proteolytic properties of endothelial cells may facilitate the design of angiostatic agents capable of inhibiting inappropriate blood vessel growth in a variety of clinical settings. Identification of additional inhibitors of angiogenesis is potentially of great importance, because different anti-angiogenic agents may act through diverse mechanisms and may therefore achieve maximum therapeutic effect when administered in appropriate combinations.

Stimulation of angiogenesis, on the other hand, has been shown to accelerate the healing of wounds and peptic ulcers, and to promote growth of collateral vessels in ischemic diseases. Coronary atherosclerosis, peripheral arterial occlusion and cerebral vascular insufficiency are among the most common causes of morbidity and mortality in Western societies. Under normal conditions, ischemia caused by obstruction of an artery stimulates the development of collateral vessels, which may however often be insufficient to maintain normal tissue perfusion. This situation may result in coronary infarction, stroke, or gangrene of the extremities. Recent work has demonstrated that administration of angiogenic factors can enhance the growth of collateral vessels in animal models of myocardial, peripheral, and cerebral arterial insufficiency (reviewed by Höckel et al., 1993; Symes and Sniderman, 1994). Considering that currently no effective drug therapy exists for many patients with critical leg ischemia or disabling claudication, these results suggest that "*therapeutic angiogenesis may have an immense clinical potential*" (Symes and Sniderman, 1994). We are confident that the use of in vitro systems will contribute to the development of appropriate therapeutic strategies by allowing for the identification of additional physiological stimulators of angiogenesis and of other examples of synergistic interactions, such as those we have previously shown for the combination of bFGF and VEGF (Pepper et al., 1992a) and HA oligosaccharides and VEGF (Montesano et al., 1996). Evidence from recent in vivo studies (Asahara et al., 1995) suggests that in situations where stimulation of angiogenesis is desired, the benefit derived from coaddition of two agents whose interaction is synergistic would be greater than that derived from the addition of one of these agents alone.

Methods

Extraction of type I collagen from rat tail tendons. Collagen is solubilized from rat tail tendons essentially as described by Strom and Michalopoulos (1982) and Dharmsathaphorn and Madara (1990):

1. Tails from 10–20 rats are stored frozen.
2. After thawing, the tail skin is incised longitudinally and stripped. The end tailbone is grasped with forceps and twisted to break the connection with adjoining caudal vertebrae, then pulled away together with the attached tendons. These are cut off from the bone, placed in an open 10 cm dish and weighed (aim for 6–7 g).
3. The tendons are dried and sterilized by placing under UV light for 48 hours. Thereafter, all manipulations are done at 4°C under sterile conditions, i.e., in the cold room or on ice under a sterile hood using sterile glassware which is stored at 4°C.
4. The tendons are placed in glacial acetic acid (diluted 1/1000 in twice-distilled water) at 4°C for 24–48 hours with constant stirring (1 g of tendons in 300 ml acetic acid).
5. The viscous collagen solution with residual undissolved material is centrifuged at 2000 g for 30 min at 4°C. The pellet is discarded and the supernatant centrifuged at 4°C for 1 hour at 16000 g. The supernatant is kept and collagen not used immediately is stored at –20°C in 50 ml aliquots.
6. The collagen solution is dialysed for at least 24 hours at 4°C (in the cold room) against 10× diluted minimal essential medium (MEM) without bicarbonate (1 l for 100 ml collagen).
7. The dialysed collagen solution is aliquoted into sterile glass bottles, which are stored at 4°C until use.

Preparation of three-dimensional collagen gels. Gels of reconstituted collagen fibrils are prepared essentially as described (Montesano et al., 1983) by quickly mixing 8 volumes of cold collagen solution (approximately 1.5 mg/ml) with 1 volume of 10× MEM without bicarbonate and 1 volume of sodium bicarbonate (11.76 mg/ml) on ice. This mixture is quickly dispersed into 35 mm tissue culture dishes (1000–1500 μl) or into 16-mm wells (400 μl) and allowed to gel at 37°C for 10 min.

Preparation of three-dimensional fibrin gels. Fibrin gels are prepared essentially as described (Montesano et al., 1985). Bovine fibrinogen (Calbiochem) is dissolved immediately before use at 37°C in calcium- and magnesium-free MEM to obtain a final protein concentration of 2.5 mg/ml. Clotting is initiated by adding 1/10 v/v of $CaCl_2$ (2 mg/ml) and 25 U/ml of thrombin (Sigma). The mixture is immediately transferred into tissue culture wells and allowed to gel for 2 minutes at room temperature.

Collagen (or fibrin) gel invasion assay. Endothelial cells are seeded onto collagen or fibrin gels at 2.5–5 × 10^4 cells/well in 500 μl complete medium (for bovine adrenal cortex microvascular endothelial cells, we use α-MEM with 5% donor calf serum). Medium is changed every 2–3 days, and when the cells reach confluence (after 3–6 days), the serum concentration is reduced to 2% and treatments are begun. Medium and compounds are changed every 2–3 days. Experiments on fibrin gels are usually performed in the presence of the serine protease inhibitor Trasylol (Bayer, 200 KIU/ml), since our previous observations demonstrated that protection of the fibrin matrix from endothelial cell fibrinolytic activity was an absolute requirement for the preservation of an intact three-dimensional matrix into which stimulated cells could migrate; in the absence of Trasylol, the underlying matrix is completely lysed, thereby precluding invasion (Montesano et al., 1987).

Quantification of invasion. Endothelial cell invasion of collagen or fibrin gels is quantified as described (Pepper et al., 1992a). Three randomly selected fields measuring 1.0 mm × 1.4 mm are photographed in each well at a single level beneath the surface monolayer by phase contrast microscopy, using a Nikon Diaphot TMD inverted photomicroscope. Invasion is quantified by determining the total additive length of all cellular structures which have penetrated beneath the surface monolayer either as apparently single cells or in the form of cell cords or tubes. Results are shown as mean additive sprout length ± s.e.m. (in μm) of three photographic fields per experiment, for each of at least three experiments per condition.

Processing for light and electron microscopy. Collagen or fibrin gel cultures are fixed in situ overnight with 2.5% glutaraldehyde in 100 mM sodium cacodylate buffer (pH 7.4). After extensive washing in cacodylate buffer, the gels are removed from the wells and cut into 2 mm × 2 mm fragments. These are postfixed in 1% osmium tetroxide in Veronal acetate buffer for 45 minutes, stained en bloc with 2.5% uranyl acetate in ethanol, dehydrated in graded ethanols, and embedded in Epon 812 in flat molds (Montesano and Orci, 1985; Montesano et al., 1990). Semithin (0.5–1 μm-thick) and thin sections are cut with an LKB ultramicrotome. Semithin sections are stained with 1% Toluidine Blue and photographed under bright field illumination using a Zeiss photomicroscope II. Thin sections are stained with uranyl acetate and lead citrate and photographed in a Philips CM 10 electron microscope.

Acknowledgments

We would like to thank Dr. J.-D. Vassalli for his important contribution to our work, and Dr. L. Orci for continued support, advice, and constructive criticism. We are also grateful to J. Rial-Robert, C. Di Sanza, and M. Quayzin for excellent technical assistance and to F. Hellal for preparing the manuscript. We apologize for not having cited our many colleagues who have provided the cytokines and other potential regulators listed in Tables I and II, and wish to acknowledge their invaluable contributions. Work performed in the authors' laboratory has been supported by the Swiss National Science Foundation, the Juvenile Diabetes Foundation International and the Sir Jules Thorn Charitable Overseas Trust.

References

Alexander C.A., Werb Z. (1991): Extracellular matrix degradation. In: *Cell Biology of Extracellular Matrix,* 2nd ed. Hay E.D., ed., New York: Plenum Press, pp. 255–302.

Alexander H.R., Billingsley K.G., Block M.I., Fraker D.L. (1994): D- factor/leukemia inhibitory factor: evidence for its role as a mediator in acute and chronic inflammatory disease. *Cytokine* 6:589–596.

Anand-Apte B., Pepper M.S., Bao L., Smith R.C., Voest E., Iwata K., Montesano R., Olsen B., Murphy G., Apte S.S., Zetter B. (1997): Inhibition of angiogenesis and tumor growth by tissue inhibitor of metalloproteinase-3 (TIMP-3), a matrix bound TIMP. *Invest Ophtalmol Vis Sci* 38:817–823.

Andrade S.P., Fan T-P.D., Lewis G.P. (1987): Quantitative in vivo studies on angiogenesis in a rat sponge model. *Br J Exp Path* 68:755–766.

Aruffo A., Stamenkovic I., Mulnick M., Underhill C.B., Seed B. (1990): CD44 is the principal cell surface receptor for hyaluronate. *Cell* 61:1303–1313.

Asahara T., Bauters C., Zheng L.P., Takeshita S., Bunting S., Ferrara N., Symes J.F., Isner J-M.

(1995): Synergistic effects of vascular endothelial growth factor and basic fibroblast growth factor on angiogenesis in vivo. *Circulation* 92 (Suppl II):II-365–II-371.

Ausprunk D.H. (1986): Distribution of hyaluronic acid and sulfated glycosaminoglycans during blood-vessel development in the chick chorioallantoic membrane. *Am J Anat* 177: 313–331.

Ausprunk D.H., Folkman J. (1977): Migration and proliferation of endothelial cells in preformed and newly formed blood vessels during angiogenesis. *Microvasc Res* 14:53–65.

Ausprunk D.H., Boudreau C.L., Nelson D.A. (1981): Proteoglycans in the microvasculature. II. Histochemical localization in proliferating capillaries of the rabbit cornea. *Am J Pathol* 103:367–375.

Baird A., Durkin T. (1986): Inhibition of endothelial cell proliferation by type β-transforming growth factor: interactions with acidic and basic fibroblast growth factors. *Biochem Biophys Res Commun* 138:476–482.

Baird A., Klagsbrun M. (1991): The fibroblast growth factor family. *Cancer Cells* 3:239–243.

Banerjee S.D., Toole B.P. (1991): Monoclonal antibody to chick embryo hyaluronan-binding protein: changes in distribution of binding during early brain development. *Dev Biol* 146: 186–197.

Banerjee S.D., Toole B.P. (1992): Hyaluronan-binding protein in endothelial cell morphogenesis. *J Cell Biol* 119:643–652.

Basilico C., Moscatelli D. (1992): The FGF family of growth factors and oncogenes. *Adv Cancer Res* 59:115–165.

Battegay E., Rupp J., Iruela-Arispe L., Sage E.H., Pech M. (1994): PDGF-BB modulates endothelial proliferation and angiogenesis in vitro via PDGFβ-receptors. *J Cell Biol* 125:917–928.

Bautch V.L., Toda S., Hassell J.A., Hanahan D. (1987): Endothelial cell tumors develop in transgenic mice carrying polyoma virus middle T oncogene. *Cell* 51:529–538.

Bertrand P., Girard N., Delpech B., Duval C., D'Anjour J., Dance J.P. (1992): Hyaluronan (hyaluronic acid) and hyaluronectin in the extracellular matrix of human breast carcinomas. *Int J Cancer* 52:1–6.

Boudreau N., Turley E.A., Rabinovitch M. (1991): Fibronectin, hyaluronan and hyaluronan binding protein contribute to increased ductus arteriosus smooth muscle cell migration. *Dev Biol* 143:235–247.

Bourguignon L.Y.W., Lokeshwar V.B., He J., Chen X., Bourguignon G.J. (1992): A CD44-like endothelial cell transmembrane glycoprotein (GP 116) interacts with extracellular matrix and ankyrin. *Mol Cell Biol* 12:4464–4471.

Breier G., Albrecht U., Sterrer S., Risau W. (1992): Expression of vascular endothelial growth factor during embryonic angiogenesis and endothelial cell differentiation. *Development* 114:521–532.

Canfield A.E., Schor A.M. (1995): Evidence that tenascin and thrombospondin-1 modulate sprouting of endothelial cells. *J Cell Sci* 108:797–809.

Culty M., Miyake K., Kincaide P.W., Silorski E., Butcher E., Underhill A.M. (1990): The hyaluronate receptor is a member of the CD44 (H- CAM) family of cell surface glycoproteins. *J Cell Biol* 111:2765–2774.

Deroanne C.F., Colige A.C., Nusgens B.V., Lapiere C.M. (1996): Modulation of expression and assembly of vinculin during in vitro fibrillar collagen-induced angiogenesis and its reversal. *Exp Cell Res* 224:215–223.

Dharmsathaporn K., Madara J.L. (1990): Established intestinal cell lines as model systems for electrolyte transport studies. *Meth Enzymol* 192:354–389.

Dvorak H.F., Brown L.F., Detmar M., Dvorak A.M. (1995): Vascular permeability factor/vascular endothelial growth factor, microvascular hyperpermeability and angiogenesis. *Am J Pathol* 146:1029–1039.

Faham S., Hileman R.E., Fromm J.R., Lindhart R.J., Rees D.C. (1996): Heparin structure and interactions with basic fibroblast growth factor. *Science* 271:1116–1120.

Fajardo L.F., Kowalski J., Kwan H.H., Prionas S.D., Allison A.C. (1988): The disc angiogenesis system. *Lab Invest* 58:718–724.

Feinberg R.N., Beebe D.L. (1983): Hyaluronate in vasculogenesis. Science 220:1177–1179.

Fenderson B.A., Stamenkovic I., Aruffo A. (1993): Localization of hyaluronan in mouse embryos during implantation, gastrulation and organogenesis. *Differentiation* 54:85–98.

Ferrara N. (1995): The role of vascular endothelial growth factor in pathological angiogenesis. *Breast Cancer Res Treat* 36:127–137.

Ferrara N., Houck K., Jackeman L., Leung D.W. (1992a): Molecular and biological properties of the vascular endothelial growth factor family of proteins. *Endocrine Rev* 13:18–35.

Ferrara N., Winer J., Henzel W.J. (1992b): Pituitary follicular cells secrete an inhibitor of aortic endothelial cell growth: identification as leukemia inhibitory factor. *Proc Natl Acad Sci USA* 89:698–702.

Fidler I.J., Ellis L.M. (1994): The implications of angiogenesis for the biology and therapy of cancer metastasis. *Cell* 79:185–188.

Fisher C., Gilberston-Beadling S., Powers E.A., Petzold G., Poorman R., Mitchell M.A. (1994): Interstitial collagenase is required for angiogenesis in vitro. *Dev Biol* 162:499–510.

Flaumenhaft R., Abe M., Mignatti P., Rifkin D.B. (1992): Basic fibroblast growth factor-induced activation of latent transforming growth factor beta in endothelial cells: regulation of plasminogen activator activity. *J Cell Biol* 118:901–909.

Folkman J. (1971): Tumor angiogenesis: therapeutic implications. *N Engl J Med* 285:1182–1186.

Folkman J. (1995a): Angiogenesis in cancer, vascular, rheumatoid and other diseases. *Nature Med* 1:27–31.

Folkman J. (1995b): Clinical applications of research on angiogenesis. *N Engl J Med* 333: 1757–1763.

Folkman J., Haudenschild C. (1980): Angiogenesis in vitro. *Nature* 288:551–555.

Folkman J., Klagsbrun M. (1987): Angiogenic factors. *Science* 235:442–447.

Folkman J., Shing Y. (1992): Angiogenesis. *J Biol Chem* 267:10931–10934.

Fotsis T., Pepper M., Adlercreutz H., Fleischmann G., Hase T., Montesano R., Schweigerer L. (1993): Genistein, a dietary-derived inhibitor of in vitro angiogenesis. *Proc Natl Acad Sci USA* 90:2690–2694.

Fotsis T., Zhang Y., Pepper M.S., Adlercreutz H., Montesano R., Nawroth P.P., Schweigerer L. (1994): The endogenous oestrogen metabolite 2-methoxyoestradiol inhibits angiogenesis and suppresses tumor growth. *Nature* 368:237–239.

Fotsis T., Pepper M.S., Aktas E., Rasku S., Adlercreutz H., Wähälä K., Montesano R., Schweigerer L. (1997): Flavonoids, dietary-derived inhibitors of cell proliferation and in vitro angiogenesis. *Cancer Res* 57:2916–2921.

Fràter-Schröder M., Müller G., Birchmeier W., Böhlen P. (1986): Transforming growth factor-beta inhibits endothelial cell proliferation. *Biochem Biophys Res Commun* 137:295–302.

Gajdusek C.M., Luo Z., Mayberg M.R. (1993): Basic fibroblast growth factor and transforming growth factor beta-1: synergistic mediators of angiogenesis in vitro. *J Cell Physiol* 157: 133–144.

Gimbrone M.A., Jr, Cotran R.S., Leapman S.B., Folkman J. (1974): Tumor growth and neovascularization: an experimental model using the rabbit cornea. *J Natl Cancer Inst* 52: 413–427.

Goto M., Goto K., Weindel K., Folkman J. (1993): Synergistic effects of vascular endothelial growth factor and basic fibroblast growth factor on the proliferation and cord formation of bovine capillary endothelial cells within collagen gels. *Lab Invest* 69:508–517.

Grant D.S., Tashiro K.I., Segui-Real B., Yamada Y., Martin G.R., Kleinman H.K. (1989): Two different laminin domains mediate the differentiation of human endothelial cells into capillary-like structures in vitro. *Cell* 58:933–943.

Gross J.L., Moscatelli D., Jaffe E.A., Rifkin D.B. (1982): Plasminogen activator and collagenase production by cultured capillary endothelial cells. *J Cell Biol* 95:974–981.

Hall C.L., Wang C., Lange L.A., Turley E.A. (1994): Hyaluronan and the hyaluronan receptor RHAMM promote focal adhesion turnover and transient tyrosine kinase activity. *J Cell Biol* 126:575–588.

Hardwick C., Hoare K., Owens R., Hohn H.P., Hook M., Moore D., Cripps V., Austen L., Nance D.M., Turley E.A. (1992): Molecular cloning of a novel hyaluronan receptor that mediates tumor cell motility. *J Cell Biol* 117:1343–1350.

Heimark R.L., Twardzik D.R., Schwartz S.M. (1986): Inhibition of endothelial regeneration by type-beta transforming growth factor from platelets. *Science* 233:1078–1080.

Hilton D.J. (1992): LIF: lots of interesting functions. *Trends Biochem Sci* 17:72–76.

Höckel M., Schlenger K., Doctrow S., Kissel T., Vaupel P. (1993): Therapeutic angiogenesis. *Arch Surg* 128:423–429.

Ingber D.E., Folkman J. (1989): Mechano-chemical switching between growth and differentiation during growth factor-stimulated angiogenesis in vitro: role of the extracellular matrix. *J Cell Biol* 109:317–330.

Iozzo R. (1985): Proteoglycans: structure, function, and role in neoplasia. *Lab Invest* 53:373–396.

Iruela-Arispe M.L., Sage E.H. (1993): Endothelial cells exhibiting angiogenesis in vitro proliferate in response to TGF-β1. *J Cell Biochem* 52:414–430.

Iruela-Arispe M., Hasselaar P., Sage H. (1991): Differential expression of extracellular proteins is correlated with angiogenesis in vitro. *Lab Invest* 64:174–186.

Kim K.J., Li B., Winer J., Armanini M., Gillett N., Phillips H.S., Ferrara N. (1993): Inhibition of vascular endothelial growth factor induced angiogenesis suppresses tumor growth in vivo. *Nature* 362:841–844.

Klagsbrun M., D'Amore P. (1991): Regulators of angiogenesis. *Annu Rev Physiol* 53:217–239.

Klagsbrun M., Soker S. (1993): VEGF/VPF: the angiogenic factor found? *Curr Biol* 3:699–702.

Klagsbrun M., Knighton D., Folkman J. (1976): Tumor angiogenesis activity in cells grown in tissue culture. *Cancer Res* 36:110–114.

Knudson C.B., Knudson W. (1993): Hyaluronan-binding proteins in development, tissue homeostasis, and disease. *FASEB J* 7:1233–1241.

Knudson W., Biswas C., Li X.Q., Nemee R.E., Toole B.P. (1989): The role and regulation of tumor-associated hyaluronan. In: *The Biology of Hyaluronan, Ciba Foundation Symposium,* Vol. 143. Evered D., Whelau J., eds., Chichester: John Wiley and Sons, pp. 150–169.

Kubota Y., Kleinman H.K., Martin G.R., Lawley T.J. (1988): Role of laminin and basement membrane in the morphological differentiation of human endothelial cells into capillary-like structures. *J Cell Biol* 107:1589–1598.

Kujawa M.J., Carrino D.A., Caplan A.I. (1986a): Substrate-bonded hyaluronic acid exhibits a size-dependent stimulation of chondrogenic differentiation of stage 24 limb mesenchymal cells in culture. *Dev Biol* 144:519–528.

Kujawa M., Pechak D.G., Fiszman M.Y., Caplan A.I. (1986b): Hyaluronic acid bonded to cell culture surfaces inhibits the program of myogenesis. *Dev Biol* 113:10–16.

Laaroubi K., Delbé J., Vacherot P., Desgranges P., Tardieu M., Jaye M., Barritault D., Courty J. (1994): Mitogenic and in vitro angiogenic activity of human recombinant heparin affin regulatory peptide. *Growth Factors* 10:89–98.

Lane T.F., Iruela-Arispe M.L., Johnson R.S., Sage E.H. (1994): SPARC is a source of copper-binding peptides that stimulate angiogenesis. *J Cell Biol* 125:929–943.

Laurent T.C., Fraser J.R.E. (1992): Hyaluronan. *FASEB J* 6:2397–2404.

Lees V.C., Fan T-P.D., West D.C. (1995): Angiogenesis in a delayed revascularization model is accelerated by angiogenic oligosaccharides of hyaluronan. *Lab Invest* 73:259–266.

Leibovich S.J., Polverini J., Shepard H.M., Wiseman D.M., Shively V., Nuseir N. (1987): Macrophage-induced angiogenesis is mediated by tumour necrosis factor-α. *Nature* 329:630–632.

Liotta L.A., Steeg P.S., Stetler-Stevenson W.G. (1991): Cancer metastasis and angiogenesis: an imbalance of positive and negative regulation. *Cell* 64:327–366.

Liu D., Pearlman E., Diaconu E., Guo K., Mori H., Haqqi T., Markowitz S., Willson G., Sy M-S. (1996): Expression of hyaluronidase by tumor cells induces angiogenesis in vivo. *Proc Natl Acad Sci USA* 93:7832–7837.

Lokeshwar V.B., Lokeshwar B.L., Pham H.T., Block N.L. (1996): Association of elevated levels of hyaluronidase, a matrix-degrading enzyme, with prostate cancer progression. *Cancer Res* 56:651–657.

Maciag T., Kadish J., Wilkins L., Stemerman M.B., Weinstein R. (1982): Organizational behavior of human umbilical vein endothelial cells. *J Cell Biol* 94:511–520.

Madri J.A., Williams S.K. (1983): Capillary endothelial cell cultures: phenotypic modulation by matrix components. *J Cell Biol* 97:153–165.

Madri J.A., Pratt B.M. (1988): Angiogenesis. In: *The Molecular and Cellular Biology of Wound Repair.* Clark R.A.F., Henson P.M., eds., New York: Plenum Press, pp. 337–358.

Madri J.A., Pratt B.M., Tucker A.M. (1988): Phenotypic modulation of endothelial cells by transforming growth factor-β depends on the composition and organization of the extracellular matrix. *J Cell Biol* 106:1357–1384.

Mandriota S.J., Pepper M.S. (1997): Vascular endothelial growth factor-induced in vitro angiogenesis and plasminogen activator expression are dependent on endogenous basic fibroblast growth factor. *J. Cell Sci* 110:2293–2302.

Mandriota S.J., Seghezzi G., Vassalli J-D., Ferrara N., Wasi S., Mazzieri R., Mignatti P., Pepper M.S. (1995): Vascular endothelial growth factor increases urokinase receptor expression in vascular endothelial cells. *J Biol Chem* 270:9709–9716.

Mandriota S.J., Menoud P-A., Pepper M.S. (1996): Transforming browth factor β1 down-regulates vascular endothelial growth factor receptor-2/flk-1 expression in vascular endothelial cells. *J Biol Chem* 271:11500–11505.

Mawatari M., Okamura K., Matsuda T., Hamanaka R., Mizoguchi H., Higashio K., Kohno K., Kuwano M. (1991): Tumor necrosis factor and epidermal growth factor modulate migration of human microvascular endothelial cells and production of tissue-type plasminogen activator and its inhibitor. *Exp Cell Res* 192:574–580.

Merwin J.R., Anderson J.M., Kocher O., van Itallie C.M., Madri J.A. (1990): Transforming growth factor-β1 modulates extracellular matrix organization and cell–cell junctional complex formation during in vitro angiogenesis. *J Cell Physiol* 142:117–128.

Mignatti P., Rifkin D.B. (1996): Plasminogen activators and angiogenesis. In: *Current Topics in Microbiology and Immunology, Vol. 213-I: Attempts to Understand Metastasis Formation.* Günthert U, Birchmeier W, eds., Berlin and Heidelberg: Springer Verlag, pp. 31–49.

Mignatti P., Tsuboi R., Robbins E., Rifkin D.B. (1989): In vitro angiogenesis on the human amniotic membrane: requirements for basic fibroblast growth factor-induced proteases. *J Cell Biol* 108:671–682.

Millauer B., Wizigman-Voos S., Schnürch H., Martinez R., Møller N.P.H., Risau W., Ullrich A.

(1993): High affinity VEGF binding and developmental expression suggest Flk-1 as a major regulator of vasculogenesis and angiogenesis. *Cell* 72:835–846.

Millauer B., Shawver L.K., Plate K.H., Risau W., Ullrich A. (1994): Glioblastoma growth inhibited in vivo by a dominant negative Flk-1 mutant. *Nature* 367:576–579.

Miyake K., Underhill C.B., Lesley J., Kincaide P.W. (1990): Hyaluronate can function as a cell adhesion molecule and CD44 participates in hyaluronate recognition. *J Exp Med* 172: 69–75.

Montesano R., Orci L. (1985): Tumor-promoting phorbol esters induce angiogenesis in vitro. *Cell* 42:469–477.

Montesano R., Orci L. (1987): Phorbol esters induce angiogenesis in vitro from large vessel endothelial cells. *J Cell Physiol* 130:284–291.

Montesano R., Orci L., Vassalli P. (1983): In vitro rapid organization of endothelial cells into capillary-like networks is promoted by collagen matrices. *J Cell Biol* 97:1648–1652.

Montesano R., Mouron P., Orci L. (1985): Vascular outgrowths from tissue explants embedded in fibrin or collagen gels: a simple in vitro model of angiogenesis. *Cell Biol Int Rep* 9:869–875.

Montesano R., Vassalli J-D., Baird A., Guillemin R., Orci L. (1986): Basic fibroblast growth factor induces angiogenesis in vitro. *Proc Natl Acad Sci USA* 83:7297–7301.

Montesano R., Pepper M.S., Vassalli J-D., Orci L. (1987): Phorbol ester induces cultured endothelial cells to invade a fibrin matrix in the presence of fibrinolytic inhibitors. *J Cell Physiol* 132:509–516.

Montesano R., Pepper M.S., Belin D., Vassalli J-D., Orci L. (1988): Induction of angiogenesis in vitro by vanadate, an inhibitor of phosphotyrosine phosphatases. *J Cell Physiol* 134: 460–466.

Montesano R., Pepper M.S., Möhle-Steinlein U., Risau W., Wagner E.F., Orci L. (1990): Increased proteolytic activity is responsible for the aberrant morphogenetic behavior of endothelial cells expressing the middle T oncogene. *Cell* 62:435–445.

Montesano R., Pepper M.S., Orci L. (1993): Paracrine induction of angiogenesis in vitro by Swiss 3T3 fibroblasts. *J Cell Sci* 105:1013–1024.

Montesano R., Vassalli J-D., Orci L., Pepper M.S. (1994): The role of growth factors and extracellular matrix in angiogenesis and epithelial morphogenesis. In: *Frontiers in Endocrinology. Vol. 6: Developmental Endocrinology.* Sizonenko P.C., Aubert M.L., Vassalli J.-D., eds., Rome: Ares-Serono Symposia Publications, pp. 43–66.

Montesano R., Kumar S., Orci L., Pepper M.S. (1996): Synergistic effect of hyaluronan oligosaccharides and vascular endothelial growth factor on angiogenesis in vitro. *Lab Invest* 75:249–262.

Moscatelli D., Presta M., Rifkin D.B. (1986): Purification of a factor from human placenta that stimulates capillary endothelial cell protease production, DNA synthesis and migration. *Proc Natl Acad Sci USA* 83:2091–2095.

Moses M.A., Sudhalter J., Langer R. (1990): Identification of an inhibitor of neovascularization from cartilage. *Science* 248:1408–1460.

Müller G., Behrens J., Nussbaumer U., Böhlen P., Birchmeier W. (1987): Inhibitory action of transforming growth factor-β on endothelial cells. *Proc Natl Acad Sci USA* 84:5600–5604.

Nathan C., Sporn M. (1991): Cytokines in context. *J Cell Biol* 113:981–986.

Nehls V., Drenckhahn D. (1995): A novel, microcarrier-based in vitro assay for rapid and reliable quantification of three-dimensional cell migration and angiogenesis. *Microvasc Res* 50:311–322.

Neufeld G., Tessler S., Gitay-Goren H., Cohen T., Levi B-Z. (1994): Vascular endothelial growth factor and its receptors. *Progr Growth Factor Res* 5:89–97.

Nicosia R.F., Ottinetti A. (1990): Growth of microvessels in serum-free matrix culture of rat aorta. *Lab Invest* 63:115–122.

Nicosia R.F., Bonanno E., Smith M., Yurchenco P. (1994): Modulation of angiogenesis in vitro by laminin-entactin complex. *Dev Biol* 164:197–206.

Okamura K., Morimoto A., Hamanaka R., Ono M., Kohno K., Uchida Y., Kuwano M. (1992): A model system for tumor angiogenesis: involvement of transforming growth factor-α in tube formation of human microvascular endothelial cells induced by esophageal cancer cells. *Biochem Biophys Res Commun* 186:1471–1479.

Ornitz D.M., Herr A.B., Nilsson M., Westman J., Svahn C-M., Waksman G. (1995): FGF binding and FGF receptor activation by synthetic heparan-derived di- and trisaccharides. *Science* 268:432–436.

Paku S., Paweletz N. (1991): First steps of tumor-related angiogenesis. *Lab Invest* 65:334–346.

Pardenaud L., Yassine F., Dieterlen-Lièvre F. (1989): Relationship between vasculogenesis, angiogenesis and haemopoiesis during avian ontogeny. *Development* 105:437–485.

Passaniti A., Taylor R.M., Pili R., Guo Y., Long P.V., Haney J.A., Pauly R.R., Grant D.S., Martin G.R. (1992): A simple, quantitative method for assessing angiogenesis and antiangiogenic agents using reconstituted basement membrane, heparin, and fibroblast growth factor. *Lab Invest* 67:519–528.

Pepper M.S., Montesano R. (1990): Proteolytic balance and capillary morphogenesis. *Cell Diff Dev* 32:319–328.

Pepper M.S., Vassalli J-D., Montesano R., Orci L. (1987): Urokinase-type plasminogen activator is induced in migrating capillary endothelial cells. *J Cell Biol* 105:2535–2541.

Pepper M.S., Belin D., Montesano R., Orci L., Vassalli J-D. (1990): Transforming growth factor-beta 1 modulates basic fibroblast growth factor-induced proteolytic and angiogenic properties of endothelial cells in vitro. *J Cell Biol* 111:743–755.

Pepper M.S., Montesano R., Vassalli J-D., Orci L. (1991a): Chondrocytes inhibit endothelial sprout formation in vitro: evidence for the involvement of a transforming growth factor-beta. *J Cell Physiol* 146:170–179.

Pepper M.S., Montesano R., Orci L., Vassalli J-D. (1991b): Plasminogen activator-inhibitor-1 is induced in microvascular endothelial cells by a chondrocyte-derived transforming growth factor-beta. *Biochem Biophys Res Commun* 176:633–638.

Pepper M.S., Ferrara N., Orci L., Montesano R. (1992a): Potent synergism between vascular endothelial growth factor and basic fibroblast growth factor in the induction of angiogenesis in vitro. *Biochem Biophys Res Commun* 189:824–831.

Pepper M.S., Sappino A-P., Montesano R., Orci L., Vassalli J-D. (1992b): Plasminogen activator inhibitor-1 is induced in migrating endothelial cells. *J Cell Physiol* 153:129–139.

Pepper M.S., Sappino A-P., Stocklin R., Montesano R., Orci L., Vassalli J-D. (1993a): Upregulation of urokinase receptor expression on migrating endothelial cells. *J Cell Biol* 122:673–684.

Pepper M.S., Vassalli J-D., Orci L., Montesano R. (1993b): Biphasic effect of transforming growth factor-beta-1 on in vitro angiogenesis. *Exp Cell Res* 204:356–363.

Pepper M.S., Vassalli J-D., Orci L., Montesano R. (1994a): Angiogenesis in vitro: cytokine interactions and balanced extracellular proteolysis. In: *Angiogenesis. Molecular Biology, Clinical Aspects.* Maragoudakis M.E., Gullino P.M., Lelkes P.I., eds., New York: Plenum Press, pp. 149–170.

Pepper M.S., Wasi S., Ferrara N., Orci L., Montesano R. (1994b): In vitro angiogenic and proteolytic properties of bovine lymphatic endothelial cells. *Exp Cell Res* 210:298–305.

Pepper M.S., Vassalli J-D., Wilks J.W., Schweigerer L., Orci L., Montesano R. (1994c): Modulation of microvascular endothelial cell proteolytic properties by inhibitors of angiogenesis. *J Cell Biochem* 55:419–434.

Pepper M.S., Ferrara N., Orci L., Montesano R. (1995): Leukemia inhibitory factor (LIF) is a potent inhibitor of in vitro angiogenesis. *J Cell Sci* 108:73–83.

Pepper M.S., Mandriota S.J., Vassalli J-D., Orci L., Montesano R. (1996a): Angiogenesis-regulating cytokines: activities and interactions. In: *Current Topics in Microbiology and Immunology, Vol. 213-II: Attempts to Understand Metastasis Formation.* Günthert U., Birchmeier W., eds., Berlin and Heidelberg: Springer-Verlag, pp. 31–67.

Pepper M.S., Montesano R., Mandriota S.J., Orci L., Vassalli J-D. (1996b): Angiogenesis: a paradigm for balanced extracellular proteolysis during cell migration and morphogenesis. *Enzyme Protein* 49:138–162.

Pepper M.S., Tacchini-Cottier F., Sabapathy T.K., Montesano R., Wagner E.F. (1997): Endothelial cells transformed by polyoma virus middle-T oncogene: a model for hemangiomas and other vascular tumors. In: *Tumor Angiogenesis.* Lewis C.E., Bicknell R., Ferrara N., eds., Oxford: Oxford University Press pp. 310–331.

Peters K.G., De Vries C., Williams L.T. (1993): Vascular endothelial growth factor receptor expression during embryogenesis and tissue repair suggests a role in endothelial differentiation and blood vessel growth. *Proc Natl Acad Sci USA* 90:8915–8919.

Phillips G.D., Whitehead R.A., Knighton D.R. (1992): Inhibition by methylprednisolone acetate suggests an indirect mechanism for TGF-β induced angiogenesis. *Growth Factors* 6:77–84.

Ponting J., Kumar S., Pye D. (1993): Localization of hyaluronan and hyaluronectin in normal and tumour breast tissues. *Int J Oncol* 2:889–893.

Rak J.W., Hegmann E.J., Kerbel R.S. (1993): The role of angiogenesis in tumor progression and metastasis. *Adv Mol Cell Biol* 7:205–251.

Risau W., Sariola A., Zerwes H-G., Sasse J., Ekblom P., Kemler R., Doetschman T. (1988): Vasculogenesis and angiogenesis in embryonic stem cell derived embryoid bodies. *Development* 102:471–478.

Roberts A.B., Sporn M.B., Assoian R.K., Smith J.M., Roche N.S., Wakefield L.M., Heine U.I., Liotta L.A., Falanga V., Kehrl J.H., Fauci A.S. (1986): Transforming growth factor type β: rapid induction of fibrosis and angiogenesis in vivo and stimulation of collagen formation in vitro. *Proc Natl Acad Sci USA* 83:4167–4171.

Rooney P., Kumar S., Ponting J., Wang M. (1995): The role of hyaluronan in tumor neovascularization. *Int J Cancer* 60:632–636.

Rooney P., Kumar P., Ponting J., Kumar S. (1996): The role of collagens and proteoglycans in tumor angiogenesis. In: *Tumor Angiogenesis.* Bicknell R., Lewis C.E., Ferrara N., eds., Oxford: Oxford University Press, 141–151.

Saksela O., Moscatelli D., Rifkin D.B. (1987): The opposing effects of basic fibroblast growth factor and transforming growth factor beta on the regulation of plasminogen activator activity in capillary endothelial cells. *J Cell Biol* 105:957–963.

Sakuda H., Nakashima Y., Kuriyama S., Sueishi K. (1992): Media conditioned by smooth muscle cells cultured in a variety of hypoxic environments stimulates in vitro angiogenesis. A relationship to transforming growth factor-β1. *Am J Pathol* 141:1507–1516.

Sato Y., Rifkin D.B. (1988): Autocrine activities of basic fibroblast growth factor: regulation of endothelial cell movement, plasminogen activator synthesis, and DNA synthesis. *J Cell Biol* 107:119–1205.

Sato Y., Okamura K., Morimoto A., Hamanaka R., Hamaguchi K., Shimada T., Ono M., Kohno K., Sakata T., Kuwano M. (1993): Indispensable role of tissue-type plasminogen activator in growth factor-dependent tube formation of human microvascular endothelial cells in vitro. *Exp Cell Res* 204:223–229.

Sattar A., Kumar S., West D.C. (1992): Does hyaluronan have a role in endothelial cell proliferation of the synovium? *Semin Arthr Rheum* 22:37–43.

Sattar A., Rooney P., Kumar S., Pye D., West D.C., Scott I., Ledger P. (1994): Application of angiogenic oligosaccharides of hyaluronan increase blood vessel numbers in skin. *J Invest Dermatol* 103:573–579.

Savani R.C., Wang C., Yang B., Zhang S., Kinsella M.G., Wight T.N., Stern R., Nance D.M., Turley E.A. (1995): Migration of bovine aortic smooth muscle cells after wounding injury. The role of hyaluronan and RHAMM. *J Clin Invest* 95:1158–1168.

Schlessinger J., Lax I., Lemmon M. (1995): Regulation of growth factor activation by proteoglycans: what is the role of the low affinity receptors? *Cell* 83:357–360.

Sherman L., Sleeman J., Herrlich P., Ponta H. (1994): Hyaluronate receptors: key players in growth, differentiation, migration and tumor progression. *Curr Opin Cell Biol* 6:726–733.

Shweiki D., Itin A., Soffer D., Keshet E. (1992): Vascular endothelial growth factor induced by hypoxia may mediate hypoxia-mediated angiogenesis. *Nature* 359:843–845.

Shweiki D., Itin A., Neufeld G., Gitay-Goren H., Keshet E. (1993): Patterns of expression of vascular endothelial growth factor (VEGF) and VEGF receptors in mice suggest a role in hormonally-regulated angiogenesis. *J Clin Invest* 91:2235–2243.

Slevin M.A., Gaffney J., Kumar S. (1996): Hyaluronan induced proliferation of bovine aortic endothelial cells requires activation of MAP kinase (submitted).

Stahl N., Yancopoulos G.D. (1993): The alphas, betas and kinases of cytokine receptor complexes. *Cell* 74:587–590.

Stamenkovic I., Aruffo A., Amiot M., Seed B. (1991): The hematopoietic and epithelial forms of CD44 are distinct polypeptides with different adhesion potentials for hyaluronate-bearing cells. *EMBO J* 10:343–348.

Strom S.C., Michalopoulos G. (1982): Collagen as a substrate for cell growth and differentiation. *Meth Enzymol* 82:544–555.

Symes J.F., Sniderman A.D. (1994): Angiogenesis: potential therapy for ischaemic disease. *Curr Opin Lipidol* 5:305–312.

Taga T., Kishimoto T. (1993): Cytokine receptors and signal transduction. *FASEB J* 7:3387–3396.

Tessler S., Rockwell P., Hicklin D., Cohen T., Levi B-Z., Witte L., Lemischka I.R., Neufeld G. (1994): Heparin modulates the interaction of $VEGF_{165}$ with soluble and cell associated Flk-1 receptors. *J Biol Chem* 269:12456–12461.

Thomas K.A. (1996): Vascular endothelial growth factor, a potent and selective angiogenic factor. *J Biol Chem* 271:603–606.

Toole B.P. (1990): Hyaluronan and its binding proteins, the hyaladherins. *Curr Opin Cell Biol* 2:839–844.

Toole B.P. (1991): Proteoglycans and hyaluronan in morphogenesis and differentiation. In: *Cell Biology of Extracellular Matrix,* 2nd ed. Hay ED, ed., New York: Plenum Press, pp. 305–341.

Toole B.P., Biswas C., Gross J. (1979): Hyaluronate and invasiveness of the rabbit V2 carcinoma. *Proc Natl Acad Sci USA* 76:6199–6203.

Turley E.A., Tretiak M. (1985): Glycosaminoglycans produced by murine melanoma variants in vivo and in vitro. *Cancer Res* 45:5098–5105.

Turley E.A., Austen L., Vandeligt K., Clary C. (1991): Hyaluronan and a cell-associated hyaluronan binding protein regulate the locomotion of Ras-transformed cells. *J Cell Biol* 112: 1041–1047.

Vassalli J-D., Sappino A-P., Belin D. (1991): The plasminogen activator/plasmin system. *J Clin Invest* 88:1067–1072.

Vernon R.B., Sage E.H. (1995): Between molecules and morphology: extracellular matrix and creation of vascular form. *Am J Pathol* 147:873–833.

Vernon R.B., Lara S.L., Drake C.J., Iruela-Arispe M.L., Angello J.C., Little C.D., Wight T.N., Sage E.H. (1995): Organized type I collagen influences endothelial patterns during "spontaneous angiogenesis in vitro" : planar cultures as models of vascular development. *In Vitro Cell Dev Biol* 31:120–131.

Wahl S.M., Hunt D.A., Wakefield L.A., McCartney-Francis N., Wahl L.M., Roberts A.B., Sporn M.B. (1987): Transforming growth factor type β induces monocyte chemotaxis and growth factor production. *Proc Natl Acad Sci USA* 84:5788–5792.

Wang D-Y., Kao C-H., Yang V.C., Chen J-K. (1994): Glycosaminoglycans enhance phorbol ester-induced proteolytic activity and angiogenesis in vitro. *In Vitro Cell Dev Biol* 30A:777–782.

Weigel P.H., Fuller G.M., LeBoeuf R.D. (1986): A model for the role of hyaluronic acid and fibrin in the early events during the inflammatory response and wound healing. *J Theor Biol* 119:219–234.

West D.C., Kumar S. (1988): Endothelial cell proliferation and diabetic retinopathy. *Lancet* 1:715–716.

West D.C., Kumar S. (1989): The effects of hyaluronate and its oligosaccharides on endothelial cell proliferation and monolayer integrity. *Exp Cell Res* 183:179–196.

West D.C., Hampson I.N., Arnold F., Kumar S. (1985): Angiogenesis induced by degradation products of hyaluronic acid. *Science* 228:1324–1326.

Williams R.L., Risau W., Zerwes H.G., Drexler H., Aguzzi A., Wagner E.F. (1989): Endothelioma cells expressing the polyoma middle T oncogene induce hemangiomas by host cell recruitment. *Cell* 57:1053–1063.

Wiseman D.M., Polverini P.J., Kamp D.W., Leibovich S.J. (1988): Transforming growth factor beta (TGFβ) is chemotactic for human monocytes and induces their expression of angiogenic activity. *Biochem Biophys Res Commun* 157:793–800.

Yang E.Y., Moses H.L. (1990): Transforming growth factor β1-induced changes in cell migration, proliferation, and angiogenesis in the chicken chorioallantoic membrane. *J Cell Biol* 111:731–741.

Yasunaga C., Nakashima Y., Sueishi K. (1989): A role of fibrinolytic activity in angiogenesis. Quantitative assay using in vitro method. *Lab Invest* 61:698–704.

Yoneda M., Yamagata M., Sakaru S., Kimata K. (1988): Hyaluronic acid modulates proliferation of mouse dermal fibroblasts in culture. *J Cell Sci* 90:265–273.

Yoshida A., Anand-Apte B., Zetter B.R. (1996): Differential endothelial migration and proliferation to basic fibroblast growth factor and vascular endothelial growth factor. *Growth Factors* 13:57–64.

2.2

The Rat Aorta Model of Angiogenesis and Its Applications

Roberto Francesco Nicosia

Introduction

Research in angiogenesis has flourished in recent years thanks to ground-breaking discoveries in cell and molecular biology. Experimental studies have opened the field to clinical applications raising new hopes for patients with angiogenesis-dependent disorders. Several clinical trials with inhibitors of angiogenesis are now in progress while scores of academic laboratories and pharmaceutical companies are trying to identify key molecular events in the angiogenic process for targeted therapeutic intervention (Folkman, 1996).

Pharmacologic inhibition of angiogenesis may prolong the life of cancer patients, prevent diabetic blindness, cause regression of hemangiomas, alleviate the crippling complications of rheumatoid arthritis, and prevent the life-threatening swelling of atherosclerotic plaques caused by rupture of neovessels in coronary arteries. Conversely, stimulation of angiogenesis may facilitate wound healing and the development of collateral circulation (Folkman, 1995).

The study of angiogenic mechanisms and the search for angiogenesis agonists and antagonists has created a need in both academic institutions and pharmaceutical companies for sensitive and specific assays of angiogenesis. A variety of in vivo and in vitro models of angiogenesis have been developed which have greatly contributed to the remarkable advances in this field. Commonly used in vivo models include the rabbit cornea, the chorioallantoic membrane of the chick embryo and the hamster cheek pouch. In vitro models with isolated endothelial cell lines have been used to study selected aspects of the angiogenic process such as endothelial migration, proliferation, proteolytic digestion of the extracellular matrix (ECM) and capillary tube formation (Auerbach et al., 1991).

In our laboratory we have found that angiogenesis can be studied in vitro by culturing explants of rat aorta in gels of biological matrices (Nicosia et al., 1982; Nicosia and Ottinetti, 1990b). The rat aorta model can be used to evaluate various aspects of the angiogenic process including neovessel–cancer interactions, paracrine regulation of angiogenesis by the vessel wall, modulation of angiogenesis by the ECM, stimulation of microvessel formation by angiogenic factors, and inhibition of angiogenesis. This chapter reviews our experience with the rat aorta model and its multiple applications.

The Early Experience with Rat Aorta Cultures and the Development of the Serum-Free Model

In 1982, we reported that explants of rat aorta embedded in plasma clots and cultured in presence of serum generated luxuriant outgrowths of branching microvessels (Nicosia et al., 1982). We subsequently observed that the aorta-derived microvessels enhanced the spread of cancer cells in three-dimensional culture (Nicosia et al., 1983). Cancer cells co-cultured with the aortic explants invaded the plasma clot using the neovascular networks as a trellis for attachment, migration and proliferation. This was the first demonstration that neovascular outgrowths enhanced the aggressiveness of a malignant neoplasm in the absence of blood flow. We hypothesized that microvessels promoted the spread of cancer cells through juxtacrine or paracrine mechanisms mediated by ECM molecules and growth factors. The study of angiogenesis and angiogenesis-dependent cancer spread in this model was, however, complicated by the confounding stimulatory effects of plasma clot and serum. In addition, labor intensive histologic examination of the cultures was needed to quantitate neovascular growth.

These limitations were overcome by the serendipitous discovery that serum and plasma clot were not required for the angiogenic response of the rat aorta (Nicosia and Ottinetti, 1990b). We made this observation while studying the effect on rat aortic angiogenesis of the molecules fibronectin and thrombospondin-1. Since these molecules are present at high concentration in serum, it became necessary to culture the rat aorta without serum. To our surprise, aortic explants cultured in collagen or fibrin gels under serum-free conditions generated a self-limited angiogenic response, which was easily quantifiable by direct visual examination of the living cultures. Similar observations were made concurrently by Kawasaki and co-workers (Kawasaki et al., 1989).

In the following years, the serum-free rat aorta model proved to be a sensitive and reproducible assay of angiogenesis. The next section contains a detailed description of the methodology that we are currently using to prepare and analyze the rat aorta cultures. This is followed by an overview of rat aortic angiogenesis and its regulation by angiogenic factors, ECM molecules and angiogenesis inhibitors.

The Serum-Free Rat Aorta Model: Methodology

Excision of the Aorta

The main steps in the preparation of the aortic cultures are illustrated in Figure 1. Excision of the aorta is performed in a designated area of the laboratory which is clean and dust-free. Two- to three-month-old Fischer 344 male rats are sacrificed by a lethal intraperitoneal injection of sodium pentobarbital or by CO_2 asphyxiation. Immediately after the animal has died, the thoracic and abdominal skin are shaved with a hair clipper. The rat is then laid on a rectangular block of styrofoam while its legs are extended and pinned down with needles. The skin is wetted and sterilized with 70% ethanol. A Y-shaped incision is employed to cut the thoracic and abdominal skin. After the skin has been dissected from the underlying muscle layer, the abdominal cavity is opened using a cross-shaped cut. The sternal plate and the attached diaphragm are then cut with scissors. The xyphoid process of the sternum is clamped with a hemostat and the whole sternal plate is folded over the right side of the animal to expose the thoracic cavity. The intestines, the stomach, the spleen and the liver are displaced to the right side. The diaphragm is then sectioned in a ventral/dorsal direction paying attention not to cut the diaphragmatic vessels. The thoracic

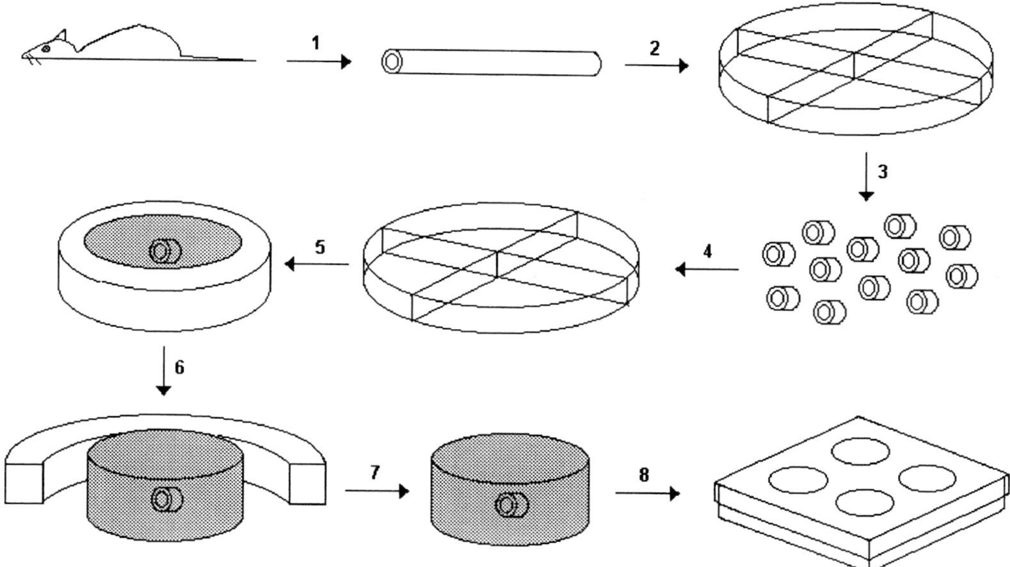

Fig. 1 Schematic drawing showing the main steps in the preparation of serum-free collagen gel cultures of rat aorta. The thoracic aorta of a 2- to 3-month-old male Fischer 344 rat is excised (1) and transferred to a compartmentalized Felsen dish where it is washed and cleaned of surrounding fibroadipose tissue (2). The aorta is cross-sectioned into 1- to 2-mm-long rings (3) which are washed (4) and embedded in collagen gels (5). Each collagen gel culture (collagen is shown as a gray matrix) is prepared in a cylidrical culture well made of agarose. After the collagen has gelled, the agarose of the culture well is cut open (6). The gel is separated form the agarose (7) and transferred to an 18-mm well of a 4-well culture dish, where it floats in serum-free MCDB-131 growth medium (8). Collagen gel cultures of rat aorta are kept in a humidified incubator at 35.5°C.

aorta, which is now visible along the vertebral column, is ligated with 3-0 silk sutures both proximally, below the aortic arch, and distally, above the diaphragm. The sutures are passed around the aorta after creating an opening with fine curved microdissection forceps (Fine Science Tools Inc., Foster City, CA) between the aorta and the vertebral column. While it is held by the distal suture, the aorta is excised from the posterior mediastinum and transferred into a compartmentalized Felsen dish. Transfer is carried out by cutting the aorta just below the suture. The Felsen dish has four compartments each containing 4 ml of serum-free MCDB 131 growth medium (Knedler and Ham, 1987; Clonetics, San Diego, CA). During excision, attention is paid not to stretch the aorta and not to cut the surrounding veins, to avoid bleeding.

Preparation of the Aortic Rings

Dissection of the aorta including preparation of the aortic rings is carried out under an American Optical dissecting microscope in a tissue culture room with HEPA-filtered air. The periaortic fibroadipose tissue is dissected with Noyes scissors (Fine Science Tools Inc.) and curved microdissection forceps. Care is taken not to stretch, cut, or crush the aortic wall. Intraluminal blood clots are removed with forceps. Small periaortic interstitial hemorrhages, which might occur during excision of the aorta from the animal, are removed by cutting the hemorrhagic adventitial tissue away from the aortic wall. While it is cleaned, the aorta is transferred to successive compartments of the Felsen dish. When it is in the fourth compartment of the dish, the aorta is cross-sectioned into 1- to 2-mm-long rings with Noyes scissors. The proximal and distal 1-mm-long segments of the aorta, which are used to hold the explant during dissection, are discarded. The aortic rings are then cleaned of residual blood by sequential transfer in 12 consecutive baths of serum-free medium (three Felsen dishes). The microdissection forceps are used to transfer the rings. After each transfer, the aortic rings are washed by gently shaking the dish a few times. Care is taken during each transfer not to crush or otherwise mechanically damage the rings.

The procedures for washing the aortic rings and preparing the collagen gel cultures (see below) are performed in a black wooden hood closed on top by a transparent glass. The glass protects the cultures from possible dust and allows direct vertical view from above. This arrangement greatly facilitates the handling of the rat aorta which is otherwise difficult in a laminar flow hood. Alternatively, the procedure can be carried out in an open tissue culture hood with positive air flow. However, if this type of hood is used, it is important to avoid air drying of the endothelium which can occur if the aorta is kept for too long out of the growth medium.

Preparation of Collagen Gel Cultures

Collagen gel cultures of rat aorta are prepared in agarose culture wells. The same procedure is used to prepare fibrin gel cultures (Nicosia and Ottinetti, 1990b). Here we describe the method that we routinely use for the collagen gel cultures. Agarose culture wells are prepared before sacrificing the animal by placing rings of agarose on the bottom of culture dishes. A sterile 1.5% solution of agarose (type VIA, Sigma Chemical Company, St. Louis, MO) is poured into 100 × 150 culture dishes (35 ml/dish; Falcon, Lincoln Park, NJ) and allowed to gel in a laminar flow hood. Dishes with solidified agarose can be stored up to 4 weeks in an airtight container at 4°C. Agarose rings are prepared before each experiment by punching two concentric circles in the agarose with nylon punchers having diameters of 10 and 17 mm, respectively. The punchers have a thin wall (1/3 mm) and a sharp cutting rim. The excess agarose inside and outside the rings is removed with a bent glass pipette. The agarose rings are then transferred with a bent spatula to clean 100 × 150 dishes and gently tapped from above with the spatula to insure adherence to the bottom of the dish. This method produces a cylindrical culture well with a plastic bottom and an agarose wall.

Four drops of collagen solution are placed with a transfer pipette on the bottom of each agarose culture well. Collagen is prepared by mixing on ice 8 volumes of 1 mg/ml collagen in 1/10 Minimum Essential Medium (MEM), pH 4.0, with a freshly made solution containing 1 volume of 10× MEM and one volume of 23.4 mg/ml Na HCO_3. Collagen is purified from the rat tail as reported (Elsdale and Bard, 1972). The dishes holding the agarose wells are then transferred into a humidified incubator for 5 min at 35.5–37°C. After the bottom collagen has gelled, each aortic ring is placed on the agarose at the edge of the culture well. Using a transfer pipette (Fisher, Malvern, PA), the well is completely filled with collagen solution and the aortic ring is displaced to the bottom of the well. Fine microdissection forceps are used to position the ring on the bottom collagen gel so that the profiles of the two cut edges of the explant are clearly visible. Then the agarose wells are incubated for 30 min at 35.5–37°C. After the collagen has gelled, the agarose of each culture well is cut open with a scalpel blade and separated from the gel with a bent spatula. Each collagen gel is then transferred with the spatula to an 18-mm well containing 0.5 ml of serum-free medium (4-well NUNC dish, Interlab, Thousand Oaks, CA). The collagen gels float in serum-free medium which can be supplemented with angiogenic agonists or antagonists. Each experimental group includes three to four cultures. The cultures are kept in a humidified CO_2 incubator at 35.5°C. Growth medium is changed three times a week starting from day 3 of culture.

Quantitation of Angiogenesis

The angiogenic response of the rat aorta can be quantitated by visual counts (Nicosia and Ottinetti, 1990b) or by computer-assisted image analysis (Nicosia et al., 1993; Nissanov et al., 1995). For visual quantitation, curves of microvascular growth are generated by counting the number of microvessels daily or every other day. Cultures are examined and scored under bright-field microscopy using an inverted Leitz microscope equipped with 2× and 4× objectives. Phase contrast microscopy is not recommended for quantitation, because the three-dimensional outgrowths cannot be visualized adequately at low magnification with this method. Optimal contrast and depth of field are obtained by closing the iris diaphragm of the condenser. The following criteria are used for counting microvessels: (a) microvessels are distinguished from fibroblast-like cells based on their thickness and cohesive pattern of growth; (b) the branching of one microvessel generates two new microvessels, which are added to the count; (c) each loop is counted as two microvessels because it originates from two converging microvessels. The time required for scoring each culture by a trained observer ranges from less than 1 minute to 5 minutes depending on the number of microvessels.

Image analysis can be used to measure the length of microvessels directly from living cultures with a personal computer. Microvessels are visualized under a 2× or 4× objective with a digital videocamera mounted on a Leitz Laborlux microscope. Length is measured by manually tracing individual microvessels with a computer mouse. The data are recorded and analyzed with Bioquant IV image analysis software (Nicosia et al., 1993).

Nissanov et al. have recently developed an Apple computer-based fully automated method to quantify the aortic outgrowth (Nissanov et al., 1995). An image-processing algorithm segments the vessels from gray scale images. A digital filter separates the images into vascular and nonvascular (fibroblast-like cells) compartments based on object size and shape. Quantification relies on the identification of vessels intersecting a closed transect set at a fixed distance from the aortic explant. Correlation between computer-assisted quantification and visual microvessel count is high. The application can process approximately 30 images/hour and is particularly useful for pharmacologic screening studies which require quantitation of a large number of cultures.

Morphologic Studies

Living Cultures

Angiogenesis in the serum-free rat aorta model is a self-limited process triggered by the injury of the dissection procedure. The aortic outgrowth is composed of three main cell types: fibroblast-like cells, endothelial cells, and pericytes (Nicosia and Ottinetti, 1990b; Nicosia and Villaschi, 1995). The first cells to migrate out of the explants are fibroblast-like cells which appear in the gel at day 2 of culture. Endothelial cells sprout from the cut edges of the explant at day 3–4.

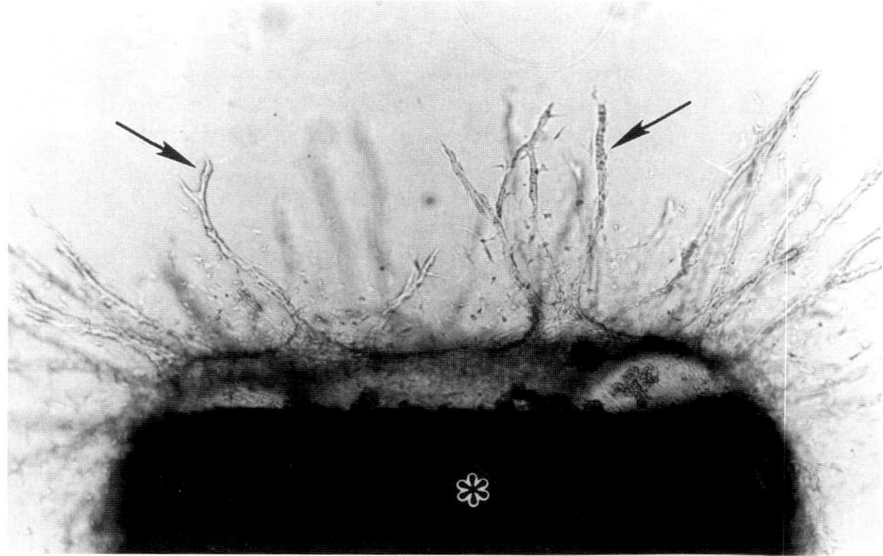

Fig. 2 Serum-free collagen gel culture of rat aorta. The aortic ring (asterisk) appears with its luminal axis parallel to the plane of the culture dish. The aortic lumen is therefore not visible. The vascular outgrowth originates from the cut edge of the explant. Newly formed microvessels are marked by arrows. Magnification: ×40. Reprinted with permission from John Wiley & Sons, Inc.

Endothelial sprouts are recognizable by their thickness and cohesive growth pattern (Fig. 2). By contrast, fibroblast-like cells are thin, isolated and finely tapered. During the second week, microvessels elongate, branch, anastomose, and eventually stop growing. Subsequent remodeling of the vascular outgrowth results in regression of the small branches which retract into the main stems of the larger microvessels.

During the first days of culture, the neovascular sprouts are composed primarily of endothelial cells. As the microvessels mature, pericytes emerge from the explant and migrate along the abluminal aspect of the sprouting endothelium. Pericytes migrate and proliferate increasing progressively in number after microvessels have stopped growing. Thus, as they regress and remodel, microvessels become surrounded by a large number of pericytes.

Time-lapse videomicroscopy can be used to study the behavior of the endothelium and pericytes (Nicosia and Villaschi, 1995). For this type of study, collagen gel cultures are transferred with a spatula into a T-25 flask containing serum-free medium. After equilibrating the flask atmosphere with the air of a CO_2 incubator, the cap is screwed on tightly and the flask is transferred to the heated stage of an Olympus IMT-2 microscope equipped with phase contrast optics, a videocamera, a JVC time-lapse videocassette recorder and a high resolution monitor.

Examination of videotapes recorded over the various stages of angiogenesis demonstrates distinct behaviors of migration and proliferation by endothelial cells and pericytes. The sprouting endothelium of immature microvessels advances by piercing the collagen fibers with its elongated cytoplasm. The tapered tip of the sprouting endothelium penetrates the gel with a rhythmic seething motion characterized by alternating phases of activity and rest. Because they are connected by junctions, endothelial cells form continuous cords and cannot be distinguished as individual cells.

Pericytes migrate and proliferate individually at the abluminal surface of the endothelium. As the microvessels grow and eventually mature, pericytes migrate from the root to the tip of the microvessels, crawling along the endothelium, which they use as a surface for attachment, proliferation, and contact guidance. Pericytes continuously contract exhibiting a variety of shapes ranging from flat, to conical, and arcuate, as described for pericytes in vivo. During the late stages of maturation of the microvessels, pericytes develop extensive dendritic cytoplasmic processes. Regression of microvessels is at least in part due to endothelial retraction. This process results in the reabsorption of smaller branches into the main stalks of the microvessels. Mature microvessels become thicker not only because they are surrounded by pericytes but also because they are composed of a larger number of endothelial cells per cross-sectional area.

Immunohistochemical and Ultrastructural Studies

The cultures can be fixed in formalin for light microscopic and immunohistochemical studies or in a glutaraldehyde-containing fixative for ultrastructural analysis (Nicosia et al., 1982; Nicosia and Ottinetti, 1990b). For paraffin embedding, the collagen gels are processed manually with conventional methods for dehydration, paraffin infiltration, and embedding. Formalin-fixed or unfixed collagen gels can be also embedded in Optimal Cutting Temperature (OCT) compound (Baxter, McGaw Park, IL), and snap frozen in cold isopentane for cryosectioning (Nicosia and Villaschi, 1995). This procedure, which may be used for lectin or immunohistochemical studies, requires pre-treatment of the collagen gels with graded dilutions of OCT compound in phosphate buffered saline. Infiltration with OCT compound greatly facilitates sectioning with the cryostat.

Endothelial cells express factor VIII-related antigen (FVIII-RAg) and stain with the Griffonia Simplicifolia isolectin B4 which typically binds to rodent endothelial cells but not to smooth muscle cells, pericytes, or fibroblasts (Nicosia and Ottinetti, 1990b; Nicosia et al., 1994a). Pericytes are negative for FVIII-RAg and positive for α-smooth muscle actin. Fibroblast-like cells are negative for FVIII-RAg. Approximately 10–15% of these cells become α-smooth muscle actin-positive in older cultures. Endothelial cells, smooth muscle cells, and fibroblast-like cells are all positive for vimentin (Nicosia and Villaschi, 1995).

By electron microscopy, the neovascular outgrowths appear as either solid endothelial cords or patent microvessels. Solid cords are composed of plump endothelial cells connected by junctions and have either no lumina or slitlike luminal spaces. Patent microvessels are composed of polarized endothelial cells resting on a discontinuous basal lamina. Prominent pinocytotic activity is observed at both the luminal and abluminal surfaces of the endothelium. The endothelium is surrounded by a discontinuous layer of pericytes with characteristic electron dense cytoplasm and abundant cisternae of rough endoplasmic reticulum. Pericytes exhibit abundant heterochromatin, and peripherally condensed cytoplasmic filaments. They share the basal lamina with the endothelium which they envelop with thin and long cytoplasmic processes (Nicosia et al., 1982; Nicosia and Ottinetti, 1990b; Nicosia and Villaschi, 1995).

Origin of Endothelial Cells and Pericytes

The majority of microvessels arise from the injured endothelium at the cut edges of the aortic explants. Vasa vasorum, if still present after the dissection, may participate in the angiogenic response (Diglio et al., 1989). However, the aorta of young rats has very few vasa vasorum and the great majority of periaortic microvessels are removed during the dissection procedure.

Serial histologic sections demonstrate that the endothelium of the aortic intima migrates and sprouts into the collagen. The endothelium responds to the combined effects of injury and collagen overlay by undergoing a complex morphogenetic rearrangement which we have documented with light and electron microscopic studies (Nicosia and Ottinetti, 1990b; Nicosia et al., 1992). A schematic representation of the morphogenetic changes resulting in microvessel formation at the collagen/intimal interface is shown in Figure 3. These changes are particularly prominent in the intimal endothelium near the openings of the intercostal arteries.

Injured endothelial cells of the aortic intima become first plump and activated. They then lose their characteristic monolayer arrangement and crawl over each other forming tubelike structures which coalesce generating an endothelial-lined space between the aortic intima and the collagen gel. Lumina form by an extracellular mechanism mediated by the development of junctions between adjacent cells and by an intracellular mechanism, based on intracytoplasmic vacuolization. The aortic portion of the endothelial sheet that lines the newly developed space between the

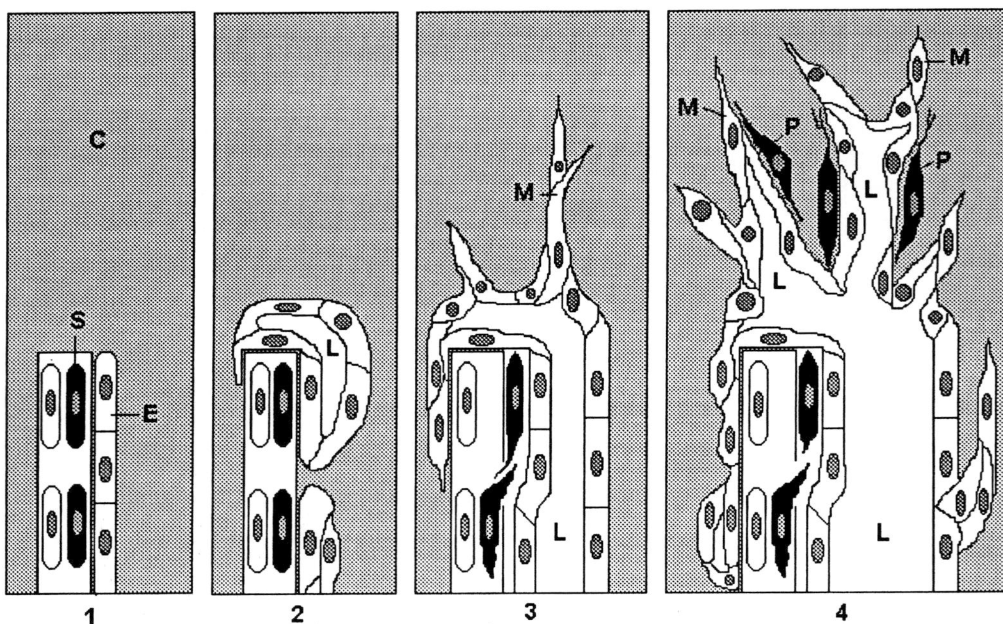

Fig. 3 Sequential stages (panels 1–4) in the angiogenic response of the rat aorta in serum-free collagen gel culture. The drawing shows a longitudinal section of the cut edge of an aortic ring. For simplicity, the aortic wall is represented by intimal endothelial cells (E) and two layers of smooth muscle cells (S). Endothelial cells penetrate the collagen gel (C) forming microvessels (M). During their morphogenetic response, endothelial cells migrate along the cut edge of the aorta and over the collagen gel forming the lining of a chamber which communicates with the lumina (L) of the microvessels. Sprouting endothelial cells recruit pericytes (P) from a subpopulation of smooth muscle cells located in the intimal/subintimal layers of the rat aorta (shown as black cells).

aorta and the collagen gel migrates over the cut edges of the aorta onto the adventitia, which in late cultures becomes covered by endothelial cells. At the same time, the endothelium that covers the gel remodels and penetrates the underlying collagen generating microvessels. Pericytes arise from a subpopulation of smooth muscle cells located in the intima or the subintimal layers of the media (Villaschi et al., 1994; Nicosia and Villaschi, 1995). These cells have a distinct endothelial tropism and probably respond to paracrine cues generated by the endothelium by migrating and proliferating along the microvessels into the collagen gel.

The vasoformative properties of the rat aortic endothelium can be further demonstrated either by culturing everted aortas in collagen gels or by studying the behavior of isolated endothelial cells (Nicosia et al., 1992). The evertion procedure sequesters adventitial cells and vasa vasorum inside the aortic tube leaving only the intimal endothelium exposed to the surrounding collagen. Endothelial cells above the internal elastic lamina respond to the injury of the aortic evertion and to the exposure of their apical surface to collagen fibrils (Jackson and Jenkins, 1991) by forming branching microvessels. Similarly, endothelial cells isolated form the rat aorta rapidly reorganize into networks of microvessels when sandwiched between two layers of collagen and sprout into collagen when treated with fibroblast conditioned medium (Nicosia et al., 1994a).

Endogenous Regulation of Angiogenesis by Growth Factors

When we first reported that rat aortic rings generated microvessels in serum-free culture, we proposed that the system was regulated by endogenous growth factors released by the explants in response to the injury of the dissection procedure (Nicosia and Ottinetti, 1990b). In a later study, immunohistochemical stain of the aorta revealed that basic fibroblast growth factor (bFGF), a potent angiogenic factor, was stored in the cytoplasm of endothelial and smooth muscle cells. The injury of the dissection procedure caused a release of bFGF from its storage sites. bFGF levels in aorta-conditioned medium were highest during the first days of culture and gradually decreased over time becoming undetectable when microvessels stopped growing (Villaschi and Nicosia, 1993).

Injured rat aortic explants secrete also vascular endothelial growth factor (VEGF, Nicosia et al., 1996) and platelet-derived growth factor (PDGF, unpublished observations). Secretion of VEGF and PDGF is highest during the early stages of angiogenesis and decreases over time, as seen for bFGF. Thus the rat aorta responds to the injury of the dissection procedure by releasing a variety of angiogenic factors.

The role of endogenous growth factors in the angiogenic response of the rat aorta can be demonstrated by treating collagen gel cultures with anti-bFGF or anti-VEGF neutralizing antibodies, both of which have significant anti-angiogenic activity (Nicosia et al., 1996; Villaschi and Nicosia, 1993). No studies have yet been performed with anti-PDGF antibodies to evaluate the role of endogenous PDGF.

The importance of endogenous growth factors can be further demonstrated by comparing the angiogenic response of freshly cut explants, which secrete measurable amounts of bFGF, VEGF, and PDGF, to that of quiescent explants, which produce low levels of these factors. Quiescence is obtained by pre-incubating aortic rings in suspension culture for 10–14 days at 35.5°C. Each aortic ring is housed in an 18 mm agarose-coated well (4-well Nunc dish) containing 0.5 ml of serum-free MCDB 131. The wells are coated with agarose to avoid attachment of the rings and growth of cells shed from the explants. After the pre-incubation period, during which the growth medium is changed four times, the aortic rings are rinsed with serum-free medium, embedded in collagen gels, and cultured as described for the freshly cut explants. Quiescent aortic rings, which produce low amounts of angiogenic factors, have a markedly reduced angiogenic response and generate only a few short microvessels (Nicosia et al., 1996). Addition of exogenous growth factors to the growth medium restores the angiogenic response to values observed in cultures of freshly cut aortic rings (Fig. 4; see below).

Effect of Exogenous Angiogenic Factors

The rat aorta model can be used to test the angiogenic activity of growth factors (Table I). We initially found that exogenous bFGF induced a dose-dependent increase in the number and length of microvessels (Villaschi and Nicosia, 1993). In a subsequent study, we found that the angiogenic response of the rat aorta was stimulated by VEGF, natural PDGF, recombinant PDGF AA and BB, and insulin-like growth factor-1 (IGF-1) (Nicosia et al., 1994c). Stimulatory effects were characterized by an increase in the number and length of the microvessels. Maximal stimulation was obtained with VEGF and PDGF BB. The angiogenic effect of PDGF and its recombinant forms was preceded and accompanied by an increase in the number and migratory activity of fibroblast-like cells. A similar, but less pronounced phenomenon was observed with IGF-1. At variance with PDGF, VEGF produced an outgrowth composed primarily of microvessels.

Since fibroblasts produce angiogenic factors (Sato et al., 1991), it is likely that the PDGF effect is indirect. This possibility is supported by evidence in the literature indicating that large vessel endothelial cells have no PDGF receptors and that fibroblasts stimulated by PDGF can in turn promote angiogenesis (Sato et al., 1993; Villaschi and Nicosia, 1994b). The VEGF effect is

Table I Soluble factors and extracellular matrix molecules that promote or inhibit angiogenesis in the rat aorta model

PROMOTERS		INHIBITORS	
SOLUBLE FACTORS	ECM MOLECULES	SOLUBLE FACTORS	ECM MOLECULES
bFGF	FIBRIN	HYDROCORTISONE	LAMININ-ENTACTIN**
VEGF	FIBRONECTIN*	CIS-HYDROXY-PROLINE	TYPE IV COLLAGEN**
PDGF	LAMININ-ENTACTIN**	GRGDS PEPTIDE	
PDGF BB	LAMININ	ANTI-bFGF ANTIBODY	
PDGF AA	ENTACTIN*	ANTI-VEGF ANTIBODY	
IGF-1	TYPE IV COLLAGEN**	TGF-β1	
	TSP-1	IL-1α	

Promoters stimulate proliferation and/or elongation of microvessels
*Fibronectin and entactin promote selective elongation of microvessels
**The laminin-entactin complex and type IV collagen have inhibitory effects at high concentration
Abbreviations: ECM, extracellular matrix; bFGF, basic fibroblast growth factor, VEGF, vascular endothelial growth factor, PDGF, platelet-derived growth factor, IGF-1, insulin-like growth factor-1; GRGDS, Glycine-Arginine-Glycine-Aspartic Acid-Serine; TSP-1, thrombospondin-1; TGF-β1, transforming growth factor-β1; IL-1α, interleukin-1α.

consistent with evidence in the literature that this molecule is a potent angiogenic factor with endothelial target specificity (Leung et al., 1989). Thus, factors that have emerged in recent years as potent regulators of blood vessel growth such as VEGF and bFGF are very active in the system. The model can be also used to study growth factors which, like PDGF, probably stimulate angiogenesis indirectly through the activation of nonendothelial cells such as fibroblasts.

Exogenous growth factors can be tested also in cultures of quiescent explants (Fig. 4). Because of the low levels of endogenous growth factors that they produce, quiescent explants are particularly sensitive to stimulation by exogenous bFGF, PDGF, or VEGF (Nicosia et al., 1996). As a result, the percentage stimulation of angiogenesis by exogenous factors is even greater in these cultures than in cultures of freshly cut explants. IGF-1, at variance with bFGF, PDGF, and VEGF, does not stimulate the angiogenic response of quiescent explants (unpublished observations). This result, taken together with data from cultures of freshly cut explants, suggests that IGF-1 is capable of enhancing the efficiency of an existing angiogenic response but is unable to trigger a new one.

Modulation of Angiogenesis by Extracellular Matrix Molecules

Effect of Fibrin, Interstitial Collagen, and Basement Membrane Molecules

At the time we first described the rat aorta model, evidence from other laboratories indicated that the molecules of the ECM had the capacity to regulate endothelial adhesion, migration, proliferation, and morphogenesis (Madri and Williams, 1983; Montesano et al., 1983). Since these processes are critical for vessel formation, we set out to investigate the composition and distribution of the microvascular ECM at different stages of rat aortic angiogenesis (Nicosia and Madri, 1987). These studies predated the development of the serum-free model and were carried out with serum-stimulated plasma clot cultures. Using light and electron immunohistochemical methods, we found that the ECM of aorta-derived microvessels underwent complex maturational changes. The provisional ECM of immature microvessels was primarily composed of a fibrillary network of fibronectin and type V collagen and of patchy amorphous deposits of laminin and type IV collagen. As microvessels matured and acquired a patent lumen they became surrounded by a continuous layer of laminin and type IV collagen, by the interstitial collagen types I and III, and by type V collagen. Perivascular deposition of collagen was promoted by the addition of ascorbic acid to the culture medium (Nicosia et al., 1991). Ascorbic acid deficiency caused a reduction of collagen production which in turn resulted in a marked dilatation of the lumen of the microvessels. Vascular dilatation was not seen in cultures supplemented with ascorbic acid. The importance of collagen in vessel formation was confirmed by the observations that angiogenesis was inhibited by collagen synthesis inhibitors (Ingber and Folkman, 1988; Nicosia et al., 1991) and that capillary tube formation was rapidly induced by exposing isolated endothelial cells to interstitial collagen (Jackson and Jenkins, 1991; Montesano et al., 1983; Nicosia et al., 1994a).

To better understand the effect of the ECM on vessel growth we studied whether the angiogenic response of the rat aorta was modified by culturing the aortic explants in different types of biomatrix gels. Rat aortic explants were cultured in plasma clot, fibrin, interstitial collagen, or Matrigel, a basement membrane-like matrix, in the presence of serum (Nicosia and Ottinetti, 1990a). Maximal growth was obtained in plasma clot whereas Matrigel appeared to

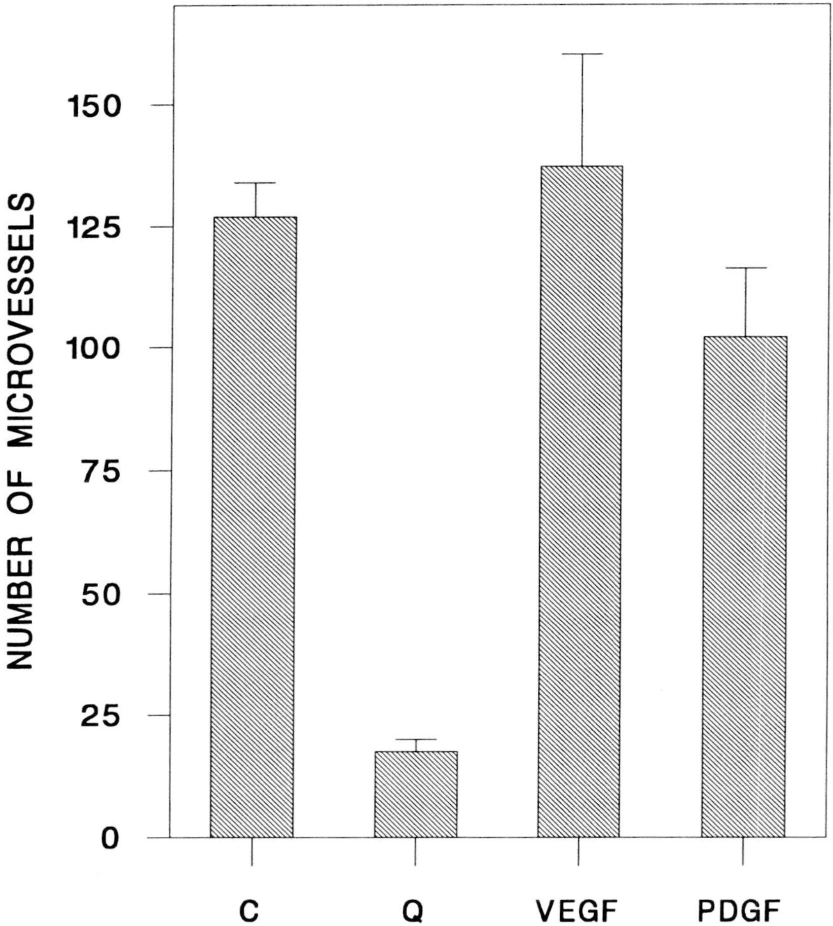

Fig. 4 Angiogenic response of freshly cut rat aortic rings (C), untreated quiescent rings (Q), or quiescent rings stimulated with either 20 ng/ml PDGF BB or 10 ng/ml VEGF, in serum-free collagen gel culture. Shown here is the maximal number of microvessels counted in each set of cultures. Note: quiescent rings have a much lower capacity to generate microvessels than freshly cut rings. Treatment with growth factors restores the angiogenic response of quiescent aortic rings to the values recorded in cultures of freshly cut explants. N = 4. Error bar represents the standard error of the mean.

interfere with angiogenesis causing a reduction in the number of vessels and a decrease in DNA synthesis. The stimulatory effect of the plasma clot was in part due to fibrin which promoted more vessel growth than Matrigel. Interstitial collagen exhibited intermediate effects between those of fibrin and Matrigel.

The effect of Matrigel was paradoxical since basement membrane molecules are normally produced by migrating and proliferating cells during the angiogenic process. One possible explanation for this unexpected result was that the endothelial cells of immature sprouts, when cultured in Matrigel, were forced to migrate into a matrix that contained concentrations of basement membrane molecules that are normally present in the basement membrane of quiescent vessels, which do not proliferate. Was Matrigel inhibiting endothelial cell proliferation by inducing differentiation at a time when endothelial cells had to divide in order to form new vessels? This interpretation was suggested by the observation that isolated endothelial cells cultured on Matrigel ceased to proliferate and reorganized into networks of capillary-like structures (Kubota et al., 1988; Nicosia et al., 1990a). Conversely, fibrin and interstitial collagen, which are not components of the basement membrane, were functioning as permissive substrates allowing new vessels to surround themselves with basement membrane molecules according to a developmentally regulated program of synthesis and secretion. If this hypothesis was correct, what role in angiogenesis might we expect from the individual components of the basement membrane and what was the significance of the quantitative and qualitative changes that we and others had detected in the microvascular basement membrane at different stages of angiogenesis?

To answer these questions it became critical to evaluate the effect of purified basement membrane molecules in a chemically defined culture environment. In fact, serum contains fibronectin and other adhesive proteins, which might have masked the effect of the molecules being tested. Matrigel on the other hand is contaminated with growth factors (Vukicevic et al., 1992), which might have affected the angiogenic response of the rat aorta. The serum-free rat aorta model (Nicosia and Ottinetti, 1990b) helped us overcome these technical difficulties.

Serum-free collagen gel cultures of rat aorta were treated with increasing concentrations of the basement membrane molecules fibronectin (Nicosia et al., 1993), laminin (Nicosia et al., 1994b), or type IV collagen (unpublished observations), which were added to the solution of interstitial collagen before gelation (Table I). Laminin was used in its purified form or as a laminin–entactin complex. Entactin is a basement membrane protein which co-purifies with laminin under nondenaturing conditions. Fibronectin, laminin, entactin, and type IV collagen, all caused a dose-dependent elongation of the neovessels.

The mechanisms responsible for the ECM-mediated microvascular elongation were studied in cultures treated with fibronectin (Nicosia et al., 1993). Histologic studies demonstrated that the microvessels of fibronectin-treated cultures were longer because they were composed of an increased number of cells. The increase in cell number did not appear to be the result of increased cell proliferation since fibronectin-treated cultures had the same DNA synthesis and mitotic activity as untreated control cultures. On this basis we hypothesized that microvascular elongation was mediated by an adhesion-dependent migratory recruitment of endothelial cells by fibronectin. This interpretation confirmed in a three-dimensional model the observations by others that fibronectin stimulates endothelial cell migration (Bowersox and Sorgente, 1982).

The laminin–entactin complex and type IV collagen had also a dose dependent bimodal effect on the number of microvessels, which was probably due in part to an interaction of these molecules with the endogenous growth factors released by the aortic explants (Nicosia et al., 1994b). Laminin-entactin and type IV collagen showed no significant effects at low concentration, promoted vascular proliferation at intermediate concentration, and had anti-angiogenic activity at high concentration. The inhibitory effect was not due to toxicity since the few vessels that formed in these cultures were highly stable and survived much longer than the microvessels of the untreated cultures. The effects observed when laminin–entactin was used at high concentration were consistent with the partial inhibitory activity of Matrigel which we had found in the earlier studies with serum-stimulated cultures. Matrigel, in fact, contains large amounts of laminin and type IV collagen, as seen in mature basement membranes.

Based on these results we now hypothesize that during the early stages of angiogenesis, endothelial cells stimulated by growth factors released by the aortic explant sprout into the collagen gel where they are induced to form tubes by interstitial collagen fibrils. During this phase, the endothelial cells secrete a provisional matrix which promotes microvascular sprouting and elongation. Immature microvessels, however, are prone to regression because they are not yet stabilized by a well formed basement membrane. The gradual accumulation of basement membrane molecules promotes endothelial proliferation when laminin and type IV collagen are secreted at intermediate concentrations that are permissive for growth. The formation of a mature basement membrane containing high concentration of laminin and type IV collagen results in stabilization of the microvessels which differentiate and cease to proliferate.

Effect of Thrombospondin-1

Thrombospondin-1 (TSP-1) is a glycoprotein secreted by a variety of cell types including vascular cells, which is incorporated in the ECM where it regulates cell–matrix and cell–cell interactions. The role of TSP-1 in angiogenesis is, however, unclear, since different groups have reported either stimulatory (Ben Ezra et al., 1993) or inhibitory (Good et al., 1990; Iruela-Arispe et al., 1990) effects. Our serum-free model gave us the opportunity to test TSP-1 under chemically-defined culture conditions. TSP-1 incorporated in collagen or fibrin gels promoted a dose-dependent growth of microvessels and fibroblast-like cells (Nicosia and Tuszynski, 1994). TSP-1 stimulated angiogenesis indirectly by promoting the growth of fibroblast-like cells which in turn secreted heparin-binding angiogenic factor(s). Interestingly, TSP-1 has inhibitory effects on isolated endothelial cells (Bavagandoss and Wilks, 1990). This effect may, however, be offset when endothelial cells are exposed to TSP-1 in the presence of fibroblasts which are stimulated by this molecule. Thus, the function of TSP-1 appears to vary depending on the cellular context in which this molecule is presented to the endothelial cells. The manner by which TSP-1 is delivered is also important since TSP-1 has multiple binding domains for both ECM molecules (Nicosia and Tuszynski, 1994) and angiogenic factors such as bFGF (Taraboletti et al., 1992). TSP-1 may sequester bFGF in the ECM impeding its diffusion toward the endothelium. Alternatively, TSP-1 may localize bFGF at sites of endothelial sprouting, thereby promoting angiogenesis. The latter condition is likely to occur in the rat aorta model because TSP-1 is incorporated in the same matrix that contains endothelial cells and bFGF released by the aorta. Assays such as the cornea model, which require diffusion of exogenous bFGF from a polymeric implant, may give opposite results because TSP-1, by binding to bFGF, is likely to limit the diffusion of this growth factor toward the distant endothelium of the limbal vessels. These considerations suggest that the rat aorta model may be more suitable than the cornea model for testing the effect on angiogenesis of ECM molecules, which diffuse poorly, and function primarily as solid-phase substrates.

Effect of Angiogenesis Inhibitors

The angiogenic response of the rat aorta can be inhibited by molecules that interfere with endothelial cell adhesion, migration and proliferation (Table I). Since ECM molecules are critical for these processes, substances that inhibit ECM production and cell–matrix interactions, have significant anti-angiogenic activity. For example, cis-hydroxyproline blocks angiogenesis by interfering with collagen metabolism (Ingber et al., 1988; Nicosia et al., 1991). The potent anti-angiogenic effect of hydrocortisone (Nicosia and Ottinetti, 1990b) is at least in part due to the disruptive effect of this molecule on the microvascular basement membrane (Ingber et al., 1986).

Of particular interest is the anti-angiogenic activity of RGD-containing peptides. While we were studying the mechanisms of angiogenesis in the rat aorta model, Pierschbacher and

Ruoslahti discovered that the amino acid sequence Arginine-Glycine-Aspartic Acid (RGD) was critical for the binding of fibronectin to its integrin receptors on the cell surface (Pierschbacher and Ruoslahti, 1984). The same group demonstrated that synthetic peptides containing the RGD sequence interfered with cell adhesion causing endothelial cells to round up and detach from a fibronectin substrate (Hayman et al., 1985). On this basis we evaluated the effect of an RGD-containing peptide on the angiogenic response of the rat aorta. Addition of the pentapeptide GRGDS (Gly-Arg-Gly-Asp-Ser) to the culture medium caused a marked inhibition of angiogenesis whereas the control peptide GRGES (Gly-Arg-Gly-Glu-Ser), which lacked the RGD sequence, had no effect (Nicosia and Bonanno, 1991). GRGDS induced also regression of neovessels formed in fibronectin-treated cultures, thereby abolishing the fibronectin effect on microvascular elongation. The subsequent report that a neutralizing antibody against the vitronectin receptor (Brooks et al., 1994) inhibits angiogenesis corroborated our hypothesis that angiogenesis is an RGD-sensitive process. The vitronectin receptor, which is expressed by sprouting endothelial cells and binds to fibronectin and other RGD-containing ECM molecules (Brooks et al., 1994), is sensitive to inhibition by RGD-containing peptides.

Although endothelial migration is probably sufficient for the formation of microvessels by the aortic explants, endothelial proliferation is needed for an optimal response. Thymidine uptake studies demonstrate that endothelial cells during rat aortic angiogenesis are actively engaged in DNA synthesis (Kawasaki et al., 1989; Nicosia et al., 1982; Nicosia and Villaschi, 1995). Treatment of cultures with colchicine to block dividing cells in mitosis demonstrates mitotic figures by the sprouting endothelium in histologic sections (Nicosia et al., 1993). Inhibition of mitotic activity in living cultures with mitomycin C causes a marked reduction in the angiogenic response (Nicosia et al., 1993).

Transforming growth factor-β1 (TGF-β1) and interleukin-1α (IL-1α) have antiangiogenic effects in the rat aorta model (Table I). These findings are consistent with reports that TGF-β1 and IL-1α inhibit endothelial cell proliferation in vitro (Cozzolino et al., 1990; Ray Chaudhury and D'Amore, 1991). Thus, TGF-β1, which also inhibits endothelial cell migration and promotes endothelial cell differentiation (RayChaudhury and D'Amore, 1991), behaves as an inhibitor when added to a developmental system such as the rat aorta model, which requires endothelial de-differentiation, migration, and proliferation for an adequate angiogenic response. The role of endogenous TGF-β1, however, remains to be determined. Endogenous TGF-β1 may promote the differentiation and stabilization of the microvessels when endothelial cells establish contact with pericytes, during the late stages of the angiogenic process (Nicosia and Villaschi, 1995; RayChaudury and D'Amore, 1991). It is also possible that TGF-β1 may have angiogenic activity at low concentration and anti-angiogenic activity at high concentration, as proposed by others (Pepper et al., 1993). In this instance, since low amounts of active TGF-β1 may already be present in the rat aortic cultures, addition of exogenous TGF-β1 is likely to generate only an inhibitory effect. Thus, more studies are needed to clarify how endogenous TGF-β1 functions in the rat aorta model.

Advantages of the Serum-Free Rat Aorta Model

The rat aorta model combines advantages of both in vivo and in vitro models of angiogenesis. Angiogenesis occurs in a chemically defined culture environment which can be easily adapted to different experimental conditions. Because the endothelial and nonendothelial cells of the aortic outgrowth have not been modified by repeated passages in culture, the newly formed microvessels reproduce in vitro the characteristics of microvessels formed during angiogenesis in vivo. The interactions between endothelial cells, pericytes, and fibroblast-like cells as well as the temporal relation between the growth of these cell types and the angiogenic process can be easily monitored and analyzed. Inflammatory complications, which may affect the interpretation of in vivo assays, are eliminated because there is no inflammation in the system. Both the soluble phase (growth medium) and the solid phase (collagen or fibrin gel) in which microvessels grow can be modified to test the activity of growth factors and ECM molecules. The effect of angiogenesis agonists and antagonists can be evaluated in the absence of serum molecules which may otherwise bind, inactivate, or simulate the action of the substances being tested. Angiogenesis can be quantitated reproducibly by either visual counts or image analysis. Finally, animal sacrifice is reduced to a minimum since many assays can be prepared from one aorta, thereby decreasing the number of animals needed for each study.

Limitation of the Rat Aorta Model

Direct quantitation of angiogenesis from the living cultures is limited when stimulation of angiogenesis results in the formation of more than 250 microvessels. For example, aortic explants embedded in fibrin gel or plasma clot and treated with 20% fetal bovine serum generate richly vascularized outgrowths surrounded by numerous fibroblasts (Nicosia et al., 1982; Nicosia and Ottinetti, 1990a). The margin of error for the observer who is counting microvessels becomes too high due to the three-dimensional complexity of the vascular network and the masking effect of the surrounding fibroblasts. Quantitation in these cultures is accomplished by counting the number of microvessels in histologic sections taken from the central portion of the gels (Nicosia and Ottinetti, 1990a). Histologic quantitation may be needed also in serum-free culture when two or more angiogenic agonists are tested together for additive or synergistic effects.

Conclusions and Future Directions

The rat aorta model of angiogenesis has been a powerful tool in our research armamentarium for many years. The observation that a blood vessel such as the rat aorta is capable of generating new vessels in the absence of exogenous stimuli such as serum, inflammation, or cancer has raised challenging questions which may hold the clue for a better understanding of the angiogenic process and its mechanisms.

Injury is clearly the trigger that activates the angiogenic response of the rat aorta. But, how does injury regulate the expression of angiogenic factors, ECM molecules, proteolytic enzymes, and cell membrane receptors, which have been implicated in the rat aorta model and in other systems? The spatiotemporal staggering that characterizes the appearance of different cell types and the self-limited nature of the angiogenic response suggest a cascade of gene and cellular activation. Do fibroblast-like cells, which are the first cells to migrate out of the explants send chemotactic signals to the endothelial cells that follow? Do endothelial cells in turn generate molecular cues to attract pericytes around the microvessels? Which signals do endothelial cells send to each other to generate anastomoses between microvessels? What molecular and cellular mechanisms turn angiogenesis off? Why do newly formed microvessels regress?

Our studies to date have implicated several growth factors and ECM molecules in the system. We have also found that fibroblasts play an important role in promoting the formation and survival of microvessels (Villaschi and Nicosia, 1994b). Future studies with the rat aorta model will help us define the sequence of molecular and cellular events required for the angiogenic process. The identification of key molecular checkpoints in this sequence is likely to contribute to the development of new drugs for the treatment of both angiogenesis-dependent disorders and conditions caused by inadequate neovascularization.

References

Auerbach R., Auerbach W., Polakowski I. (1991): Assays of angiogenesis: a review. Pharmac Ther 51:1–11.

Bagavandoss P., Wilks J.W. (1990): Specific inhibition of endothelial cell proliferation by thrombospondin. Biochem Biophys Res. Commun 170:867–872.

Ben Ezra D., Griffin B.W., Maftzir G., Aharonov O. (1993): Thrombospondin and in vivo angiogenesis induced by basic fibroblast growth factor or lipopolysaccharide. Invest Ophthalmol Sci 34:3601–3608.

Bowersox J.C., Sorgente N. (1982): Chemotaxis of aortic endothelial cells in response to fibronectin. Cancer Res 42:2547–2551.

Brooks P.C., Clark R.A., Cheresh D.A. (1994): Requirement of vascular integrin avb3 for angiogenesis. Science 264:569–571.

Cozzolino F., Torcia M., Aldinucci D., Ziche M., Almerigogna F., Bani D., Stern D.M. (1990): Interleukin 1 as autocrine regulator of endothelial cell growth. Proc Natl Acad Sci USA 87:6487–6491.

Diglio C.A., Grammas P., Giacomelli F., Wiener J. (1989): Angiogenesis in rat aorta ring explant cultures. Lab Invest 60:523–531.

Elsdale T., Bard J. (1972): Collagen substrata for studies of cell behavior. J Cell Biol 54:626–637.

Folkman J. (1995): Clinical applications of research on angiogenesis. New Eng J Med 333(26): 1757–1763.

Folkman J. (1996): Fighting cancer by attacking its blood supply: What you need to know about cancer. Scientific Am 275:150–154.

Good D.J., Polverini P.J., Rastinejad F., LeBeau M.M., Lemons R.S., Frazier W.A., Bouck N.P. (1990): A tumor-suppressor dependent inhibitor of angiogenesis is immunologically and functionally indistinguishable from a fragment of thrombospondin. Proc Natl Acad Sci USA 87:6624–6628.

Hayman E.G., Pierschbacher M.D., Ruoslahti E. (1985): Detachment of cells from culture substrate with soluble fibronectin peptides. J Cell Biol 100:1948–1954.

Ingber D., Folkman J. (1988): Inhibition of angiogenesis through modulation of collagen metabolism. Lab Invest 59:44–51.

Ingber D.E., Madri J.A., Folkman J. (1986): A possible mechanism for inhibition of angiogenesis by angiostatic steroids: induction of capillary basement membrane dissolution. Endocrinology 119:1768–1774.

Iruela-Arispe M.L., Bornstein P., Sage H. (1990): Thrombospondin exerts an anti-angiogenic effect on cord formation by endothelial cells in vitro. Proc Natl Acad Sci USA 88: 5026–5030.

Jackson C.J., Jenkins K.L. (1991): Type I collagen fibrils promote rapid vascular tube formation upon contact with the apical side of cultured endothelium. Exp Cell Res 192:319–323.

Kawasaki S., Mori M., Awai M. (1989) Capillary growth of rat aortic segments cultured in collagen without serum. Acta Pathol Japonica 39(11):712–718.

Knedler A., Ham R. (1987): Optimized medium for clonal growth of human microvascular endothelial cells with minimal serum. In Vitro Cell Dev Biol 23:481–491.

Kubota Y., Kleinman H.K., Martin G.R., et al. (1988): Role of laminin and basement membrane in the morphological differentiation of human endothelial cells into capillary-like structures. J Cell Biol 107:1589–1598.

Leung D.W., Cachianes G., Kuang W.J., Goeddel D.V., Ferrara N. (1989): Vascular endothelial growth factor is a secreted angiogenic mitogen. Science 246:1306–1309.

Madri J.A., Williams S.K. (1983): Capillary endothelial cell cultures: phenotypic modulation by matrix components. J Cell Biol 97:153–165.

Montesano R., Orci L., Vassalli P. (1983): In vitro rapid organization of endothelial cells into capillary-like networks is promoted by collagen matrices. J Cell Biol 97:1648–1652.

Nicosia R.F., Tchao R., Leighton J. (1982): Histotypic angiogenesis in vitro: light microscopic, ultrastructural, and radioautographic studies. In Vitro 18:538–549.

Nicosia R.F., Tchao R., Leighton J. (1983): Angiogenesis-dependent tumor spread in reinforced fibrin clot culture. Cancer Res 43:2159–2166.

Nicosia R.F., Madri, J.A. (1987): The microvascular extracellular matrix: developmental changes during angiogenesis in the aortic-ring plasma clot model. Am J Pathol 128:78–90.

Nicosia R.F., Ottinetti A. (1990a): Modulation of microvascular growth and morphogenesis by reconstituted basement membrane-like gel in three-dimensional culture of rat aorta: a comparative study of angiogenesis in matrigel, collagen, fibrin, and plasma clot. In Vitro Cell Dev Biol 26:119–128.

Nicosia R.F., Ottinetti A. (1990b): Growth of microvessels in serum-free matrix culture of rat aorta: a quantitative assay of angiogenesis in vitro. Lab Invest 63:115–122.

Nicosia R.F., Bonanno E. (1991): Inhibition of angiogenesis in vitro by arg-gly-asp-containing synthetic peptide. Am J Pathol 138:829–833.

Nicosia R.F., Belser P., Bonanno E., Diven J. (1991): Regulation of angiogenesis in vitro by collagen metabolism. In Vitro Cell Dev Biol 27A:961–966.

Nicosia R.F., Bonanno E., Villaschi S. (1992): Large-vessel endothelium switches to a microvascular phenotype during angiogenesis in collagen gel culture of rat aorta. Atherosclerosis 95:191–199.

Nicosia R.F., Bonanno E., Smith M. (1993): Fibronectin promotes the elongation of microvessels during angiogenesis in vitro. J Cell Physiol 154:654–661.

Nicosia R.F., Tuszynski G. (1994): Matrix-bound thrombospondin promotes angiogenesis in vitro. J Cell Biol 124:183–193.

Nicosia R.F., Villaschi S., Smith M. (1994a): Isolation and characterization of vasoformative endothelial cells from the rat aorta. In Vitro Cell Dev Biol 30A:394–399.

Nicosia R.F., Bonanno E., Yurchenco P. (1994b): Modulation of angiogenesis in vitro by laminin-entactin complex. Dev Biol 164:197–206.

Nicosia R.F., Nicosia S.V., Smith M. (1994c): Vascular endothelial growth factor, platelet derived growth factor, and insulin-like growth factor-1 promote rat aortic angiogenesis in vitro. Am J Pathol 145:1023–1029.

Nicosia R.F., Villaschi S. (1995): Rat aortic smooth muscle cells become pericytes during angiogenesis in vitro. Lab Invest 73(5):658–666.

Nicosia R.F., Lin Y.J., Hazelton D., Quian X. (1996): Role of vascular endothelial growth factor in the rat aorta model of angiogenesis. J Vasc Res (Abstracts IX International Vascular Biology Meeting) 33(S1):73.

Nissanov J., Tuman R.W., Gruver L.M., Fortunato J.M. (1995): Automatic vessel segmentation and quantification of the rat aortic ring assay of angiogenesis. Lab Invest 73:734–739.

Pepper M.S., Vassalli J.D., Orci L., Montesano R. (1993): Biphasic effect of transforming growth factor beta 1 on in vitro angiogenesis. Exp Cell Res 204:356–363.

Piershbacher M.D., Ruoslahti E. (1984): Cell attachment activity can be duplicated by small synthetic fragments of the molecule. Nature 309:30–33.

RayChaudhury A., D'Amore P.A. (1991): Endothelial cell regulation by transforming growth factor beta. J Cell Biochem 47:224–229.

Sato N., Tsuroka N., Yamamoto M., Nishihara T., Goto T. (1991): Identification of non-heparin binding endothelial cell growth factor from rat myofibroblasts. EXS 61:179–187.

Sato N., Beitz J.G., Kato J., Yamamoto M., Clark J.W., Calabresi P., Frackelton R., Jr. (1993): Platelet-derived growth factor indirectly stimulates angiogenesis in vitro. Am J Pathol 4:1119–1130.

Taraboletti G., Belotti D., Giavazzi R. (1992): Thrombospondin modulates basic fibroblast growth factor activities on endothelial cells. EXS 61:210–213.

Villaschi S., Nicosia R.F. (1993): Angiogenic role of basic fibroblast growth factor released by rat aorta after injury. Am J Pathol 143:182–190.

Villaschi S., Nicosia R.F., Smith M. (1994a): Isolation of a morphologically and functionally distinct muscle cell type from the intimal aspect of the normal rat aorta. Evidence for smooth muscle cell heterogeneity. In Vitro Cell Dev Biol 30A:589–595.

Villaschi S., Nicosia R.F. (1994b): Paracrine interactions between fibroblasts and endothelial cells in a serum-free co-culture model: modulation of angiogenesis and collagen gel contraction. Lab Invest 71:291–299.

Vukicevic S., Kleinman H.K., Luyten F.P., Roberts A.B., Roche N.S., Reddi A.H. (1992): Identification of multiple active growth factors in basement membrane Matrigel suggests caution in interpretation of cellular activity related to extracellular matrix components. Exp Cell Res 202:1–8.

Acknowledgments

My thanks go to the many co-workers and collaborators who over the years have contributed to the studies presented in this chapter and whose names are cited in the references. I am particularly grateful to Dr. Joseph Leighton, my mentor, whose seminal work with three-dimensional tissue culture models has laid the foundation for our progress in the field of angiogenesis. I gratefully acknowledge also the support from the W.W. Smith Charitable Trust and the Heart Lung and Blood Institute (HL52585).

2.3

In Vitro Coculture Models to Study Vessel Formation and Function

Karen K. Hirschi and Patricia A. D'Amore

The vasculature is one of the first organ systems to develop and provides a means of distributing nutrients and oxygen, as well as removing waste products in the developing embryo. Hence, blood vessel formation is vital to the establishment and maintenance of all tissues of the body. The primitive vascular network is formed almost entirely from mesoderm during vasculogenesis, which begins with the differentiation of angioblasts into tubelike endothelial structures (Doetschman et al., 1985; Coffin and Poole, 1988; Noden, 1989).

Mural cells, smooth muscle cells (SMC) and pericytes, are associated with forming vessels at later stages of development (Nakamura, 1988; Hungerford et al., 1996; Topouzis and Majesky, 1996) and are thought to be recruited by the endothelium. Endothelial cells (EC) synthesize and secrete soluble, diffusible factors such as PDGF (Westermark et al., 1990), bFGF (Montesano et al., 1986), and HB-EGF (Higashiyama et al., 1993), which may act as chemoattractants and mitogens for SMC and fibroblasts. The distribution of PDGF receptors and ligands in the developing vessels of the placenta suggests that EC-derived PDGF may play a role in mesenchymal recruitment. Specifically, EC in the developing vessels were found to produce PDGF-B, whereas surrounding SMC and mesenchymal cells express the PDGF-β receptor (Holmgren, 1993).

Although data suggesting a role for PDGF have been obtained from in situ developmental models, it is not possible using most in vivo systems to dissect and analyze the exact cell–cell interactions that occur during vessel assembly, or in the intact vessel. Furthermore, although much information has been gained from investigating cellular behavior in solo cultures of EC or SMC, heterotypic interactions, representative of a true physiological state, can only be studied in coculture models. Therefore, we and others have created in vitro models to more directly examine the paracrine interactions between EC and SMC or mesenchymal cells.

Interactions between EC and mesenchymal cells will be discussed first, in the context of vessel formation. Subsequently, interactions between EC and mature mural cells or astrocytes will be discussed with regard to vessel function and stability. We will focus on two-dimensional coculture systems; three-dimensional systems are addressed in another chapter.

EC-Mesenchymal Cell Interactions

To elucidate interactions that may occur during the formation of vessels, we have recently created coculture systems using EC and undifferentiated mesenchymal cells (Hirschi et al., submitted). Using these systems, we can examine the effects of heterotypic cell culture on processes of migration, proliferation, and differentiation, in order to determine the mechanisms and regulatory factors that influence the various aspects of vessel assembly.

As previously stated, observations of the developing vasculature suggest that vessel assembly begins with the formation of EC tubes. We hypothesize that EC of forming vessels recruit undifferentiated mesenchymal cells and direct their differentiation into SMC or pericytes, which form the surrounding medial layers of the vessels. Initially, we examined the ability of EC to recruit mesenchymal cells using an under agarose assay system (depicted in Fig. 1) in which we cocultured EC from bovine aortas or rodent microvessels and 10T1/2 cells, multipotent murine embryonic cells, which serve as presumptive SMC/pericyte precursors. The 10T1/2 cells have been used in other systems as precursors to adipocytes, skeletal myocytes, and osteoblasts, which demonstrates their multipotent nature. The EC and 10T1/2 cells were cocultured in 5 mm wells created ~2 mm apart in 1% agarose and incubated for up to 48 hr. In this system, the 10T1/2 cells migrated directionally toward the EC. Furthermore, as cellular morphology can be readily monitored in this system, we observed a dramatic shape change in the 10T1/2 cells that were on the side of the well closest to the EC, from flat and spread-shaped to elongated with many pseudopodia.

Another advantage of the under agarose assay system is that neutralizing reagents can be directly incorporated into the agarose to determine the precise soluble regulators influencing cellular behavior. We therefore used this same assay to explore which EC-derived soluble factor(s) mediated the 10T1/2 cell recruitment. We incorporated into the agarose neutralizing antibodies to various candidate chemoattractants. We found that neutralization of PDGF-B completely blocked the EC-directed 10T1/2 cell recruitment, whereas inhibition of PDGF-A or bFGF had no effect (Hirschi et al., submitted).

Further studies using the same under agarose coculture assay revealed that the EC not only induced directed migration of the 10T1/2 cells, but stimulated their proliferation, as well. The cocultures were labeled with BrdU, which is incorporated into replicating cells and can be detected with a specific antibody. Results of these studies indicated that the labeling index of 10T1/2 cells on the side of the agarose well closest to the EC increased by twofold, in contrast to the proliferation rate of the 10T1/2 cells on the opposite side of the well, which was unchanged. Again, neutralizing antisera to PDGF-B specifically suppressed the EC-induced 10T1/2 cell proliferation, whereas antibodies to PDGF-A and bFGF had no effect (Rohovsky et al., submitted). These studies implicate PDGF-B in EC recruitment of mesenchymal cells. These in vitro findings are corroborated by recent in vivo studies, which document the absence of pericytes in mice lacking the PDGF-B ligand (Lindahl et al., 1997).

In a separate series of studies, we tested the possibility that EC could influence the differentiation of the recruited 10T1/2 toward a SMC/pericyte lineage. Using the under agarose assay, or a direct 1:1 coculture of EC and 10T1/2 cells, we measured the effect of EC-10T1/2 cell coculture on 10T1/2 cell expression of proteins reflective of a SMC/pericyte phenotype, including α-SM-actin, SM myosin heavy chain, calponin, and SM22α. Immunohistochemical and Western blot analyses revealed an upregulation of the SM-specific proteins in 10T1/2 cells cocultured with EC. Furthermore, there was a dramatic change in the shape of the 10T1/2 cells from flat and spread to elongated and spindle-shaped with pseudopodia, reminiscent of the morphology of vascular SMC in primary culture. These phenotypic changes appeared to be contact depen-

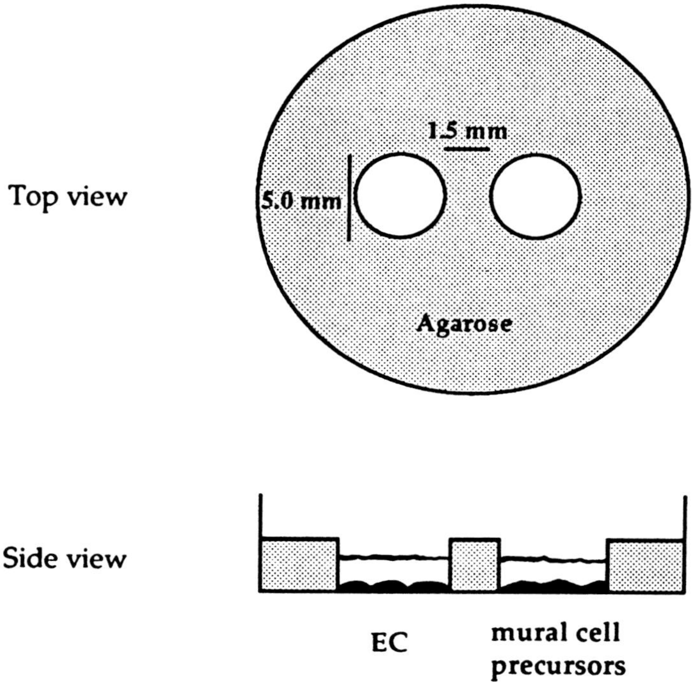

Fig. 1 Under Agarose Assay System. EC and mural cell precursors are plated into two 5-mm wells created 1–2 mm apart in 1% agarose/1% BSA in DMEM. Cells are cocultured for various time points, fixed in 4% paraformaldehyde and stained with Coomassie blue for migration assays or with primary antibodies against SM-specific markers for analyses of protein expression.

dent. A representative result is shown in Figure 2 in which 10T1/2 cells cultured alone (A) or in contact with rat microvascular EC (B) were immunostained for αSM-actin protein.

Previous studies from our laboratory demonstrate that cocultures of EC and mural cells lead to the activation of endogenous transforming growth factor-beta (TGF-β) (Antonelli-Orlidge et al., 1989). Activated TGF-β has been shown to influence cellular differentiation, growth, and extracellular matrix production in a number of systems. We therefore investigated the possibility that TGF-β might mediate the induction of 10T1/2 cells toward a SMC lineage. We first tested the effects of exogenous TGF-β on 10T1/2 cells, which were treated with 1 ng/ml TGF-β1 for 24 hr. Indeed, TGF-β1 produced an upregulation of SM-specific proteins and morphological changes similar to those seen in 10T1/2 cells cocultured with EC.

To more directly determine the role of TGF-β in the coculture, EC and 10T1/2 cells were plated in the under agarose assay and incubated in the presence of neutralizing antisera against TGF-β1, 2, and 3 for 3–6 days, allowing the two cell types to contact. In these experiments, the presence of the neutralizing antisera against TGF-β suppressed the upregulation of αSM-actin and SM-myosin, and inhibited the shape change in the 10T1/2 cells cocultured with EC. These studies suggest a role for TGF-β in the process of vessel assembly and are consistent with the observation that mice lacking TGF-β1 exhibit defective vasculogenesis (Dickson et al., 1995).

Fig. 2 EC-10T1/2 Cell Coculture. 10T1/2 cells were cultured alone (A) or with primary rat microvascular EC prepared in our laboratory from rat adipose tissue (B) in 2% CS/DMEM for 48 hr. Cells were subsequently fixed and stained for αSM-actn, which is greatly increased in 10T1/2 cells contacting EC. 10T1/2 cells in coculture with EC also underwent a dramatic shape change from flat and spread to elongated with many pseudopodia.

The growth of both EC and 10T1/2 cells in coculture was also investigated (Rohovsky et al., submitted). We hypothesize that, upon contact, both cell types would exhibit growth inhibition necessary for the formation of an intact vessel. To examine this possibility, EC and 10T1/2 cells were prelabeled with different fluorescent dyes, cocultured for 3–7 days, and then separated and counted by fluorescence-activated cell sorting. The growth of both EC and 10T1/2 cells was inhibited by 30–50%. Interestingly, the growth inhibition observed in these cocultures appears not to be due to TGF-β, which we had previously shown to influence the differentiation of the 10T1/2 cells cocultured with EC. Instead, we suspect that another yet uncharacterized agent must be mediating the inhibition. Thus, although TGF-β appears to play a role in the differentiation of mesenchymal cells, it does not appear to be the sole mediator.

Our studies demonstrate that 10T1/2 cells can be induced to become SMC-like by coculture with EC in vitro. We also confirmed that 10T1/2 cells could become SMC in a more complex in vivo environment. To accomplish this, 10T1/2 cells were prelabeled with a permanent fluorescent dye, seeded into a collagen matrix and placed either onto the developing chick chorioallantoic membrane of 10-day-old fertilized chick embryos or implanted subcutaneously into the backs of 6- to 8-wk-old C57 mice. In both cases, the 10T1/2 cells became incorporated into the walls of newly developing blood vessels and stained positively for αSM-actin, SM-myosin, and calponin. These in vivo studies lend credibility to our novel in vitro model of SMC differentiation and, together, they will prove useful in evaluating further the roles of PDGF-B and TGF-β, as well as other factors involved in vessel assembly.

Importantly, other investigators have demonstrated a similar effect of mesenchymal cells on EC differentiation and tube formation in three-dimensional coculture model systems. Kuzuya and Kinsella (1994) found that conditioned media from human skin fibroblasts induces migration and capillary-like tube formation of EC in three-dimensional collagen gels or in Matrigel. Furthermore, fibroblasts were found to stabilize EC tubes in collagen gels and to increase the deposition of subendothelial extracellular matrix (Villaschi and Nicosia, 1994).

EC-Mural Cell Interactions

Growth Control. The most extensive studies of the interactions between EC and mature SMC or pericytes have focused on growth control and have shed light on the establishment and maintenance of cellular quiescence in intact vessels. Coculture of pericytes with EC were found to inhibit endothelial growth (Orlidge and D'Amore, 1987) and migration (Sato and Rifkin, 1989). The inhibition of EC growth was shown to be dependent on contact between pericytes and EC and due to the activation of TGF-β1 (Antonelli-Orlidge et al., 1989; Sato et al., 1990). The activation of TGF-β1 in cocultures of EC and SMC is suspected to occur via plasmin. It can be inhibited by mannose-6-phosphate (Dennis and Rifkin, 1991), which presumably competes for the binding of latent TGF-β1 to the surface of SMC and, thereby, prevents its accessibility to plasmin cleavage and, hence, its activation (reviewed in Nunes et al., 1996).

EC-mural cell cocultures have also revealed an effect of EC on the growth of SMC or pericytes. Media conditioned by EC vary in their effect on mural cell growth, depending on the density of the conditioning EC (Dodge et al., 1993). Media collected from sparse EC is stimulatory for SMC or pericytes whereas media collected from post confluent EC is inhibitory. Similar inhibitory activity is observed when EC and SMC are cultured on opposite sides of a porous membrane, through which heterotypic contacts are possible (Saunders and D'Amore, 1992; Fillinger et al., 1993). These results are consistent with hypothesized developmental interactions in which growing EC synthesize and secrete factors that stimulate mural cell proliferation, whereas non-proliferating EC, such as in established vessels, contribute to SMC/pericyte growth quiescence. Presumably, continued heterotypic contact and bidirectional communication would promote growth control and maintenance of quiescent vessels. In pathological situations, such as in diabetic retinopathy, where normal heterotypic contact between EC and mural cells in disrupted secondary to "pericyte drop-out," there is uncontrolled vessel growth and impairment of retinal function (Kuwabara and Cogan, 1963).

Vessel Tone. EC-mural cell interactions are not only important in growth control and the maintenance of intact vessels, but for normal physiological vessel function, as well. The contribution of EC to SMC contractile function was demonstrated by Furchgott and Zawadzki (1980) who found that removal of EC from isolated blood vessels prevents the vasorelaxation normally observed upon administration of acetylcholine. In vitro models have since been established to study the regulation of vessel tone, which is dictated by the contractile state of SMC, mediated in response to EC-derived soluble factors.

Nitric oxide (NO) (reviewed in Busse et al., 1995) is one such diffusible effector that is constitutively expressed by EC (Rees et al., 1989) and released in response to shear stress (Griffith et al., 1987). EC-derived NO has been shown in EC-SMC cocultures to exert vasodilatory effects on SMC (Wang et al., 1996), in a myosin light chain kinase/cGMP-dependent process (Hataway et al., 1985). In vitro studies have also demonstrated an antiproliferative effect of NO on SMC (Garg and Hassid, 1989).

Conversely, endothelin-1 (ET-1) (reviewed in Masaki 1995) is a potent vasoconstrictor produced by EC in response to various vasoactive substances, including angiotensin II and epinephrine (Yanagisawa et al., 1988). The secretion of ET-1 from EC is thought to occur in a polar fashion, from the abluminal side of the cell (Wagner et al., 1992), enabling immediate access to underlying SMC or pericytes. Furthermore, SMC have been shown to regulate ET-1 release from EC in vitro; that is, ET-1 release is significantly reduced in EC cocultured with SMC (Bonin and Damon, 1994) or treated with SMC conditioned media, suggesting negative feedback regulation (Stewart et al., 1990).

ET-1 has also been shown to influence mural cell growth. In culture, EC-derived soluble factors have been shown to stimulate pericyte growth. Addition of neutralizing reagents against ET-1 were found to inhibit these effects (Yamagishi et al., 1993).

Pathological Conditions. EC-SMC coculture models have also been recently established to mimic the process of atherogenesis. In one model, SMC and EC are cultured on opposite sides of a porous filter (Axel et al., 1996). Mechanical injury of a confluent monolayer of EC leads to an increased proliferation and accumulation of SMC on the underlying surface. Furthermore, addition of LDL and monocytes to the coculture results in the formation of a plaquelike structure with a lipid core containing necrotic cells and a fibromuscular cap of αSM-actin positive cells. Such a model system appears useful for investigating potential mediators and inhibitors of various aspects of plaque formation.

The same type of transmembrane coculture system has also been used to investigate the regulation of adhesion molecules in vascular SMC and subsequent leukocyte binding (Gamble et al., 1995). Tumor necrosis factor-α was shown to induce VCAM-1 expression in SMC which, in turn, mediates T-lymphocyte binding. Coculture of SMC with human umbilical vein EC decreases VCAM-1 expression in SMC, which supports a role for heterotypic contact between EC and SMC in the prevention of atheroma.

EC-Astrocyte Interactions

Interactions between EC and abluminal SMC or pericytes are of great importance to the development and physiology of blood vessel, in general. However, the interactions between EC and astrocytes are equally important to the formation and function of vessels in neuronal tissues, such as in the retina and brain, which serve a more specialized function (for detailed discussion, see review by Garcia and D'Amore, 1997). EC-astrocyte cocultures have been created to study the establishment and regulation of the blood–brain barrier (BBB) and the blood–retinal barrier (BRB).

Increased tight junction formation and function, as well as increased gamma-glutamyl transferase (GGT) activity are characteristic of EC in brain and retinal vessels, which form a highly-selective lumenal barrier between blood flow and tissues in these regions. Establishment of the BBB and BRB in the EC are thought to be dependent on the influence of neighboring glial cells. Coculture models have demonstrated increased tight junction formation (Tao-Cheng and Brightman, 1988) and resistance (Janzer and Raff, 1987) among EC in the presence of astrocytes or astrocyte conditioned media (Rubin et al., 1991). GGT activity, which is high in brain capillaries but lost in cultured EC, is also restored in EC cocultured with astrocytes (DeBault and Cancilla, 1980; Hafney et al., 1996). Furthermore, astrocytes have been shown to alter the extracellular matrix composition of EC (Jiang et al., 1994), and the ability of EC to form capillary-like structures (Laterra et al., 1990). Conversely, our lab and others have found that astrocyte cellular morphology and growth are altered in astrocytes cocultured with EC (Garcia and D'Amore, unpublished observations; shown in Fig. 3).

Three-way communication between astrocytes, EC, and mural cells are physiologically relevant, as well, but not often modeled in culture. One study by Risau et al. (1992) found that

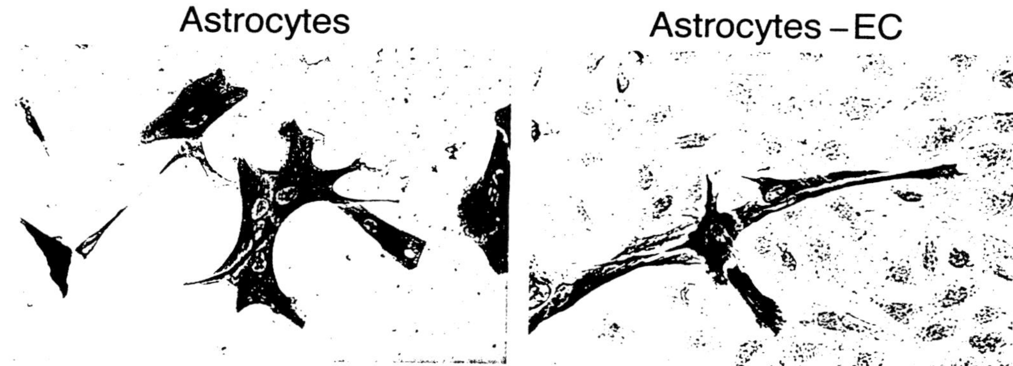

Fig. 3 EC-Astrocyte Cell Coculture. Rat astrocytes were cultured alone (A) or with rat microvascular EC (B) in 10% CS/DMEM for 48 hr. Cells were subsequently fixed and stained for GFAP. Astrocytes in coculture with EC underwent a dramatic shape change from flat to elongated with extensive processes.

coculture of EC with astrocytes resulted in increased GGT activity when compared to EC cultured alone; addition of pericytes to the EC-astrocyte cocultures further increased GGT activity.

Summary

Studies discussed herein document the utility of in vitro analysis for dissecting and analyzing the events involved in the formation of blood vessels and the regulation of their function. Complementary in vivo models also serve as powerful tools to further elucidate the roles and synergistic effects of various regulatory factors on blood vessel development and physiology. The ultimate goal of such studies is to elucidate the mechanisms underlying normal neovascularization in simple models, so that one may then determine where the processes have become dysregulated in vascular pathologies.

References

Antonelli-Orlidge A., Saunders K.B., Smith S.R., and D'Amore P.A. (1989). An activated form of transforming growth factor β is produced by cocultures of endothelial cells and pericytes. Proc Natl Acad Sci USA 86:4544–4548.

Axel D.I., Brehm B.R., Wolburg-Buchholz K., Betz E.L., Koveker G., and Karsch K.R. (1996). Induction of cell-rich and lipid-rich plaques in a transfilter coculture system with human vascular cells. J Vasc Res 33:327–339.

Bonin L.R., and Damon D.H. (1994). Vascular cell interactions modulate the expression of endothelin-1 and platelet-derived growth factor BB. Am J Physiol 267:H1698–H1706.

Busse R., Fleming I., and Schini V.B. (1995). Nitric oxide formation in the vascular wall: regulation and functional implications. Curr Topics Microbiol Immunol 196:7–18.

Coffin J.D., and Poole T.J. (1988). Embryonic vascular development: immunohistochemical identification of the origin and subsequent morphogenesis of the major vessel primordia in quail embryo. Development 102:1–14.

DeBault L.E., and Cancilla P.A. (1980). Gamma-glutamyl transpeptidase in isolated brain endothelial cells: induction by glial cells in vitro. Science 207(8):653–655.

Dennis P.A., and Rifkin D.B. (1991). Cellular activation of transforming growth factor β requires binding to the cation-independent mannose 6-phosphate/insulin-like growth factor type II receptor. Proc Natl Acad Sci USA 88:580–584.

Dickson M.C., Martin J.S., Cousins F.M., Kulkarni A.B., Karlsson S., and Akhurst R.J. (1995). Defective haematopoiesis and vasculogeneisis in transforming growth factor-β1 knock-out mice. Development 121:1845–1854.

Dodge A., Lu X., and D'Amore P. (1993). Density-dependent endothelial cell production of an inhibitor of smooth muscle cell growth. J Cell Biochem 53:21–31.

Doetschman T.C., Eistetter H., Katz M., Schmidt W., and Kemler R. (1985). The in vitro development of blastocyst-derived embryonic stem cell lines: formation of visceral yolk sac, blood islands and myocardium. J Embryol Exp Morphol 87:27–45.

Fillinger M.F., O'Connor S.E., Wagner R.J., and Cronenwett J.L. (1993). The effect of endothelial cell coculture on smooth muscle cell proliferation. J Vasc Surg 17:1058–1068.

Furchgott R.F., and Zawadzki J.V. (1980). The obligatory role of endothelium in the relaxation of arterial smooth muscle by acetylcholine. Nature 288:373–376.

Gamble J.R., Bradley S., Noack L., and Vadas M.A. (1995). TGF-beta and endothelial cells inhibit VCAM-1 expression on human vascular smooth muscle cells. Arterioscler Thromb Vasc Biol 15:949–955.

Garg U.C., and Hassid A. (1989). Nitric oxide-generating vasodilators and 8-bromo-cyclic guanosine monophosphate inhibit mitogenesis and proliferation of cultured rat vascular smooth muscle cells. J Clin Invest 83:1774–1777.

Griffith T.M., Edwards D.H., Davies R.L., Harrison T.J., and Evans K.T. (1987). EDRF coordinates the behavior of vascular resistance vessels. Nature 329:442–445.

Hafney B.E., Bourre J.-M., and Roux F. (1996). Synergistic stimulation of gamma-glutamyl transpeptidase and alkaline phosphatase activities by retinoic acid and astrogial factors inimmortalized rat brain microvessels endothelial cells. J Cell Physiol 167:454–460.

Hathaway D.R., Konicki M.V., and Coolican S.A. (1985). Phosphorylation of myosin light chain kinase from vascular smooth muscle by cAMP- and cGMP-dependent protein kinases. J Mol Cell Cardiol 17:841–850.

Higashiyama S., Abraham J.A., and Klagsbrun M. (1993). Heparin-binding EGF-like growth factor stimulation of smooth muscle cell migration: dependence of interactions with cell surface heparan sulfate. J Cell Biol 122:933–940.

Hirschi K., Rohovsky S.A., and D'Amore P.A. PDGF, TGF-β and heterotypic cell–cell interactions mediate the recruitment and differentiation of 10T1/2 cells to a smooth muscle cell fate (submitted).

Holmgren L. (1993). Potential dual roles of PDGF-B during human placental blood vessel formation. Endothelium 1:167–171.

Hungerford J.E., Owens G.K., Argraves W.S., and Little C.D. (1996). Development of the aortic vessel wall as defined by vascular smooth muscle and extracellular matrix markers. Dev Biol 178:375–392.

Janzer R.C., and Raff M.C. (1987). Astrocytes induce blood-brain barrier properties in endothelial cells. Nature 325:253–257.

Jiang B., Liou G.I., Behzadian M.A., and Caldwell R.B. (1994). Astrocytes modulate retinal vasculogenesis: effects on fibronectin expression. J Cell Sci 107:2499–2508.

Kuwabara T., and Cogan D.G. (1963). Retinal vascular patterns. VI. Mural cells of the retinal capillaries. Arch Ophthalmol 69:492–502.

Kuzuya M., and Kinsella J.L. (1994). Induction of endothelial cell differentiation in vitro by fibroblast-derived soluble factors. Exptl Cell Res 215:310–318.

Laterra J., Guerin C., and Goldstein G.W. (1990). Astrocytes induce neural microvascular endothelial cells to form capillary-like structures in vitro. J Cell Physiol 144:204–215.

Lindahl P., Johansson B.R., Leveen P., and Betsholtz C. (1997) Pericyte loss and microaneurysm formation in PDGF-B-deficient mice. Science 277:242–245.

Masaki T. (1995). Possible role of endothelin in endothelial regulation of vascular tone. Ann Rev Pharmacol Toxicol 35:235–255.

Montesano R., Vassali J.-D., Baird A., Guillemin R., and Orci L. (1986). Basic fibroblast growth factor induces angiogenesis in vitro. Proc. Natl Acad Sci USA 83:7297–7301.

Nakamura H. (1988). Electron microscopic study of the prenatal development of the thoracic aorta in the rat. Am J Anat 181:406–418.

Noden D.M. (1989). Embryonic origins and assembly of blood vessels. Am Rev Respir Dis 140: 1097–1103.

Nunes I., Munger J.S., Harpel J.G., Nagano Y., Shapiro R.L., Gleizes P.E., and Rifkin D.B. (1996). Structure and activation of the large latent transforming growth factor-beta comples. Int Obes Rel Metab Dis 3:S4–8.

Orlidge A., and D'Amore P.A. (1987). Inhibition of capillary endothelial cell growth by pericytes and smooth muscle cells. J Cell Biol 105:1455–1462.

Rees D.D., Palmer R.M.J., Hodson H.F., and Moncada S. (1989). A specific inhibitor of nitric oxide formation fro L-arginine attenuates endothelium-dependent relaxation. Br J Pharmacol 96:418–424.

Risau W., Dingler A., Albrecht U., Dehouck M.-P., and Cecchelli R. (1992). Blood–brain barrier pericytes are the main source of g-glutamyltranspeptidase activity in brain capillaries. J Neurochem 58:667–672.

Rohovsky S.A., Hirschi K.K., and D'Amore P.A. The effect of soluble factors and cell–cell interactions on proliferation in an in vitro model of vessel assembly (submitted).

Rubin L., Hall D., Porter S., Barbu K., Cannon C., Horner H., Janatpour M., Liaw C., Manning K., and Morales J. (1991). A cell culture model of the blood–brain barrier. J Cell Biol 115: 1725–1735.

Sato Y., and Rifkin D.B. (1989). Inhibition of endothelial cell movement by pericytes and smooth muscle cells: activation of a latent transforming growth factor-beta 1-like molecule by plasmin during co-culture. J Cell Biol 109:309–315.

Sato Y., Tsuboi R., Lyons R., Moses H., and Rifkin D.B. (1990). Characterization of the activation of latent TGF-β by co-cultures of endothelial cells and pericytes or smooth muscle cells: a self-regulating system. J Cell Biol 111:757–763.

Saunders K., and D'Amore P.A. (1992). An in vitro model for cell–cell interactions. In Vitro Cell Dev Biol 28A:521–528.

Stewart D.J., Langleben D., Cernacek P., and Cianflone K. (1990). Endothelin release is inhibited by coculture of endothelial cells with cells of vascular media. Am J Physiol 259: H1928–H1932.

Tao-Cheng J.H., and Brightman M.W. (1988). Tight junctions of brain endothelium in vivo are enhanced by astroglia. J Neurosci 7:3293–3299.

Topouzis S., and Majesky M.W. (1996). Smooth muscle lineage diversity in the chick embryo. Two types of aortic smooth muscle cell differ in growth and receptor-mediated transcriptional responses to transforming growth factor-β. Dev Biol 178:430–445.

Villaschi S., and Nicosia R.F. (1994). Paracrine interactions between fibroblasts and endothelial cells in a serum-free coculture model. Modulation of angiogenesis and collagen gel contraction. Lab Invest 71:291–299.

Wagner O.F., Christ G., Wojta J., Vierhapper H., Parzer S., Nowotry P.J., Schneider B., Waldhaus W., and Binder B.R. (1992). Polar secretion of endothelin-1 by cultured endothelial cells. J Biol Chem 267:16066–16068.

Wang Y., Shin W.S., and Kawaguchi H., Inukai M., Kato M., Sakamoto A., Uehara Y., Miyamoto M., Shimamoto N., and Korenaga R. (1996). Contribution of sustained Ca2+ elevation for nitric oxide production in endothelial cells and subsequent modulation of Ca2+ transient in vascular smooth muscle cells in coculture. J Biol Chem 271:5647–5655.

Westermark B., Siegbahn A., Heldin C.-H., and Claesson-Welsh L. (1990). B-type receptor for platelet-derived growth factor mediates a chemotactic response by means of ligand-induced activation of the receptor protein-tyrosine kinase. Proc Natl Acad Sci USA 87: 128–132.

Yamagishi S., Hsu C.-C., Kobayashi K., and Yamamoto H. (1993). Endothelin 1 mediates endothelial cell-dependent proliferation of vascular pericytes. Biochem Biophys Res Commun 191:840–846.

Yanagisawa M., Kurihara H., Kimura S., Tomobe Y., Kobayashi M., Mitsui Y., Yazaki Y., Goto K., and Masaki T. (1988). A novel potent vasoconstrictor peptide produced by vascular endotheial cells. Nature 3323:411–415.

Part III

Vascular Morphogenesis In Mente

Introduction: Toward the Theoretical Biology of Vascular Morphogenesis

Vladimir A. Mironov

Introduction

Since its publication in 1917, On Growth and Form, the classical work of Scottish zoologist D'Arcy Thompson, has attracted investigators from different fields and with different backgrounds to the study of biological form in mathematical terms. The five chapters of the Part III of our book are devoted to mathematical modeling and computer simulation of vascular morphogenesis. It clearly demonstrates that vascular morphogenesis is becoming mature field with established broad spectrum of hypotheses, theories, assumptions, postulates, and experimental data which are suitable for mathematical analysis and modeling as well as computer simulation. From another point of view, these chapters indicate the emergence of a new field in the study of vascular morphogenesis which we propose to call vascular morphogenesis in mente (from the Latin word mente—a mind). We think that this term will fit better than other proposed terms (in numero, in computero) with the logic of the existing classification of research based on topological principles (where was experiment performed? in vivo, in vitro, in mente). The existing mathematical models of vascular morphogenesis may be classified into at least four basic groups: (a) models based on chemical gradients of activators and inhibitors of angiogenesis (Meinhardt, 1976; Meakin, 1986); (b) models exploiting mechanical factors involved in vascular morphogenesis (Waxman et al., 1981, Manoussaki et al., 1996); (c) models analyzing the cellular behavior (Stokes and Lauffenburger, 1991; Byrne and Chaplain, 1995); (d) models analyzing a fractality of vascular networks (Bassingthwaite et al, 1994; Kiani and Hudetz, 1991; Pries et al., 1995; Sandau and Kurz, 1995). Collectively, in this book, these approaches give a whole and impressive picture of recent trends in mathematical modeling and computer simulation of vascular morphogenesis.

Who Is the Leader?

Who will develop this field of research? There are four potential scenarios: (a) biologists who will acquire necessary mathematical knowledge and computer simulation skills; (b) mathematicians and physicists who will educate themselves in biological problems; (c) both groups of scientists will effectively cooperate and perform combined research; (d) new generations of scientists with equal skill in both disciplines will be educated through specially designed courses and programs such as mathematical biology or biomedical engineering. It is interesting that this book demonstrates that all four proposed scenarios are emerging.

The requisite simplification of natural processes in mathematical models, using clearly defined but sometimes not well-sounded assumptions, have made this approach unattractive for some biologists who like to emphasize the complexity and uniqueness of the biological systems. Moreover, biologists tend to reject abstract, oversimplified, and unrealistic mathematical models. Another popular argument of experimentalists is that the enormous power of modern experimental tools moves the progress of our biological knowledge at an explosive rate. In such an environment, there is no need for mathematical theory. Mathematicians and physicists usually argue in response, that biologists undervalue mathematical methods just because they are not trained enough to understand the models and to use them correctly. They predict that new generations of biological scientists who are equally trained in both disciplines will dominate in the biological science of XXI century and that interdisciplinary training is the best way to promote broad use of mathematical methods in biology. Moreover, the growing complexity of experimental data and multidimensional, nonlinear interactions between the various parts of the developing vascular system will require mathematical modeling and computer simulation (Cox, 1992)

Validity Criteria for a Mathematical Model

According to Meinhardt (1995) (see also his chapter in this book) there are two main reasons for modeling the dynamic systems: to provide a check on whether a system is fully understood, and to make predictions, at least for the near future. It is definitely not enough to say that the proposed mathematical model allows generation of a vascular pattern which resembles the pattern of natural vascular networks. A similar pattern can be generated in several different ways (we call this a principle of equifinality), based on completely different assumptions and mechanisms. Therefore, some of these models would have no connection to naturally occurring biological processes.

The realistic model must simulate not only resulting patterns but also reconstruct realistic pathways of vascular morphogenesis. For example, free-ending dichotomic branching does not occur during natural vascular morphogenesis because vessels grow as a network of connected vessels and not as open-ended tree-like structures. Therefore, we believe that the best criteria for the validity of a proposed mathematical model is its predictive power and capacity to design model-based experiments. According to Murray (see his chapter in this book), the valid mathematical model not only gives insight as to which elements are crucial to the development of vascular pattern and helps assess the relative importance of such elements, but also predicts how the changes in these factors affect vascular pattern formation and growth. Thus, mathematical models are not only useful for the interpretation of experimental facts, but also serve as an instrument to discover the unexpected behavior of the vascular system and its elements during development. The comparison of mathematical model predictions with experimental data allows one to estimate the validity of a proposed model. It is evident that the marriage of experimentation and mathematical theory is the key to successful development of new realistic mathematical models and computer simulations of vascular morphogenesis. It is noteworthy that all of the authors of the following chapters (although to different degrees) have successfully related their models to experimental observations.

From Genes to Vascular Pattern

To understand how gene expression leads to changes in vascular pattern is a main goal of a modern developmental vascular biology. Transgenic technologies dramatically transformed the landscape of research in the field of vascular development. However, even the elegant demonstration, via sophisticated transgenic technology, that as a certain gene mutation or deletion induces a defect in the vascular morphogenesis, does not produce a complete understanding of the mechanisms of vascular morphogenesis. In order to appreciate the complexity of vascular morphogenesis, we must explore and elucidate the whole chain of events from the gene expression to the resulting vascular pattern, and identify physical driving forces. This chain includes genes–molecules–cells–vessels–networks–patterns.

Moreover, we must know how cell interactions with other cells and extracellular matrices will generate driving forces for vascular morphogenesis. Wolpert (1995) wrote that change in form is a problem of linking gene action to mechanics. The relative contribution of chemical and mechanical factors in vascular development is another longstanding and yet unsolved issue. Although interest in the mechanical factors involved in the mechanism of vascular development has gradually increased during recent years (Waxman, 1981; Manoussaki et al., 1996), the studies of the role of chemical factors are still of predominant interest to the majority of developmental vascular biologists. The chemical theory of morphogenesis postulates the existence of a so-called morphogen (Turing, 1952). The search for new genes, transcriptional factors, and molecules with angiogenic or antiangiogenic effects (chemical factors), which can serve as such a morphogen, is a top priority in the recent studies of molecular mechanisms for vascular morphogenesis.

Relative contributions of different types of cellular processes to the mechanism of vascular morphogenesis have also been gradually elucidated. It has been shown that vascular morphogenesis includes all the basic cellular morphogenetic mechanisms; cell proliferation; cell death; cell spreading, migration, and invasion; cell aggregation and sorting; and finally, cell differentiation. In addition to classical sprouting angiogenesis and embryonic vasculogenesis, several novel mechanisms of network formation and remodeling of vascular pattern were recently described. They include nonsprouting angiogenesis, splitting or intessusceptive growth vascular fusion, and recruitment of circulated cells. Mechanisms of vascular network remodeling and capillary regression are also under investigation.

Thus, vascular morphogenesis is a complex process which implicates multiple factors at different levels of structural organization, and accumulation of experimental data in this field is still in the phase of explosive growth. Keeping this in mind, it is difficult to imagine that one ideal mathematical model can combine a whole complexity of different mechanisms of vascular morphogenesis on molecular, cellular, and supracellular levels of structural organization. Therefore, a diversity of mathematical models and their focusing on different aspects or mechanisms of vascular morphogenesis in the nearest future is unavoidable. However, it is not unreasonable to think that eventually enough will be known to program a computer and simulate some important aspects of vascular development. The reader can see some impressive examples of that in the following chapters. The ideal situation will be when we collect detailed knowledge on a whole chain of events from gene expression to vascular pattern formation, and when we present this in the form of an appropriate quantitative model and/or computer simulation. Then we can "play" with the computer to see the effects of altering one gene at a time (Wolpert, 1995).

Microvascular Tissue Engineering and Mathematical Modeling

An explosive growth of tissue engineering in the last decade and the enormous significance of the emerging biomedical technology calls for a new field of research in vascular morphogenesis which we suggest calling microvascular tissue engineering. The absence of effective ways of eliciting vascularization in tissue-engineered organs is a main limitation in the engineering of complex three-dimensional organs or thick-walled tubular organs. There are three basic approaches in this field: (a) to provide the effective vascularization of tissue-engineered constructs by inducing angiogenesis in recipient tissues adjacent to the implanted construct; (b) to generate a microvascular network in tissue-engineered constructs before transplantation; (c) to use fiber meshwork of biodegradable polymers as a carrier for endothelial cells which subsequent to implantation will form a vascular channel along the cellularized scaffold in vivo. The need for theoretical work in this field is obvious. It is interesting that the researcher will be transformed from a passive observer of the natural process of vascular morphogenesis into an active designer of this biological process. Therefore, it is possible to predict that in the next 5–10 years, microvascular tissue engineering will attract the interest of researchers working in the field of mathematical modeling of vascular morphogenesis. Microvascular tissue engineering will be an exciting area for testing mathematical models of vascular morphogenesis.

Conclusion

The vascular network is one of the earliest systems to develop in the embryo, and the organization of the primary vascular network is relatively simple as compared to other more complicated tissues and organs. This explains why vascular morphogenesis has become an attractive system for mathematical modeling and computer simulation. To continue the development of theoretical biology of vascular morphogenesis, additional logical steps must be followed.

First, the mathematical models must be based on clearly formulated assumptions and defined terms. Therefore, the existing nomenclature, terminology, and definition in the area of vascular morphogenesis requires improvement and systematization. Even well-defined terms such as angiogenesis and vasculogenesis are sometimes used incorrectly. Moreover, these terms do not cover the whole spectrum of morphogenetic events which occur during vascular morphogenesis. To define the process of formation and remodeling of a vascular network, we need additional terminology. In order to improve an existing terminology, we propose to define a microvascular network formation and remodeling with the term a retegenesis (from the Latin words rete—a net, and genere—to generate). One can assume at least four mechanisms of retegenesis: (a) formation of new vessels; (b) regression of existing vessels; (c) changing interaction between existing vessels; (d) changing geometrical form and size of the vessels.

Second, conceptual bases of vascular morphogenesis must be clearly formulated, well organized, and compactly reviewed in order to be a valuable and suitable source for subsequent mathematical modeling and computer simulation. Therefore, experimental and theoretical data on the molecular and cellular aspects of vascular morphogenesis and the patterns of vascular development must be systematically collected and catalogued in an Internet database of vascular morphogenesis. It will also be extremely important to collect all proposed computer simulations of vascular morphogenesis and post them on a website.

Third, computer simulation based on quantitative models is already becoming a fundamental biological skill. Creation of virtual laboratories, some of which already exist in vascular research centers, is a critical step for the promotion of mathematical modeling and computer simulations. It is predictable that the number of computer-based simulations and experiments in mente will dramatically increase in the future. Virtual laboratories of vascular morphogenesis will be, as usual, in the biological department as cell culture laboratories are now. Vascular biologists in the future may be characterized more by their ability to use their computers for data analysis and simulation than by their ability to use traditional microscopy (Keen and Spain, 1992). We think that traditional microscopy will undergo a dramatic computerization too. Integrative studies on vascular morphogenesis combining in vivo, in vitro, and in mente approaches will dominate the field of developmental vascular biology in the next century.

References

Bassingthwaite J.B., Liebovitch L.S., West B.J. (1994) Fractal Physiology, Oxford University Press.

Byrne H.M., and Chaplain M.A.J. (1995): Mathematical models for tumor angiogenesis: numeral simulations and nonlinear wave solutions. Bull, Math. Biol. 57:461–486.

Cox E.C. (1992): Modeling and experiment in developmental biology. Curr. Opin. Genet. Dev. 2:647–650.

Keen R.E., and Spain J.D. (1993): Computer simulation in biology. A Basic introduction. New York, Wiley-Liss.

Kiani M.F., and Hudetz A.G. (1991): Computer simulation of growth of anastomosing microvascular networks. J. Theor. Biol. 150:547–560.

Manoussaki D., Lubkin S.R., Vernon R.B., and Murray J.D. (1996): A mechanical model for the formation of vascular networks in vitro. Acta Biotheoret. 44:271–282.

Meakin P. (1986): A new model for biological pattern formation. Theor. Biol. 118:101–113.

Meinhardt H. (1995): The Algorithmic Beauty of Sea Shells, Springer-Verlag.

Meinhardt H. (1976): Morphogenesis of lines and nets. Differentiation 6:117–123.

Murray C.D. (1926): The physiological principle of minimum work: I. The vascular system and the cost of blood volume. Proc. Natl. Acad. Sci USA 12:207–304.

Murray J.D (1993): Mathematical biology (2nd corrected ed.). Heidelberg New York, Springer.

Pries A.R., Secomb T.W., and Gaehtgens P. (1995): Design principles of vascular beds. Circ. Res. 77:1017–1023.

Sandau K., and Kurz H. (1995): Modeling of vascular growth processes: a stochastic biophysical approach to embryonic angiogenesis. J. Microsc. 175:215–213.

Stokes C.L., and Lauffenburger D.A. (1991): Analysis of the role of microvessel endothelial random motility and chemotaxis in angiogenesis. J. Theor. Biol. 152:377–403.

Thompson D'Arcy (1961): On growth and form. Cambridge, Cambridge University Press.

Turing A. (1952): The chemical basis of morphogenesis. Phil. Trans. B. 237:37–72.

Wolpert L. (1995): Development: is the egg computable or could we generate an angel or a dinosaur? In What is a life? The next fifty years. Murphy M.P., and O'Neill L.A.J. (Eds.), Cambridge, Cambridge University Press, pp. 57–66.

Waxman A.N. (1981): Blood vessel growth as a problem in morphogenesis. A physical theory. Microvasc. Res. 22:32–42.

3.1

Models for the Formation of Netlike Structures

Hans Meinhardt

General Models of Biological Pattern Formation

The ability to generate the complex structure of an organism in each life cycle is one of the most fascinating aspect of living beings. In their development, higher organisms grow rapidly to a size where passive diffusion becomes inappropriate to supply the tissue with oxygen, water, nutrients, and information. Nature has solved the resulting problems by the invention of complex-shaped organs that consists of long branched filaments. The blood vessels, the lymph system, the tracheae of insects, the venation of leaves, or the nervous system are examples of such distributed organs.

The pattern of blood vessels is highly complex. What type of interactions can be envisioned that generates this complex structure? In this article, I would like to discuss a model proposed more than 20 years ago, long before a molecular approach was possible. It is based on a molecularly feasible interaction. By computer simulations, it was shown that it accounts not only for basic features of the normal pattern but also for regulative phenomena as they are observed in several of the systems mentioned above. As far as possible, I will compare the model proposed with more recent observations made on the molecular-genetic level.

Central was the question of how long filaments of differentiated cells an emerge. How are their extensions oriented towards those regions that need a better oxygen supply? How is a particular element of a long filament selected to initiate locally the formation of a branch? I would like to make a case for the following assumptions

1. A local signal for filament elongation is generated by local self-enhancement and long ranging inhibition.
2. The signal causes an elongation of the filament either by a local change of the shape of a cell or by accretion of newly differentiated cells at the tip of an existing filament.
3. The tissue is assumed to produce a growth factor for the vessel system that is removed by the veins. Therefore, the highest concentration of the growth factor will be in regions less supplied by filaments.

To substantiate these models, it is necessary to provide molecular interactions that are able to accomplish the following steps: (i) Generation of a local signal. This is required for the localized elongation of a cell or to provide a signal for a cell that should obtain a particular differentiation. (ii) A long-term memory of a cell that it has seen a particular signal that has caused its (or its predecessors) differentiation.

Mechanisms involved in biological pattern formation are expected to contain nonlinear interactions that have positive and negative feedback loops. Our intuition for the behaviour of such processes is not reliable. Only by formulating a model in a mathematical way combined with computer simulation one can show that a model is at least free of internal contradictions and it has indeed the expected properties. This is, of course not a proof that the model is correct, but it allows one to eliminate models that are definitely insufficient.

An Essential Ingredient: Pattern Formation by Local Autocatalysis and Long-Ranging Inhibition

A most striking feature of some developmental systems is their capability to generate patterns from a more or less structureless initial situation. For the developing blood vessels, the emerging pattern of angioblast in the mesodermal mesenchyme is a good example. Nests of cells with such determination can be detected by specific staining before any morphological characteristics become obvious (Pardanaud et al., 1987). Therefore, some local signal must be generated to instruct the cells to become these precursor cells. In the following, a model will be outlined that enables de novo pattern formation in an initially almost homogeneous situation.

Formation of patterns from almost uniform initial conditions is not unique to living systems. The formation of high sand dunes or of sharply contoured rivers are examples. Common in all these inorganic pattern formations is that small deviations from a homogeneous distribution have a strong feedback such that the deviations grow further. The initiation has only a limited influence on the final pattern. Formed by erosion, the shape of a river is not predetermined by the distribution of the rain coming from diffuse clouds. In analogy, we have proposed that primary embryonic pattern formation is accomplished by the coupling of a short range self-enhancing (autocatalytic) process with a long-range reaction that acts antagonistically to the self-enhancement (Gierer and Meinhardt, 1972; Gierer, 1981; Meinhardt, 1982, 1992). A simple molecular realization would consist of an activator molecule whose autocatalysis is antagonized by a rapidly diffusing inhibitor. In this model, the range of the self-enhancing reaction determines the width of the activated region while the range of the antagonist is responsible for the minimum spacing between two maxima. Figure 1 shows computer simulations demonstrating that the model accounts for the generation of elementary patterns frequently encountered in development. For instance, a single maximum at one boundary of the field and graded concentration profile emerges if the range of the activator is comparable to the size of the field (Fig. 1a). Such a distribution is convenient to provide positional information (Wolpert, 1969). An alternative for the antagonistic reaction consist of the depletion of substrate that is required for the autocatalysis. The local self-enhancement proceeds are expense of a substrate (or cofactor) derived from larger surrounding. For the generation of net-like structures, both mechanisms will be used in conjunction.

The models require an exchange of signals between neighboring cells. Diffusion is certainly the simplest mechanism conceivable. The actual mechanism is, however, more complex, requiring the secretion of molecules and their recognition on neighboring cells by receptors on the cell surface. Therefore, if gene regulation is directly involved in the autocatalysis and a spreading autocatalytic component is required, it is expected that this results from a chain of interactions. For instance, a nucleus-restricted transcription factor controls the production of a small secreted molecule. The latter is able to diffuse between cells. Upon its binding at a receptor, a sig-

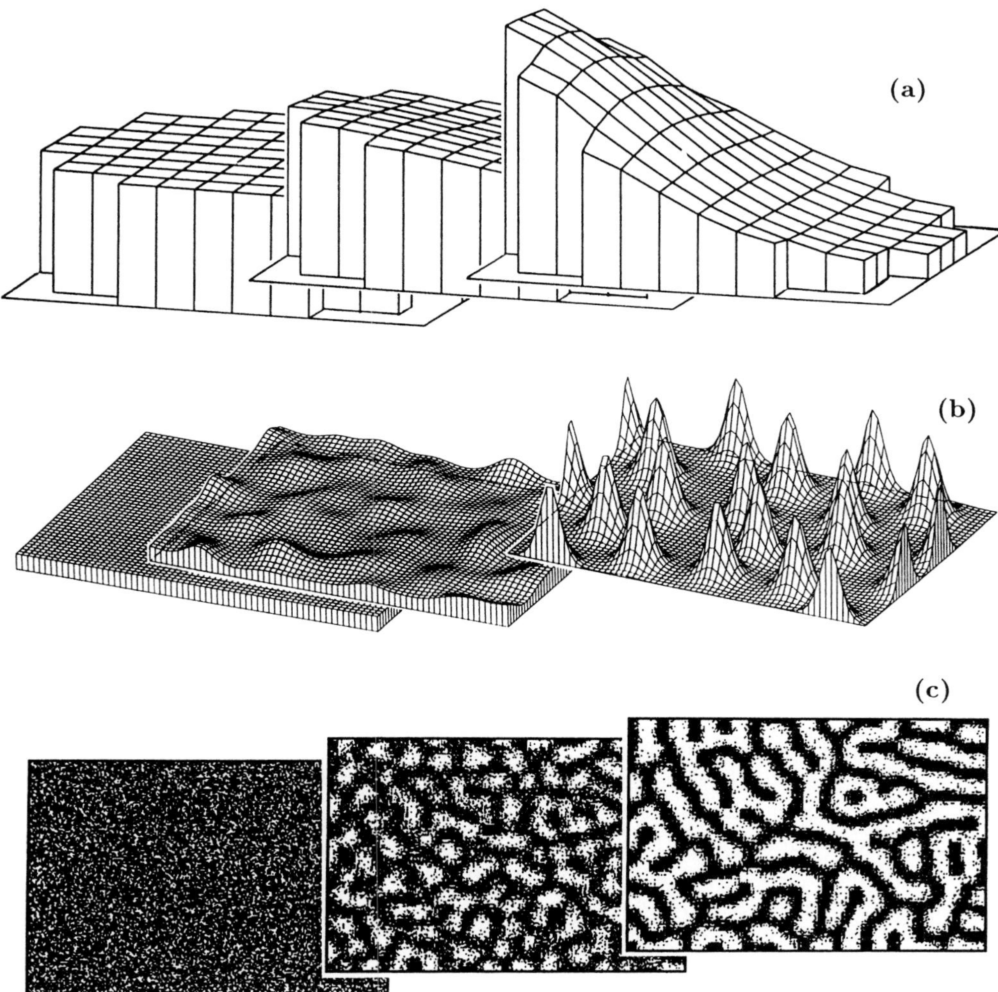

Fig. 1 Stages in the generation of elementary patterns by local self-enhancement and long-ranging inhibition. Shown are the initial, an intermediate, and the final activator distributions. (a) Monotonic gradients are formed if the range of the activator is comparable to the size of the field. The pattern orients itself along the longest extension of the field. (b) A more or less regular arrangement of peaks emerge in fields that are large compared to the range of the inhibitor. (c) Stripelike distributions result if the autocatalysis saturates (Meinhardt, 1989).

naling cascade is triggered that leads, in turn, to the activation of the transcription factor. Several system are already known in which a secreted molecule and a transcription factor form a positive loop. An example is the brachyury gene and the FGF growth factor (Schulte-Merker and Smith, 1995).

In Amphibians, blood islands are formed from the ventral portion of the mesoderm. Many molecules are already known that are involved in the dorsoventral patterning of the mesoderm. Several components have properties that are expected from the theoretical considerations mentioned above. A key gene for the formation of the Spemann organizer at the dorsal side is the gene gooscoid. In the course of time, its activation becomes more and more restricted to the dorsal side—an indication for a self-enhancing process exists. Injections of gooscoid RNA to the ventral side (where the gene is normally inactive), can initiate transcription of that gene if a certain threshold is surpassed (Niehrs et al., 1992). A BMP-3 related molecule has antagonistic properties. Although it has its highest concentration at the organizer region, an artificial overproduction leads to a switching off of all molecules characteristic for the organizer (Moos et al., 1995). This molecule has therefore properties we expect for the inhibitor. The ventral fate of the mesoderm is also accomplished by an active process. Again, a secreted molecule, BMP-4 and a transcription factor VOX constitute an self-enhancing loop (Schmidt et al., 1996; Ladher et al., 1996).

If the range of the inhibitor is much smaller than the total field, more or less regularly spaced peaks (Fig. 1b) emerge. This mode is appropriate for the initiation of repetitive structures such as feathers, bristles, leaves, stomata etc. and presumably for the angioblasts. Such a pattern shows self-regulation. For instance, if two maxima become too remote from each other due to an overall growth, the inhibitor concentration may drop to such a low level, that the autocatalysis can no longer be suppressed. This leads to the insertion of a new maximum that shapes itself due to the concomitantly produced inhibitor. Such a formation of a new signal will be an essential component in the generation of branches on existing vessels.

Stripelike distributions are a third type of pattern that can be generated by this mechanism (Fig. 1c). Such patterns emerge if the autocatalysis of the activator saturates at higher activator concentrations. Due to the upper bound of the activator concentration, the inhibitor production is also limited. Activated cells must tolerate other activated cells in their neighborhood. Stripes are the preferred pattern since in this case, each activated cell has an activated neighbor and nonactivated cells are nearby into which the inhibitor can be dumped.

Pattern formation based on autocatalysis and long-ranging inhibition has many properties that are well known from experimental systems. They are self-regulating. Removal of the activated region leads, after the decay of the remaining inhibitor, to an onset of the autocatalysis and thus to a restoration of the pattern. Pre-existing asymmetries can orient the pattern while the pattern itself is to a large extent independent of the mode of initiation.

As mentioned, the antagonistic effect can also result from the depletion of a substrate that is consumed during the autocatalytic activator production. Autocatalysis comes to rest due to a severe depletion of the necessary substrate or cofactor. While in an activator–inhibitor scheme, the antagonist has it highest concentration at the position of an activator maximum, in the activator substrate model it is the reverse (Fig. 2). Both reactions have some distinctly different properties. During growth, new maxima emerge in the activator–substrate mechanism preferentially by a split and subsequent movement of existing maxima. In contrast, in the activator–inhibitor mechanism, new maxima become inserted at a distance to existing maxima (Fig. 2). Therefore, whenever the formation of stable isolated maxima has to be described, the activator–inhibitor mechanism is more appropriate. In contrast, if a maximum has to be shifted, the activator–substrate model is more convenient. In the model for netlike structures, a combination of both mechanisms will play a decisive role.

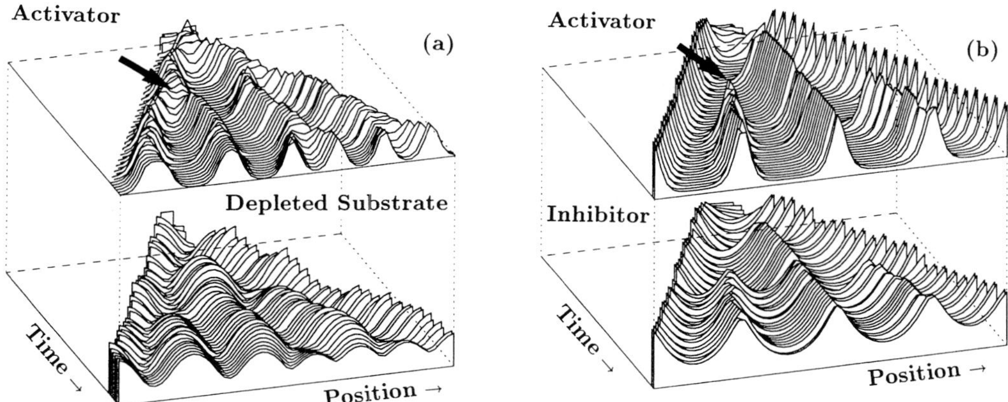

Fig. 2 Induction of new vs. the split of existing maxima: different behavior during growth. (a) In the activator–substrate model, a tendency exists to shift maxima toward higher substrate concentrations. This shift may be connected with a split of a maximum and their subsequent separation. With growth, the substrate concentration increases in the enlarging space between the activated regions since the substrate is not used up there. This can lead to a higher activator production at the side of an existing maximum, i.e., the maximum begins to wander toward higher substrate concentrations until a new optimum position is obtained. (b) In contrast, with the activator–inhibitor mechanism, new regions become activated in the enlarging interstices whenever the inhibitor becomes too low to suppress the onset of autocatalysis. One such event is marked in each figure by an arrow. If in this reaction the autocatalysis saturates, splitting and shifting can occur there too.

The generation of a stable pattern requires that the antagonistic effect reacts rapidly upon a change in the activator concentration. If this condition is not satisfied, the same interaction can generate oscillating patterns. As it will be shown later, this mode is essential for the detection of small concentration differences to orient the outgrowth of a filament in a chemotactic manner (Fig. 11).

Cells Must Remember What They Have Learned: Stable Gene Activation

The signals generated by the mechanisms discussed above are certainly only temporary. At an appropriate stage the cells have to make use of these position-specific signals by making a corresponding change of their state of determination. Afterward, the cells may maintain this determination although the evoking signal is no longer present. This requires the stable activation of genes under the influence of such "morphogenetic" signals.

According to the classical gradient concept, a morphogenetic signal can accomplish a selection between several pathways depending on its local strength. The selection of one particular gene and the suppression of the alternative genes has formal similarities with the formation of a pattern in space. In pattern formation, a substance has to be produced at a particular location but this production must be suppressed at others. Correspondingly, determination requires the activation of a particular gene and the suppression of the alternative genes that are possible in the particular developmental situation. Based on this analogy, I have proposed that stable gene activation requires autoregulation and the competition of alternative genes (Meinhardt, 1978, 1982). Gene activation is, so to say, a pattern formation among alternative pathways.

Meanwhile, many genes with autocatalytic properties (autoregulation) have been found. Examples are the genes engrailed (Condie and Brower, 1989), even-skipped (Jian et al., 1991), fushi tarazu (Schier and Gehring, 1992), and Deformed (Regulski et al., 1991). In the latter case, based on this autoregulation, a short activation of the Deformed gene under heat shock control is sufficient for a long-lasting activation of the endogenous gene (Kuziora and McGinnis, 1988). On theoretical grounds, it is expected that the autocatalysis is nonlinear. This can result from a dimerization of the activating molecules or by multiple and cooperating binding sites on the DNA. The Deformed gene is an example for the latter possibility.

Such a dynamic regulation of gene activation has many properties that are essential for the developing organism. Firstly, such a system has thresholds. The activation of a gene becomes an all-or-nothing event. If a threshold is surpassed, the answer of the cell becomes independent of the exact level of the morphogen concentration. Therefore, small deviation from a desired value are not propagated into the hierarchically next level of gene activation. Secondly, the cells obtain a long-term memory in respect of the signals a cell has seen. The signal is required only for the initiation. Due to the feedback, the gene activity is self-maintaining. Only one of the alternative genes can be active within a particular cell; the cells have to make an unequivocal choice even if the signal is sloppy.

The simplest model for the self-maintenance of a gene activity requires a gene product with a positive nonlinear feedback on its own activation (Meinhardt, 1976; Lewis et al., 1977). A saturation determines an upper bound of its activity. Such a system can remain in two states: a state with low gene activity, in which the linear decay term dominates, and in a high state, when the autoregulation comes into action. If the activation is initiated by an external signal, it flips into the high state after a sharp threshold is surpassed (Fig. 3).

An analysis of ligation experiments with early insect embryos (Sander, 1976) has revealed that the cells do not measure a particular morphogen concentration at once but that there is a stepwise promotion until the activated gene corresponds to the local morphogen concentration (Meinhardt, 1977, 1978). Meanwhile several systems with such a behavior have been described. Gurdon et al. (1995) found that a low concentration of Activin causes activation of the Xbra gene while a high concentration is required for the gene Xgsc. Both these genes remain active after removal of Activin. However, a later increase of the Activin concentration leads to a reprogramming from Xbra to Xgsc. Gurdon et al. have called this behavior a ratchet-like switching, in full agreement with the mechanism I have proposed. A similar behavior has been found in the determination of neuroblasts in Drosophila. The ventral midline cells act as an organizing region. When neuroblasts are grafted toward a more dorsal position (i.e., away from the organizer, presumably into a region of lower morphogen concentration), the neuroblast behave according to their origin, i.e., they maintain their determination. In contrast, if more dorsal neuroblasts are shifted to a more ventral position, they adapt to their new position (Udolf et al., 1995), in full agreement with the model.

Figure 4 shows a simulation of the activation of three genes under the influence of a graded morphogen distribution m. Starting with gene 1 active everywhere, those cells that are exposed to the highest m-concentration are the first to switch to gene 2 and to repress gene 1. In regions of lower m-concentrations, this step takes somewhat longer. Therefore, the pattern of gene activation is expected to be highly dynamic and to include wavelike patterns of activation and deactivation. This behavior is frequently observed in real systems (Kablar et al., 1996; Von Bub-

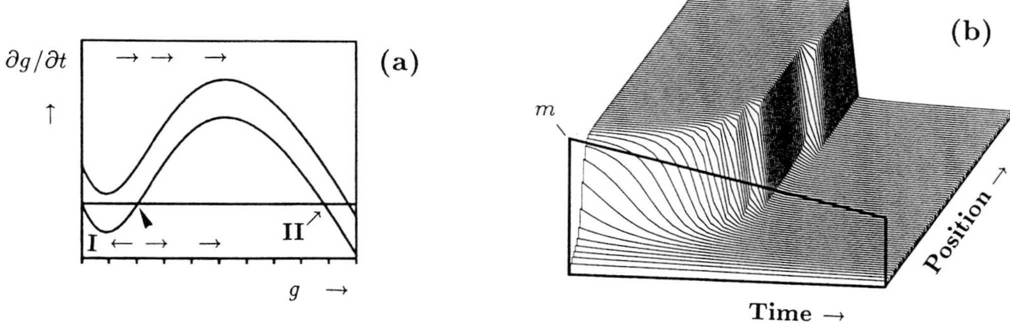

Fig. 3 Model for the switchlike behavior of a cell by saturating autoregulation (see Eq. 4). (a) Plotted is the speed of concentration change as function of the concentration. Lower curve: Without signal, three steady states ($\partial g/\partial t = 0$) exist; that at very low and that at high g concentration are stable (I and II), the third one (arrowhead) is unstable. Under the influence of an external signal the lower state becomes unstable (upper curve). The system can come to rest at the high state only. This transition is irreversible. (b) Under the influence of a graded signal (m) only those cells that are above a threshold make the transition into the high state. The transition is an all-or-nothing event. Shown is the gene activation as function of position and time in a linear array of cells

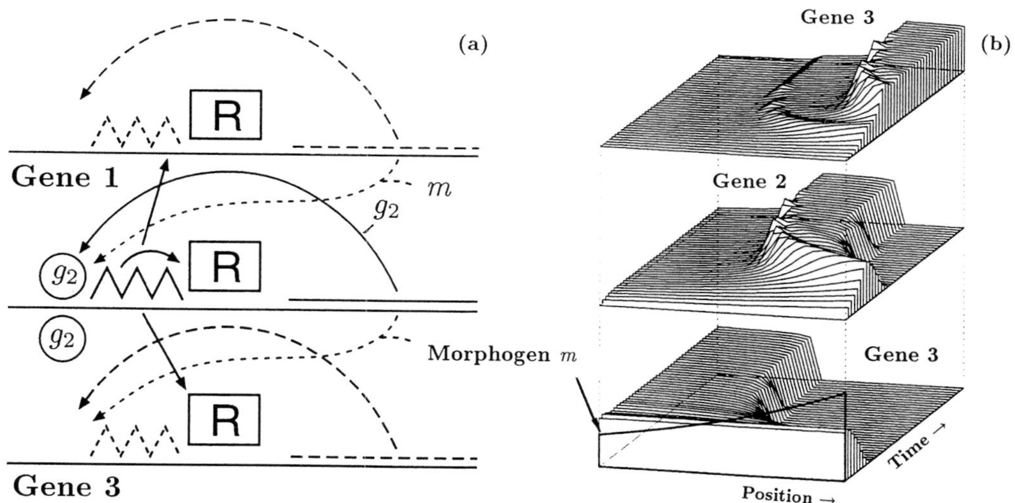

Fig. 4 Position-dependent gene activation. (a) Activation of one gene out of a set of genes. Assumed are genes 1, 2, 3 . . . whose gene products feed back on the activation of their own gene and which compete with each other either directly or via a common repressor. Only one gene of the set can remain active. (b) Particular genes become activated in a field, each covering a certain concentration range if, under the influence of a morphogen gradient m, a concentration-dependent switch from gene 1 to gene 2 etc. takes place. Shown is the activity of three genes as function of time and the morphogen distribution m. Due to the assumed autoregulation, the once activated genes remain activated independent of whether the evoking signal remains present.

noff et al., 1996). This waves sweeps over the field until the local m-concentration is too low to accomplish the switch. The same procedure repeats with the activation of gene 3, causing in addition the repression of gene 2. The result is a subdivision of the field in regions that are separated by sharp borders. In agreement with many observations, these borders sharpen in the course of time. They can become the new organizing regions for an even finer subdivision of the embryo (for models, see Meinhardt 1983a,b; for recent experimental evidence, Vincent and Lawrence, 1994; Martin, 1995).

According to the model, the stepwise irreversible promotion of the cells under the influence of the morphogen has an important biological function. During embryogenesis, the tissue usually grows. If small groups of cells act as source region of a diffusible morphogen, any growth will lead to an increase between a cell and the source region, i.e., to a decrease of the local morphogen concentration. If the cells remember the highest concentration to which they have been exposed, a once obtained subdivision will be maintained also in a growing field despite the fact of the fading morphogen concentrations.

As a possible application, Figure 5 shows a simulation that could describe the formation of blood islands. Shown is the pattern of the activator (a) and three genes under its control: a default state (b), a central core state (d, blood cells) and a narrow ring of cells that separate both populations (c), the endothelial primordia of the vessels. According to such a model, at first many cells are expected to produce a marker characteristic for endothelial cells but that, due to the further promotion, the corresponding gene becomes quiescent later on (see the activation and sub-

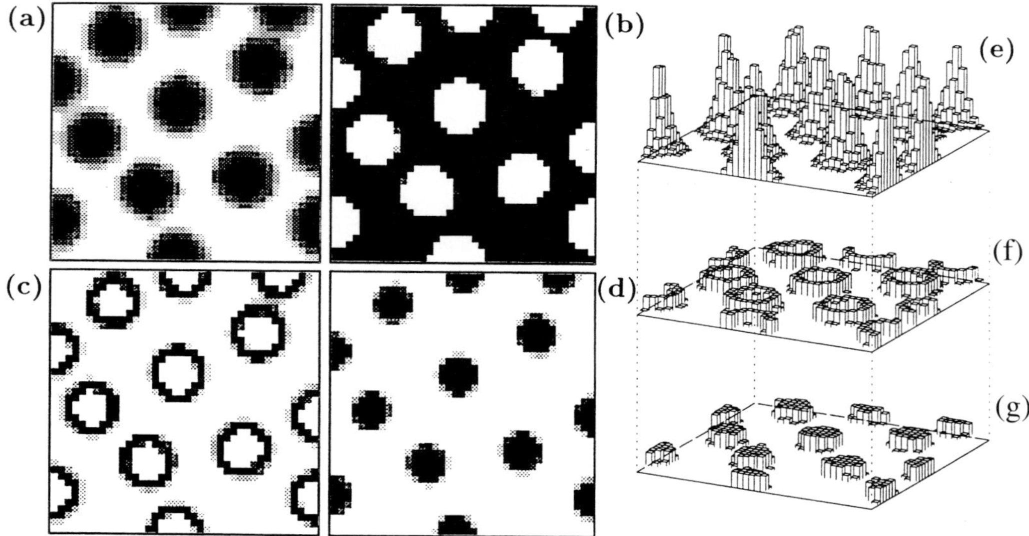

Fig. 5 A model for the formation of blood islands. Assumed is an activator–inhibitor system that generates isolated maxima (a, e; see also Fig. 1). The activator acts on two different genes in a concentration-dependent manner. Cells in the core of a maximum obtain a different determination (d,g) than those at a more peripheral position (c,f). The deterioration of the blood cells and the endothelium may be achieved in this way.

sequent down regulation of gene 2 in Fig. 4). The expression of flk-1, a marker for endothelial cells, has exactly this dynamic (Eichmann et al., 1993; Millauer et al., 1993).

In summary, the predicted principle—the maintenance of the determined state by feedback of a gene on its own activity combined with a repression of alternative genes—is presumably a general mechanism to generate stable determined cell states.

Models for Elongation and Branching of a Filament

The formation of the blood vessel system has been subdivided into two steps: "vasculogenesis" denotes the de novo formation of the rudiments during early embryonic development beginning with the blood islands mentioned above. In contrast, "angiogenesis" describes the formation of new vessels by sprouting from existing vessels (Poole and Coffin, 1989). The latter is the normal mode when the formation of new vessels is required in an adult organism, for instance, in response to an inflammation or during the vascularization of a solid tumor. This mode is also involved during embryonic development in the vascularization of the spinal cord and the brain (Nakao et al., 1988). As shown below, a similar distinction has to be made in terms of the model.

Necessary Simplifications

To make computer simulations feasible, several simplifications have to be made. Instead of a three-dimensional mass of cells, I will consider only two-dimensional fields with square-shaped elements. These restrictions exclude the consideration of several important features such as the formation of tubelike structures, the influence of the streaming of the blood on the vessel morphology, and the shape changes of individual cells. The mechanism proposed will be explained first for the special case in which filaments are formed by ordered accretion of differentiated cells within a field of "undifferentiated" cells. A generalization will be given later.

Angiogenesis: Elongation and Branching in an Existing Netlike Structure

In the model, a crucial step in the generation of netlike structures is the formation of a local signal at a particular position that causes the elongation of the filament there. With this elongation, the position of the signal is shifted too. Therefore, long filaments are formed as a trail behind a wandering filament-inducing signal.

In the following model of filament formation, the interaction of four substances is assumed. Two of them form the local signal for elongation, one is involved in the orientation of the elongation toward a target region and the last one functions as a genetic switch and marks the differentiated cells. To allow a computation of the development in time, in the following a set of differential equations is provided that describes the concentration change per time unit of each of the four substances as function of the local concentrations of other substances present at a particular moment. By assuming a certain initial situation, one can calculate the concentration change in the following short time interval. By adding this change to the given concentrations, one obtains the concentrations at a somewhat later time. By a repetition of this computation the total time course can be calculated. All the simulations shown below have been made in this way. I hope that the reader does not regard equations as a burden but as a convenient shorthand to state in an unambiguous way what is assumed.

The generation of a local signal by autocatalysis and long ranging inhibition has been discussed already above (Fig. 1). A possible interaction between an activator a and its antagonist h is given by the following interaction (Gierer and Meinhardt, 1972)

$$\frac{\partial a}{\partial t} = \frac{\rho a^2 s}{h} - \mu_a a + D_a \Delta a + \sigma_a g \tag{1}$$

$$\frac{\partial h}{\partial t} = \rho a^2 s - \mu_h h + D_h \Delta h + \sigma_h g \tag{2}$$

The activator a has a nonlinear autocatalytic influence on its own production rate (ρa^2; ρ is a constant except for small random fluctuations). It also catalyzes the production of the inhibitor h. The nonlinearity is required to have a production term stronger than the first-order decay term. The condition of nonlinearity is satisfied, if two activator molecules have to form a dimeric complex to be effective. The rate of activator removal is described by $-\mu_a a$. It is proportional to the number of molecules present (like the number of individuals dying per year in a city is proportional to the number of inhabitants). $D_a \Delta a$ describes the exchange by diffusion. A small baseline activator production that is independent of the actual activator concentration ($\sigma_a g$) is necessary to initiate the system at low activator concentrations. It is required to trigger a new activation along an existing filament for branch formation. Only the differentiated cells (high g) have this basic production. Therefore, a branch can be initiated only along an existing filament.

The inhibitor is produced under control of the activator (ρa^2) but has no direct influence on its own synthesis. It decays ($-\mu_h h$), diffuses ($D_h \Delta h$), and slows down the activator production (1/h in Eq. 1). A necessary condition for pattern formation is that the diffusion rate D_h of the inhibitor is much higher than that of the activator Da. The pattern will be stable in time if $\mu_a < \mu_h$ otherwise oscillations will occur. Both oscillating and nonoscillating modes presumably play a role in the generation of a net like structure (see Fig. 11). A small baseline (activator-independent) inhibitor production ($\sigma_h g$) causes a background level of the inhibitor that increases with increasing density of the filaments. It can be used to terminate the elongation of the filaments when a certain density is achieved.

Important is that the outgrowth has an appropriate orientation toward a region that requires additional filaments. It is assumed that a growth factor s (or a "substrate") is produced by all cells which is preferentially removed by all cells belonging to the filaments ($-\nu gs$).

$$\frac{\partial s}{\partial t} = \sigma_s - \mu_s s - \nu g s + D_s \Delta s \tag{3}$$

The higher the density of the filaments the lower the average concentration of this factor will be. Its concentration will be lowest next to the filament. The maximum concentration indicates how distant other filaments are. Since this factor must be diffusible, its local concentration near the filaments is a measure how urgent the surrounding cells need the ingrowth of new vessels. Therefore, it can be used in the control of branch initiation. To achieve the correct orientation, this factor s is assumed to have an accelerating influence on the autocatalytic signal production that causes filament elongation (sa^2 in Eqs. 1 and 2). Therefore, a high activator concentration will be shifted to a position that points toward a region that needs additional filaments.

The fourth equation describes the differentiation of the cell.

$$\frac{\partial g}{\partial t} = \frac{\rho_g g^2}{1 + \kappa g^2} - \mu_g g + \sigma_g a \tag{4}$$

The saturating feedback of the gene product g on its own activity ($g^2/(1 + \kappa g^2)$; ρ_g is a constant) leads to two stable steady states, one at a high and one at a low g concentration (see Fig. 3). Due to the additional g-activation by the a-signal ($+ \sigma_g a$), a cell in the low state can surpass the threshold and switch irreversibly into the high state.

How It Works

The simulation provided in Fig. 6 shows how these elements work together. The local high activator concentration (top row, see also Fig. 1) causes the first cell to differentiate. This leads to a decrease in the s concentration in this cell (Fig. 6b, third row), causing a slowing down of the activator autocatalysis here. The surrounding cells are, due to their higher s-concentration, better in the competition to become activated. Diffusion of the activator into neighboring cells triggers a new activator maximum there. Due to mutual competition, only a restricted group of neighboring cells will develop a new maximum. The inhibitor that spreads from the newly activated cell contributes to the complete suppression of the previously active cell. The newly activated cells will subsequently differentiate too.

The next cell to be activated will be that filament-adjacent cell that has the highest s-concentration. Usually it is the cells in front of the tip of this incipient filament, because it has the least contact with the s-removing differentiated cells. By repetition of these steps—shift of the

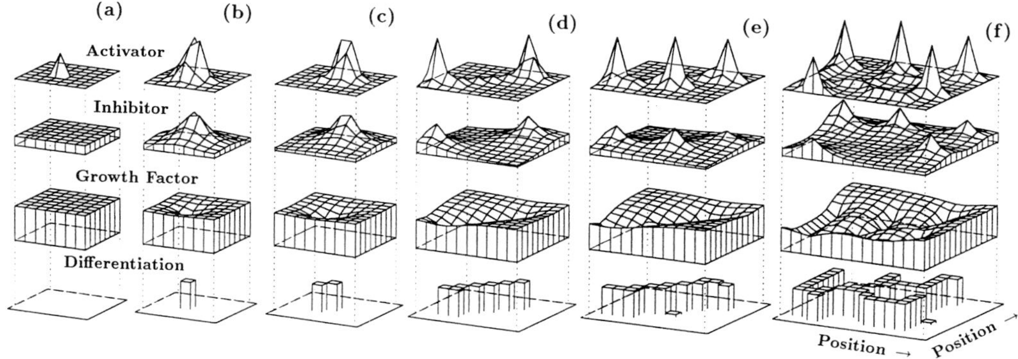

Fig. 6 Formation of a filament by differentiated cells and the initiation of a branch. Simulation is started with a single activator maximum (a, upper row; see Fig. 1) causing one cell to differentiate. The latter removes the substance s, a growth factor, from the surrounding cells (third row). Since the activator production depends on the substrate s, the maximum is shifted away from the depression in the s concentration into a neighboring cell, which differentiate too (switching g from low to high concentration). Long filaments of differentiated cells are formed behind wandering activator maxima. If the growing tips of the filaments become sufficiently remote and enough space is available, the basic activator production of the differentiated cells can trigger a new maximum (e,f). A new branch is initiated in this way.

signal in front of the tip, differentiation of that cell and shift again—long filaments of differentiated cells are formed (Fig. 6d). Figure 7 shows a similar simulation in a larger field of cells using a different mode of plotting.

As it will be shown below in detail, the netlike structures that are generated by this simple mechanism have features that resemble closely those of biological networks. For example, bifurcations and lateral branches can be formed, the density of filaments can be regulated according to local demand, the outgrowth of a filament can be oriented toward a target area, filaments of two different cell types can make connections with each other, and a damaged net can be repaired.

Formation of Lateral Branches

To form a netlike structure, individual filaments have to branch repetitively. A branch can be formed either by bifurcation at the growing tip or by the formation of a new growth point along an existing filament. According to the model, a branch is initiated when a new activator maximum is triggered along an existing filament. The inhibitor is produced mainly by the activated regions at the growing tips. As the length of a filament increases, the inhibitor concentration along the filaments may become at some positions insufficient to suppress in the filament the basic, activator-independent activator production ($\sigma_a g$ in Eq. 1). By autocatalysis, a new activator maximum will be formed. Since the concentration of s is lowest along the filaments, the activator maximum is rapidly shifted to a cell at the side and a branch is initiated (Fig. 6e,f). Figure 7 shows an example of the resulting pattern in a larger field of cells.

The concentration differences of the growth factor along filaments are expected to be small. The averaging effect resulting from its diffusion contributes to the smoothness of its distribution. Therefore, in branch formation, the problem is to keep the signal sharply localized although a relatively large region is exposed to a comparable concentration of the growth factor. Any model assuming branch initiation whenever the growth factor surpasses a certain threshold is certainly insufficient since this would occur simultaneously in very extended regions. The involvement of a pattern forming system solves this problem. The signal for branching remains localized. The initiation of activator production leads, due to the concomitantly produced inhibitor, to a competition within the filament for branch formation that will be won only by a small group. The signal shapes itself. Therefore, the system is able to make a choice, if necessary, at random. However, each decision, once made, has a pronounced influence on the further pattern of branching since it changes the concentration landscape of the growth factor.

Bifurcation of Growing Filaments

In addition to lateral branching, bifurcations of growing filaments at the tip is another mode to increase the complexity of an evolving netlike structure. This requires a split of the activator maximum, followed by a shift of the separated maxima into different directions. In the model, such a split will occur if there is some saturation in the autocatalysis of the activator production. Due to such an upper limit, the activator concentration can no longer increase in peak high, it has to broaden until an equilibrium is reached between the autocatalytic and the self-produced antagonistic reaction. The center of such a broadened peak is especially prone to become deactivated, since there the concentration of the inhibitor is the highest. This will occur preferentially if much growth factor is present, driving the activator production into a saturation. The acti-

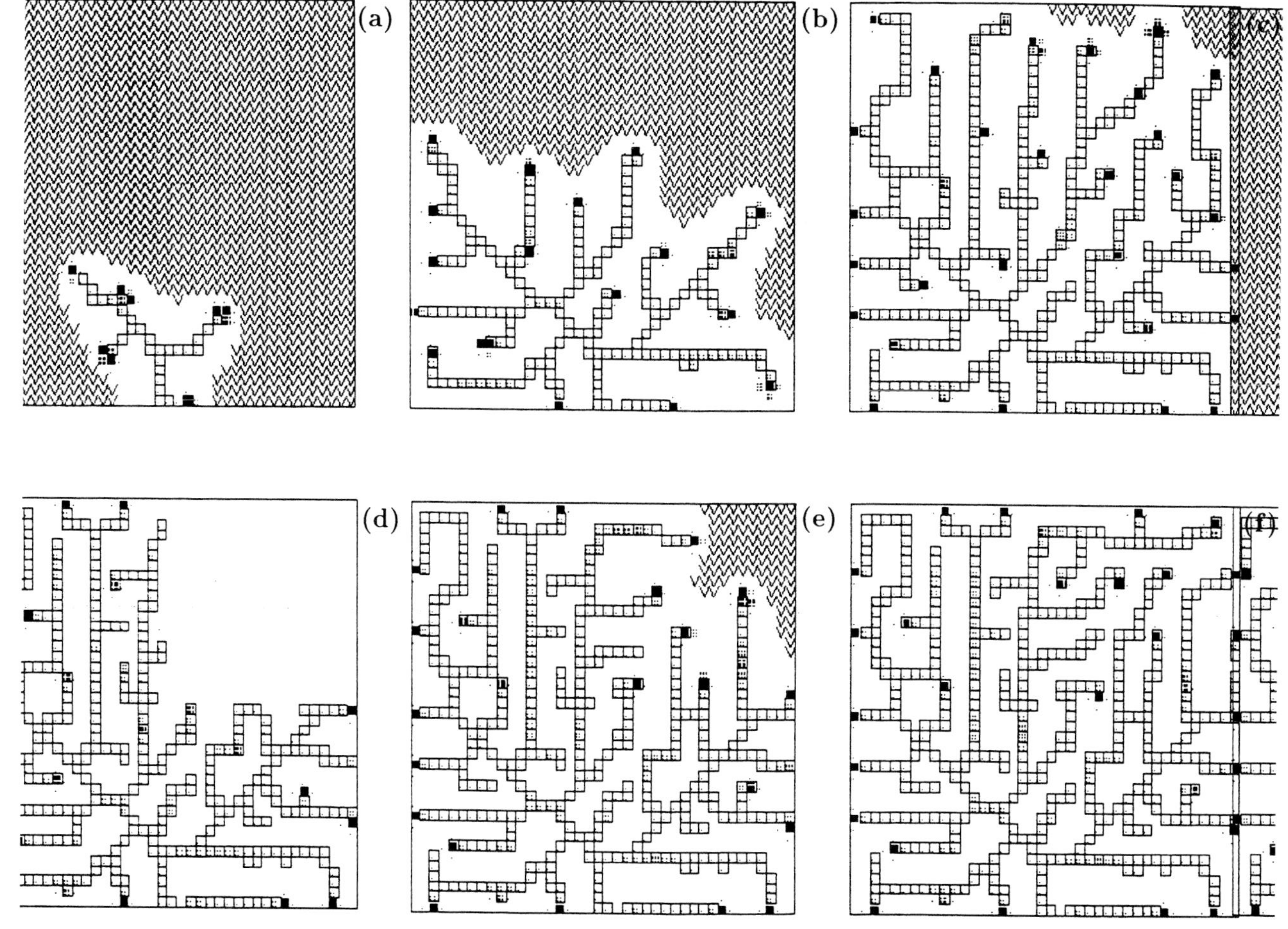

Fig. 7 Formation of netlike structure in a larger field (a–c). and its repair (d–f). After removal of the filaments in the upper right quarter, the substrate concentration recovers. New veins grow into the area. The newly formed part of the pattern looks similar but is not identical to the original net.

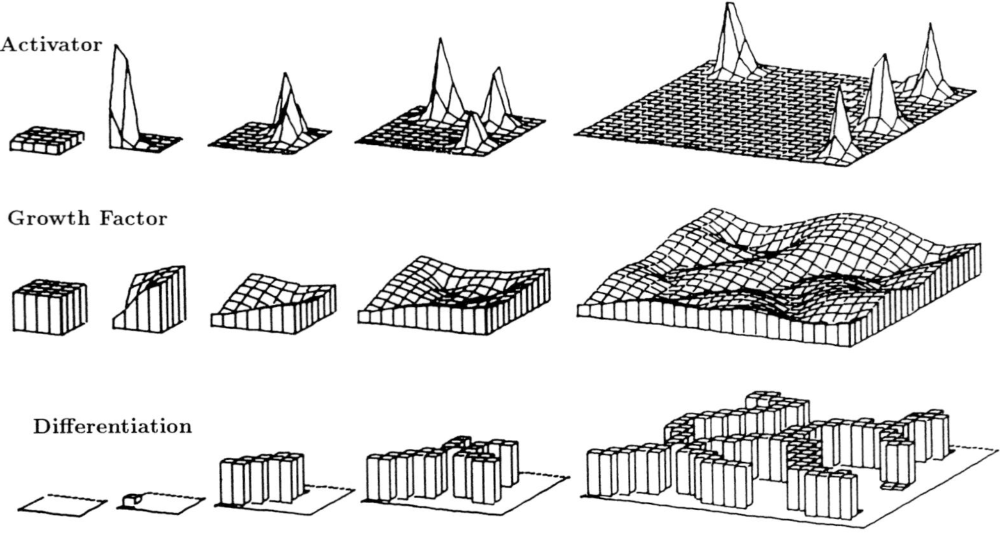

Fig. 8 A simpler model to generate netlike structures. An activator–substrate model is used that causes cell differentiation. No separate inhibitor is involved. Possible is only a splitting of the maxima at the growing tips causing bifurcations of the growing filaments. Since the substrate concentration is too low along an existing filament, no lateral branching is possible.

vator–substrate model with its inherent saturation shows this shift and split behavior very extensively (Fig. 2a).

To provide an example, in the simulation illustrated in Figure 8 an even simpler interaction has been used that is described by Eqs. (1–4). The depletion of the growth factor is used for two purposes: to keep the activated area localized and to provide the drive to shift the signal toward regions that are not yet sufficiently supplied by filaments. In this case, lateral branches are no longer possible since the concentration of the growth factor next to the filaments is too low to trigger a new signal that would lead to branch formation, but the growing tips bifurcate extensively. In the evolution of plant leaves, the transition from this dichotomous branching to the contemporary venation with lateral branching was presumable a major evolutionary step. The evolutionary old Ginkgo tree has leaves with veins that only bifurcate. In needle trees, even such branching does not occur. In terms of the model, the decisive step that enables lateral branches was a molecular separation of the two antagonistic tasks: (i) keeping the autocatalytic signal localized, and (ii) shift the localized maximum away from the differentiated filaments. Having a separate inhibitor, branch formation can be initiated along a filament although the concentration of the growth factor next to a filament is low.

Repair of a Damaged Net

Destruction of some portion of a net frequently leads to branching of nearby filaments. In this way the damage becomes repaired. After formation of a wound, new vessels are formed by sprouting. Similar repair mechanisms exist also in the formation of other netlike structures, for instance the tracheal system of insects (Wiggelsworth, 1959). The proposed model reproduces this regenerative capability. In an area without filaments, the growth factor s is no longer removed. The increasing s-concentration (and/or the absence of an inhibitory effect spreading from existing filaments) attracts new branches (Fig. 7) and the damage becomes repaired. The repaired pattern is similar but not identical to the original pattern.

Limitation of Maximum Net Density

Since any new branch can give rise to other branches, the density of filaments will increase. In the model, the elongation of the filaments can come to rest in two ways. With an increasing density of the net, the concentration of the growth factor s declines. Since s acts as a cofactor in the generation of the signal, from a certain density onward, the activator concentration is too low to accomplish a further elongation. Other way round, the more growth factor is produced by a particular tissue, the denser the final web of filaments will be. This is illustrated in Figure 9.

Alternatively, the final net density can be controlled by an activator-independent inhibitor production of the differentiated cells ($\sigma_h g$ in Eq. 2). This creates a background inhibitor concentration proportional to the local net density. Filament elongation or the formation of new branches will cease if activator production becomes choked by the rising background level of the inhibitor. Tissues with a high level of this background inhibitor production can remain free of vessel ingrowth. In the simulation illustrated in Fig. 9 the filament-free region has this origin.

The actual mechanism by which the limitation of the net density occurs depends on the details of the interactions. If only the activator but not the inhibitor production depends on s, the production rate of s has a very strong influence on the final density. In contrast, if production of both the activator and the inhibitor depend on s (as given in Eqs. 1 and 2), the absolute activator concentration becomes over a wide range independent of the s-concentration and the influence of the s-production is much weaker. Then, the baseline inhibitor production of the differentiated cells may play a decisive role in limiting the final density of the net. An advantage of the latter scheme is that range of allowed growth factor concentrations is much larger, providing a better dynamic regulation in different regimes.

Finding Target Regions and the Connections of Veins and Arteries

According to the theory proposed here, elongation proceeds in the direction of the highest concentration of s. In the case of homogeneous s-production, an approximately uniform density of filamentous structures will emerge. If, however, the factor s is produced only or preferentially in a particular area, the filament will follow the resulting s-gradient.

The system of blood vessels consists of two components: the veins and arteries. They are intimately connected by small capillaries. In the model, one possibility to connect filaments of two different cell types would be that one type produces the growth factor that provides the attraction for the other. Figure 10 provides and example.

A difficult problem is the formation of anastomosis, i.e., of reconnections of filaments derived from same cell types. Since the filaments tend to grow away from existing filaments (to achieve a homogeneous coverage with filaments), the formation of reconnections between two filaments of the same type is unlikely, especially in a three-dimensional space where a growing filament can circumvent an existing one. Indeed, most filaments end blindly in nets consisting of a single cell type only. The lymphatic system or the tracheal system in insects are examples. How-

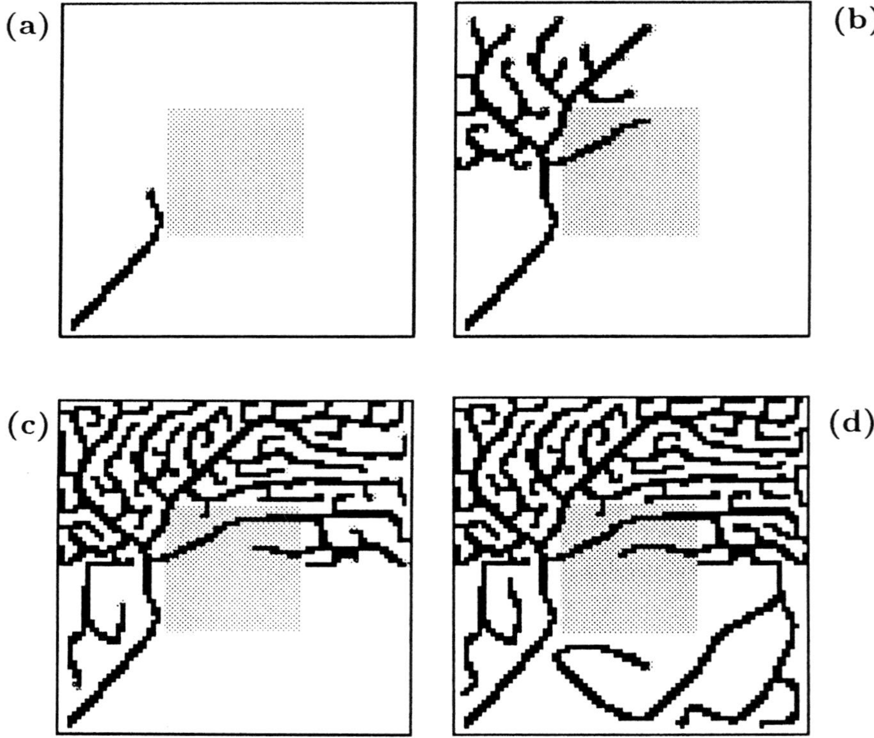

Fig. 9 Regulation of the density of a net. According to the model, the density of a net can be regulated by the concentration of the growth factor s which is produced everywhere in the tissue and which is removed by the net. A twofold higher s production in the upper half area leads to a much higher vessel density. In the center of the field, cells with an excess of inhibitor production is assumed. At this particular choice of parameters, at normal s production (lower half), no vessels extend into the inhibited area, while at the elevated s production, some may grow into that field.

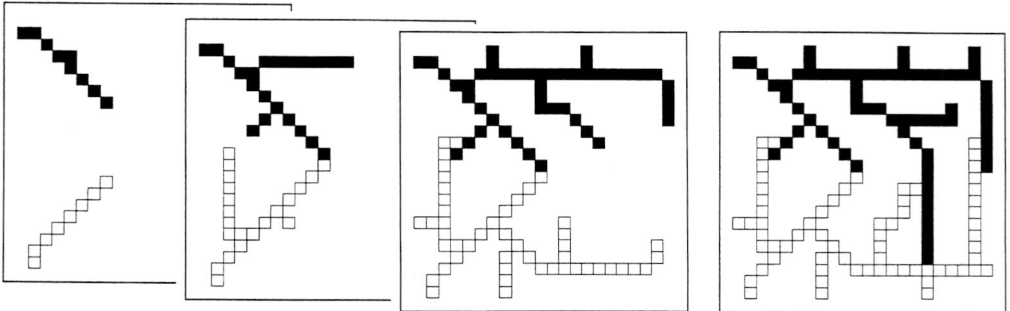

Fig. 10 Making connections between filaments of two different cell types (open and filled squares). Each cell type is assumed to produce a surplus of the growth factor required by the other. The elongation of the filaments becomes oriented toward cells of other type

ever, in leaves, anastomosis occur regularly. To my knowledge, there are no models available that describe this feature (the mechanism described in Fig. 12 produces closed loops too regularly). I will come back to that problem in the discussion of the transition from the isolated blood islands to the early plexus.

Variability of the Resulting Pattern

In the model, very complex patterns can be generated by the interactions of very few substances. These interactions can easily be encoded by genes. The question may arise as to how reproducible such patterns would be. In the model, the parameters determine only the general features of the pattern, such as average net density, the distance between branching points or the straightness of the lines. The fine details depend on external influences or even on random variations. For instance, during the initiation of a new branch, small differences determine to which side the activator maximum will escape and therefore toward which side the new branch will grow. However, once a branch has been made, let us say to the left, the resulting asymmetry strongly influences subsequent branching. Due to the presence of the new branch, the concentration of s will drop on the left side and the next branch will extend to the right, and so on. Since each decision depends to such an extent on the previous one and random fluctuations play an important role, only the overall pattern is reproducible. In the details, the pattern will be similar but not identical. The variability in the details is a prevailing element in many netlike structures. Identical twins have different patterns of blood vessels in their retinae. Leaves from the same tree are not identical, even though they are surely developed under the control of the same genetic information. Similarly, the details of neuronal branching differ among genetically identical individuals of the water flea Daphnia (Macagno et al., 1973).

Filaments Formed by Oriented Cell Division or by Extensions of Single Cells

Elongation by accretion of newly differentiated cells is only one of several possible modes of the formation of filaments that can be subsumed under the proposed theory. Indeed, in the generation of many netlike structures, the elongation of individual cells play a crucial role. The sprouting of capillaries of blood vessels is only one example, the elongation of an axon at the growth cone or of tracheae in insects (Samakovlis et al., 1996) proceeds in the same way.

For the local elongation of a cell, again a signal must be available as to where this elongation should occur. A particular surface element has to be selected while the same process must be prevented in the remaining part of the cell. For a chemotactic orientation of this elongation, the cells have to detect small concentration differences and react thereupon. Such small differences must be internally amplified to produce an unequivocal signal as to where, for instance, filopodia has to be stretched out. The mechanism of autocatalysis and long range inhibition is able to perform this detection and amplification. Since in this case pattern formation must take place within a cell or within in a part of a cell, the diffusion range of the activator must be small. This suggests that the self-enhancing process is membrane-bound. In contrast, a rapid redistribution of the antagonist (an inhibitor, for instance) is essential to suppress secondary maxima that would point in a direction opposite to the desired one. Thus, it would be advantageous if the antagonistic substance diffuses freely in the cytoplasm.

Other filaments grow by oriented cell division. This requires a similar mechanism since the polarity of a cell must be oriented. The vascular endothelial growth factor VEGF discussed below in more details is a potent mitogen for endothelial cells (Wilting et al., 1996), but whether it has a polarizing influence on the cell division is not yet known.

Use of Oscillating Pattern Forming Systems to Orient Chemotactic Cells

For a chemotactic cell, it is important that the direction of movement can be permanently readjusted. A stable pattern is inadequate since a once generated high point on the cortex of a cell would be difficult to shift. Due to the internal self-amplification, the cell would be unable to detect small changes in the environment of the cell. An adjustment in the position of the activator signal and therewith a correction in the orientation of movement would be difficult to achieve. This problem can be overcome if the activator pattern oscillates (Meinhardt and Gierer, 1974). Phases in which the system is sensitive to external signals alternate with phases of internal amplifications, in which the small external differences are converted into a strong signal for local cell elongation. In the model discussed above, oscillations will occur if the antagonist has a longer time constant than the activator. As shown in Fig. 11, after an increasing accumulation of an inhibitor(or a complete depletion of a substrate), the autocatalysis breaks down. The next autocatalytic burst has to wait until the antagonistic reaction becomes sufficiently weak. During the subsequent refractory period, an autocatalytic burst is impossible but minute external differences are sufficient for the positioning of the next sharply localized signal. Depending on the external signal, the next activation can appear at a different position on the cell cortex. Therefore, the growing filament searches for an optimum orientation. Inherently, the oscillations have been used also in the simulations in Figs. 6 and 7 since there the differentiation of the cells causes also a destabilization of the local signal allowing a new activation in the neighborhood at an optimum position.

Harrison (1910) has shown that the growth of a nerve fiber is not a continuous process but, rather, that phases of fast elongation alternate with phases of searching for a new direction. Several pseudopods may be sent out at the same time, although most of them will later be retracted. An inspection of the shapes of chemotactically-sensitive cells suggest that elongation is not a single process restricted to a particular position but that several "hot spots" like filopodia are formed and that a second competition takes place among these candidates to select the best one. This suggest that at least two antagonists are involved. A detailed discussion of the dynamic properties of systems with two or three antagonists has been provided for modeling of the pattern on tropical sea shells (Meinhardt, 1995).

Initiation of a Plexus: Formation of Avascular Regions my be an Active Process

As outlined above, the formation of blood islands in an originally uniform tissue can be explained in a straightforward manner (see Fig. 5). A problem is, however, how the early connections between these islands are formed. As mentioned, in order to generate a ramifying net that covers a field, it is necessary that the elongation of filaments occurs in a direction away from existing filaments. The question remains of how the formation of connections on the one hand and the spreading towards nonsupplied areas is reconciled.

A view to an early plexus reveals that it does not consist of fine filaments but that the endothelial cells cover a substantial fraction of the available space. The avascular areas are remarkable circular, similar to holes in swiss cheese. This suggest an alternative mechanism: a pattern forming system could exist that signals where vessels should not form. Figure 12 shows the result of a simulation under that stipulation.

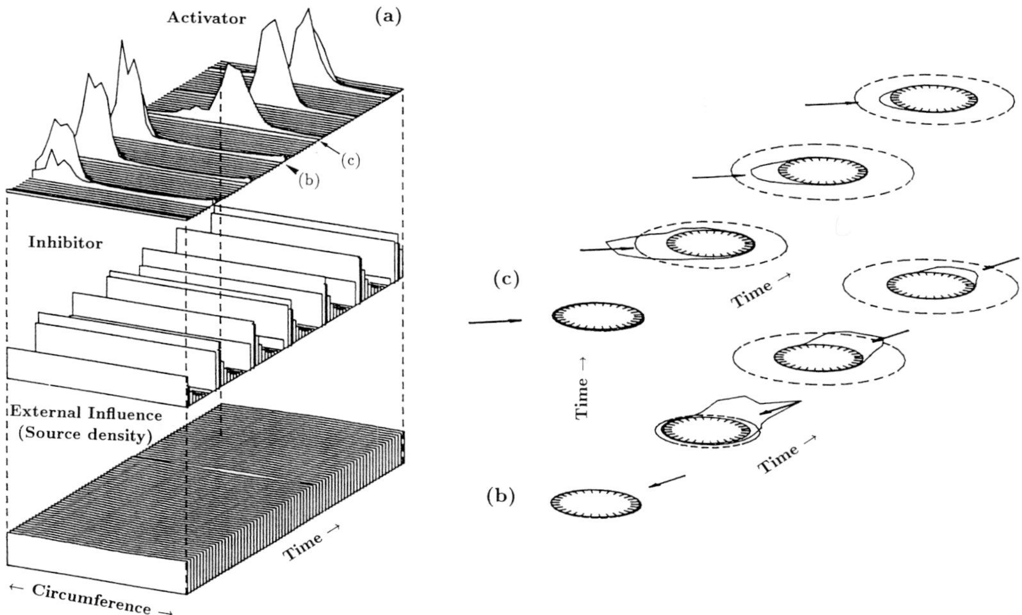

Fig. 11 The use of an oscillating activator–inhibitor system to detect small external differences to orient a chemotactic cell. For computational simplicity simulations have been made on the circumference of a circle instead of a sphere. The inhibitor is assumed to be redistributed very quickly within the cell. (a) To allow a space-time plot, the circle is cut up. The left and right elements are adjacent neighbors in reality. Due to the oscillation, the maximum appears at the position where the external signal is slightly above average. After a change of the external influence, the activator maximum is formed again at the optimum position. The system can detect and adapt to differences which are so small that they can hardly be seen by eye. This is possible even if this signal carries in addition some noise. (b,c) Activator (solid line) and inhibitor distribution during the stages marked in (a) by the arrowhead (b) and arrow (c). The long arrow indicates the position of the maximum external influence. Reorientation and phase shift between the activator and inhibitor maximum are clearly visible.

A pattern consisting of fine filaments forming closed loops are not rare. The wings of dragon flies, appearing like a precious artwork, are another example. In terms of the models, a corresponding pattern can be generated by the superposition of two systems. One system generates fine stripelike activation pattern. The second forms patches that causes filament-free regions (Koch and Meinhardt, 1994). Examples for the corresponding elementary pattern have been provided already in Figure 1a and c. A simulation of the combined system is shown in Figure 12. Assumed is an activator that removes a necessary cofactor during its autocatalysis. Such a factor has its highest concentration at the largest distance from the activator maxima. These high rims have a positive feedback on the stripe-forming system. Therefore, the stripes will appears at the maximum distance from the patches. Such a pattern has size-regulating properties. The meshsize remains nearly constant during growth since, after sufficient growth, existing maxima split into two and shift into the enlarging interstices. A new high rimlike profile appears between the now

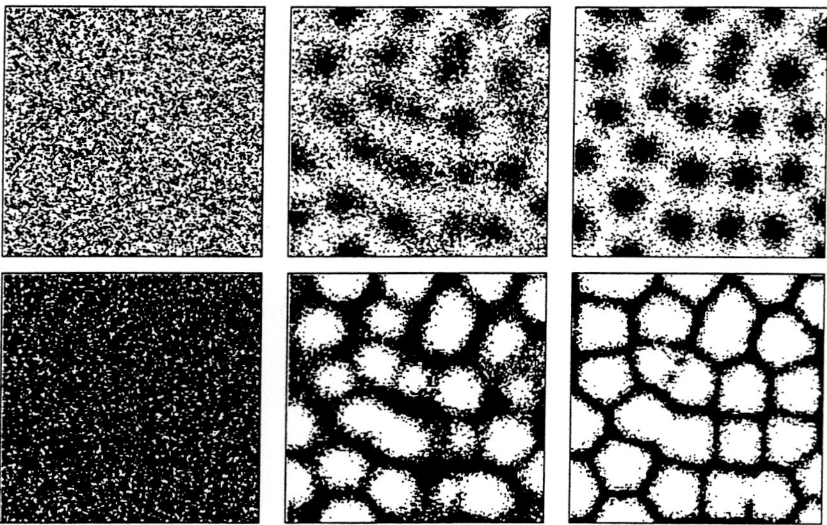

Fig. 12 Initiation of a net. Model: assumed is a patch-forming (top) and a stripe-forming system (bottom). The initial, an intermediate and the final state is shown. The patches dictate where no filament should be formed. The width of the filaments is regulated by the properties of the stripe-forming system. Closed loops of filaments emerge from an originally uniform situation.

separated maxima, causing a new filamentlike activation that connects existing filaments (Koch and Meinhardt, 1994). Although such a model may appear artificial, there are observations that can be integrated in such a model in a straightforward manner. Two modes of vascular sprouting have been described: In one mode, sprouts extend from two preexisting capillaries toward each other and fuse, forming in this way a new vessel. In a second mode, called intussuceptive growth, small but growing avascular islands arises within broader capillaries, reducing in this way large spaces within the capillary (Risau and Flamme, 1995; Burri and Tarek, 1990). The latter mode supports the view that a pattern forming mechanism exists that specifies the position where no capillaries should be. Under some experimental conditions, it can be the dominant mode of microvascular growth (Wilting et al., 1996). In contrast, the first mode requires a signal to initiate the two sprouting events at distant positions and perform a corresponding guiding during their further elongation in order that they meet each other. The rimlike pattern as shown in Figure 12 would be appropriate for that purpose. Thus, both the formation of vascular and of nonvascular tissue may be active and mutually-balanced processes.

Molecules Involved in Angiogenesis and Their Possible Role in Terms of the Model

Several factors are known that have a strong influence on the initiation and elaboration of blood vessels. In this section, a comparison between known substances and the expectation in terms of the theory should be given, hoping that the model is helpful for the understanding of their possible function. From the model, four essential components are employed: an activator and an inhibitor to generate a localized signal for differentiation and elongation of the filament. Both substances are expected to be synthesized preferentially at the tips of growing filaments and a baseline production throughout the whole net. The third component is a factor produced by the tissue into which vessels has to sprout, causing their attraction. The fourth is a substance required that the cells belonging to the vessel system maintain their differentiation.

While the last molecule is expected to be a protein restricted to the nucleus (like a homeobox protein), the first three substances are expected to be diffusible to allow a signaling between the cells. In the equations, diffusion has been treated as if the cells have little holes through which substances can be exchanged directly. It is well known that the exchange of signals require a more elaborate mechanism, employing secreted molecules as signals and receptors that receive it. Therefore, it should not come as a surprise if more substances are involved than suggested by the equations since any signal that spreads by diffusion requires a ligant, a receptor, and a signaling cascade toward the nucleus.

An assignment of experimentally-observed substances and those expected by the model is not trivial. Central is the self-enhancement to obtain a local signal. Two diffusible molecules are expected to be involved that have both a positive feedback on the generation of the signal (a and s in the equations). They differ in the location as to where they are produced and in the range of the spreading. In the following, these differences will be used as a guideline to attempt an assignment. One of the substances, the activator, should be produced preferentially at the tip of the growing filaments, and its range must be small, causing only an "infection" of neighboring cells. The second molecule involved in the self-enhancement should be a factor of longer range that is produced by the tissue that needs more vessels, not by the vessels itself and must have a longer range to allow the chemotactic orientation of the growing filaments.

An important molecule is vascular endothelial growth factor, VEGF (Ferrara and Henzel, 1989; Gospodarowicz et al., 1989), a glycoprotein of about 46kD. The factor is produced in different isoforms by different splicings. Some forms are exported through the cell cortex, but remain attached to the cell surface. These nondiffusible forms are presumably involved only in the communication between neighboring cells. Therefore, they are presumably to be assigned to the activator-class of substances expected in the model. On the other hand, the mRNA for the secreted form of VEGF is synthesized by organs that attract blood vessels such as the brain or the kidney (Breier et al., 1992; Millauer et al., 1993). Local overexpression by a retroviral system cause hypervascularization in the avian limb bud (Flamme et al., 1995). Exogenious addition leads to large vascular sacs that fuse in an unregulated way such that individual vessels can no longer be distinguished (Drake and Little, 1995). High amounts of this factor has been detected during embryonic and tumor angiogenesis (Plate et al., 1992). It orients chemotactically the elongation of the net (Weindel et al., 1994). Therefore, the same class of factors may give rise to the more local activator-like molecule and to the diffusible vessel-attracting molecule, depending on the actual splicing. The receptor for VEGF is flk-1 is (Millauer et al., 1993). There is a positive feedback of the VEGF onto its receptor (Wilting et al., 1996), making sure that if the signal is present, it can be also received. It would be interesting to see whether an increase of flk-1 receptor leads also to an increase of VEGF, either in a direct or indirect manner. If this would be the case, the autocatalytic character expected by the model would be proven. This positive loop must evoke an antagonistic reaction to obtain a steady state and to avoid an overall explosion.

The rapid growth of a tumor is only possible if it is extremely well nourished by a plethora of newly formed vessels (Algire and Chalkley, 1945). Obviously, a malignant tumor must be able to overcome the body's control of blood vessel formation. Thus, an understanding of the control of vessel density is of great importance. A tumor angiogenesis factor TAF (Folkman et al., 1971; Folkman, 1982) has been isolated from different tumors that induces ingrowth of vessels into normally avascular areas, such as the cornea or the epidermis. One such factor isolated from human tumor cells is a single chain peptide of about 14.4 kDa (Fett et al., 1985).

Inhibitory substances are known too. Folkman et al. (1971) have isolated a factor from avascular cartilage which suppresses the ingrowth of vessels. Moses et al. (1990) have determined the structure of this factor and demonstrated its function in in vitro assays. This factor is, however, not necessarily identical with that expected from the model. The factor has a collagenase-inhibiting activity. The degradation of collagen IV and V associated with the basement membrane is an important early step to initiate sprouting (Kalebic et al., 1983). Therefore, this inhibitor may change more the surface properties of the tissue than to act as a signaling molecule in a patterning step.

In the formation of blood islands, FGF-2 plays an important role. The addition of this factor to cultured chicken cells can initiate the formation of blood islands (Flamme and Risau, 1992), a process the can begin in cultured mouse cells spontaneously (Risau et al., 1988). In the model, the self-amplification needs a baseline level to start with. Minute differences can be decisive whether an activation occurs spontaneously or requires an elevation of the baseline level of the activator or of the cofactor. This may be responsible for the different behavior of the mouse and chicken system. FGF has a general function in the organization of mesoderm. FGF and the brachyury gene form together a positive loop (Schulte-Merker and Smith, 1995). While FGF is a secreted factor, the Brachyury gene product is a nuclear protein that does not leave the cell. With both molecules together, the activation obtains an autocatalytic and diffusible character as required in the model. A similar mechanism may be at work in the involvement of FGF in island formation.

Comparison with Other Systems Generating Netlike Structures

The tracheae of insects are another system that forms tubelike branched structures of epithelial cells that supplies tissue with oxygen (see Samakovlis et al., 1996). Elongation of the filaments proceeds there too by protrusions of single cells. Wigglesworth (1954) has shown, however, that this depends not solely on the activity of the extending cells. Ectodermal cells in front of a growing tip extend long protrusions toward a tracheole, attach to the tip and pull it toward itself. Thereafter, other more distant cells send similar processes and pull again. This requires a moving localized signal for the ectodermal cells to become active—a feature that is an integral part of the model. The molecular machineries that form the tracheal and the vessel systems have obviously a common evolutionary origin. For instance, the Drosophila FGF receptor homolog Breathless is necessary in each phase of tracheal morphogenesis and regulates the pattern of secondary and terminal branching (Lee et al., 1996).

In leaves, additional vascular elements are formed after application of auxin to an injury (Jost, 1942). Auxin is known to be actively transported from the leaves via the veins to the roots. The veins remove auxin from the surrounding tissue. This suggest a function like s in the model proposed. A molecule responsible for the maintenance of the differentiated cell state corresponding to g in the model may be the phloem-specific homeodomain gene VAHOX (Tornero et al., 1996).

Similarly, the nerve growth factor NGF (Levi-Montalcini, 1964), a factor necessary for the outgrowth of adrenergic nerves, is actively transported from filaments to the cell body (see Thoenen and Barde, 1980). The highest NGF-concentration that a particular nerve will encounter is presumably that at its end, since more surface elements are available to remove NGF along a fiber. As long as no other constraints are imposed, this would lead to linear elongation of the fiber. The valley of NGF (or a related factor), centered along each fiber, would cause fibers of the same type to keep a certain distance from each other. If higher concentrations of NGF were present in a particular area, this area would attract growing fibers. According to the model, NGF and auxin correspond to the cofactor s for the activation of fiber elongation. The model provides a rational why the best characterized components in these systems belong to the proposed shift-substance s. This substance must be constantly produced in all the cells into which the filaments should grow. In contrast, the proposed activator and inhibitor are expected to be produced only during short time intervals and at changing positions. Moreover, in vitro isolation of activated cells or an experimentally enhanced activator production would lead to an overproduction of the inhibitor and thus to a mutual downregulation of the activation. Therefore, the experimental observation of these substances is presumably much more difficult.

Conclusion

Relatively simple molecular interactions are sufficient to generate a complex netlike structure. Although many essential features are correctly described by the model, many questions are still open. The mutual connections formed between the blood islands is one of them. I hope that an interplay between experimental observations and theoretical considerations will contribute to a better understanding of this fascinating pattern-forming event.

References

Algire, G.H., and Chalkley, H.W. (1945). Vascular reaction of normal and malignant tissue in vivo. I. Vascular reactions of mice to wounds and to normal and neoplastic implants. J. Natl. Cancer Inst. 6, 73.

Breier, G., Albrecht, U., Sterrer, S., and Risau, W. (1992). Expression of vascular endothelial growth factor during embryonic angiogenesis and endothelial cell differentiation. Development 114, 521–532.

Burri, P.H., and Tarek, M.R. (1990). A novel mechanism of capillary growth in the rat pulmonary microcirculation. Anat. Rec. 228, 35–45.

Condie, J.M., and Brower, D.L. (1989). Allelic interactions at the engrailed locus of Drosophila—engrailed protein expression in imaginal disks. Dev. Biol. 135, 31–42.

Drake, C.J., and Little, C. (1995). Exogenous vascular endothelials growth factor induces malformed and hyperfused vessels during embryonic neovascularization. Proc. Natl. Acad. Sci. USA 92, 7657–7661.

Eichmann, A., Marcelle, C., Breant, C., and LeDouarin, N. (1993). To novel molecules related to the VEGF receptor are expressed in the early endothelial cells during avian embryonic development. Mech. Dev. 42, 33–48.

Ferrara, N., and Henzel, W.J. (1989). Pituitary follicular cells secrete a novel heparin-binding growth factor specific for the vascular endothelial cells. Biochem. Biophys. Res. Commun. 161, 851–858.

Fett, J.R., Strydom, D.J., Lubb, R.R., Alterman, E.M., Bethune, J.L., Riordan, J.F., and Vallee, B.L. (1985). Isolation and characterization of angiogenin, an aniogenetic protein from human carcinoma cells. Biochemistry 24, 5480–5486.

Flamme, I., and Risau, W. (1992). Induction of vasculogenesis and hematopoiesis in vitro. Development 116, 435–439.

Flamme, I., Vonreutern, M., Drexler, H.C.A., Syedali, S., and Risau, W. (1995). Overexpression of vascular endothelial growth-factor in the avian embryo induces hypervascularization and increased vascular-permeability without alterations of embryonic pattern-formation. Dev. Biol. 171, 399–414.

Folkman, J. (1982). Angiogenesis: initiation and control. Ann. NY Acad. Sci. 401, 212–227.

Folkman, J., Merler, E., Abernathy, C., and Williams, G. (1971). Isolation of a tumor factor responsible for angiogenesis. J. Exp. Med. 133, 275–288.

Gierer, A. (1981). Generation of biological patterns and form: some physical, mathematical, and logical aspects. Prog. Biophys. Molec. Biol. 37, 1–47.

Gierer, A., and Meinhardt, H. (1972). A theory of biological pattern formation. Kybernetik 12, 30–39.

Gospodarowicz, D., Abraham, J.A., and Schilling, J. (1989). Isolation an characterization of a vascular endothelial cell mitogen produced by the pituitary-derived folliculo stellate cells. Proc. Natl. Acad. Sci. USA 86, 7311–7315.

Gurdon, J.B., Mitchell, A., and Mahony, D. (1995). Direct and continuous assessment by cells of their position in a morphogen gradient. Nature 376, 520–521.

Harrison, R.G. (1910). The outgrowth of the nerve fiber as a mode of protoplasmic movements. J. Exp. Zool. 9, 787–846.

Jiang, J., Hoey, T., and Levine, M. (1991). Autoregulation of a segmentation gene in Drosophila—combinatorial interaction of the even-skipped homeo box protein with a distal enhancer element. Genes and Dev. 5, 265–277.

Jost, L. (1942). Über Gefäßbrücken. Z. Bot. 38, 161–215.

Kablar, B., Vignali, R., Menotti, L., Pannese, M., Andreazzoli, M., Polo, C., Giribaldi, M.G., Boncinelli, E., and Barsacchi, G. (1996). Xotx genes in the developing brain of Xenopuslaevis. Mech. Dev. 55, 145–158.

Kalebic, T., Garbisa, S., Glase, B., and Liotta, L.A. (1983). Basement membran collagen: degradation by migrating endothelial cells. Science 221, 281–283.

Koch, A.J., and Meinhardt, H. (1994). Biological pattern-formation—from basic mechanisms to complex structures. Rev. Mod. Phys. 66, 1481–1507.

Kuziora, M.A., and McGinnis, W. (1988). Autoregulation of a Drosophila homeotic selector gene. Cell 55, 477–480.

Ladher, R., Mohun, T.J., Smith, J.C., and Snape, A.M. (1996). Xom—a Xenopus homeobox gene that mediates the early effects of BMP-4. Development 122, 2385–2394.

Lee, E.C., Hu, X.X., Yu, S.Y., and Baker, N.E. (1996). The scabrous gene encodes a secreted glycoprotein dimer and regulates proneural development in Drosophila eyes. Molec. Cell. Biol. 16, 1179–1188.

Lee, K.J., Cox, E.C., and Goldstein, R.E. (1996). Competing patterns of signaling activity in dictyostelium—discoideum. Phys. Rev. Lett. 76, 1174–1177.

Lee, S.F., Egelhoff, T.T., Mahasneh, A., and Cote, G.P. (1996). Cloning and characterization of a dictyostelium myosin-i heavy-chain kinase activated by cdc42 and rac. J. Biol. Chem. 271, 27044–27048.

Levi-Montalcini, R. (1964). Growth control of nerve cells by a protein factor and its antiserum. Science 143, 105–110.

Lewis, J., Slack, J., and Wolpert, L. (1977). Thresholds in development. J. Theor. Biol. 65, 579–590.

Macagno, E.R., Lopresti, V., and Levinthal, C. (1973). Structure and development of neuronal connections in isogenetic organisms: variations and similarities in the optic system of Daphnia magua. Proc. Natl. Acad. Sci. 70, 57–61.

Martin, G.R. (1995). Why thumbs are up. Nature 374, 410–411.

Meinhardt, H. (1976). Morphogenesis of lines and nets. Differentiation 6, 117–123.

Meinhardt, H. (1977). A model of pattern formation in insect embryogenesis. J. Cell Sci. 23, 117–139.

Meinhardt, H. (1978). Space-dependent cell determination under the control of a morphogen gradient. J. Theor. Biol. 74, 307–321.

Meinhardt, H. (1982). Models of biological pattern formation. Academic Press, London.

Meinhardt, H. (1983a). A boundary model for pattern formation in vertebrate limbs. J. Embryol. Exp. Morphol. 76, 115–137.

Meinhardt, H. (1983b). Cell determination boundaries as organizing regions for secondary embryonic fields. Dev. Biol 96, 375–385.

Meinhardt, H. (1989). Models for positional signalling with application to the dorsoventral patterning of insects and segregation into different cell types. Development (Suppl.) 169–180.

Meinhardt, H. (1992). Pattern-formation in biology—a comparison of models and experiments. Rep. Prog. Phys. 55, 797–849.

Meinhardt, H. (1995). The algorithmic beauty of sea shells (with PC-software). Springer, Heidelberg, New York.

Meinhardt, H., and Gierer, A. (1974). Applications of a theory of biological pattern formation based on lateral inhibition. J. Cell Sci. 15, 321–346.

Millauer, B., Wizigmann-Voos, S., Schnürch, H., Martinez, R., Møller, N.P.H., Risau, W., and Ullrich, A. (1993). High affinity VEGF binding and developmental expression suggest flk-1 as a major regulator of vasculogenesis and angiogenesis. Cell 72, 835–846.

Moos, M., Wang, S.W., and Krinks, M. (1995). Anti-dorsalizing morphogenetic protein is a novel tgf-beta homolog expressed in the spemann organizer. Development 121, 4293–4301.

Moses, M.A., Sudhalter, J., and Langer, R. (1990). Identification of an inhibitor of neovascularization from cartilage. Science 248, 1408–1410.

Nakao, T., Ishizawa, A., and Ogawa, R. (1988). Observation of vascularization in the spinal cord of mouse embryos, with special reference to development of boundary membranes and perivascular spaces. Anat. Rec. 221, 663–677.

Niehrs, C., Steinbeisser, H., and De Robertis, E.M. (1994). Mesodermal patterning by a gradient of the vertebrate homeobox gene goosecoid. Science 263, 817–820.

Pardanaud, L., Altmann, C., Kitos, P., Dieterlen-Lievre, F., and Buck, C.A. (1987). Vasculogenesis in the early qual blastodisc as studied with monoclonal antibodies recognizing endothelial cells. Development 100, 339–349.

Plate, K.H., Breier, G., Weich, H.A., and Risau, W. (1992). Vascular endothelial growth factor is a potential tumor angiogenesis factor in human gliomas in vivo. Nature 359, 845–848.

Poole, T.J., and Coffin, J.D. (1989). Vasculogenesis and angiogenesis: two distinct morphogeneic mechanisms establish embryonic vascular pattern. J. Exp. Zool 251, 224–231.

Regulski, M., Dessain, S., McGinnis, N., and McGinnis, W. (1991). High-affinity binding-sites for the deformed protein are required for the function of an autoregulatory enhancer of the deformed gene. Genes Dev. 5, 278–286.

Risau, W., and Flamme, I. (1995). Vasculogenesis. Annu. Rev. Cell Dev. Biol. 11, 73–91.

Risau, W., Sariola, H., Zerwes, H.G., Sasse, J., Ekblom, P., Kemler, R., and Doetschman, T. (1988). Vasculogenesis and angiogenesis in embryonic stem cell-derived embryoid bodies. Development 102, 471–478.

Samakovlis, C., Hacohen, N., Manning, G., Sutherland, D.C., Guillemin, K., and Krasnow, M.A. (1996). Development of the Drosophila tracheal system occurs by a series of morphologically distinct but genetically coupled branching events. Development 122, 1395–1407.

Sander, K. (1976). Formation of the basic body pattern in insect embryogenesis. Adv. Insect Physiol. 12, 125–238.

Schier, A.F., and Gehring, W.J. (1992). Direct homeodomain–DNA interaction in the autoregulation of the fushi tarazu gene. Nature 356, 804–807.

Schmidt, J.E., Dassow, G.V., and Kimelmann, D. (1996). Regulation of dorsoventral patterning: the ventralizing effect of the novel homeobox gene Vox. Development 122, 1711–1721.

Schmidt, U., Beyer, C., Oestreicher, A.B., Reisert, I., Schilling, K., and Pilgrim, C. (1996). Activation of dopaminergic d-1 receptors promotes morphogenesis of developing striatal neurons. Neuroscience 74, 453–460.

Schulte-Merker, S., and Smith, J.C. (1995). Mesoderm formation in response to brachyury requires fgf signaling. Curr. Biol. 5, 62–67.

Thoenen, H., and Barde, Y.A. (1980). Physiology of the nerve growth factor. Physiol. Rev. 60, 1284–1334.

Tornero, P., Conejero, V., and Vera, P. (1996). Phloem-specific expression of a plant homeobox gene during secondary phases of vascular development. Plant J. 9, 639–648.

Udolf, G., Lüer, K., Bossing, T., and Technau, G.M. (1995). Commitment of the CNS progenitors along the dosroventral axis of Drosophila neuroectoderm. Science 269, 1278–1281.

Vincent, J.P., and Lawrence, P.A. (1994). It takes three to distalize. Nature 372, 132–133.

Von Bubnoff, A., Schmidt, J.E., and Kimelman, D. (1996). The Xenopus-laevis homeobox gene xgbx-2 is an early marker of anteroposterior patterning in the ectoderm. Mech. Dev. 54, 149–160.

Weindel, K., Martinybaron, G., Weich, H.A., and Marme, D. (1994). Mitogenic and chemotactic response of endothelial cells to human recombinant vegf (121), vegf (165), plgf-1 amd plgf-2. J. Cell. Biochem. 1994, 322–322.

Wigglesworth, V.B. (1954). Growth and regeneration in the tracheal system on an insect Rhodnius prolixus (Hemipter). Quart. J. Microsc. Sci. 95, 115–137.

Wigglesworth, V.B. (1959). The role of the epidermal cells in the "migration" of tracheoles in Rhodnius prolixus (hemipter). J. Exp. Biol. 36, 632–640.

Wilting, J., Birkenhäger, R., Eichmann, A., Kurz, H., Martiny-Baron, G., Marné, D., McCarthy, J.E.G., Christ, B., and Weich, H.A. (1996). $VEGF_{121}$ induces proliferation of vascular endothelial cells and expression of flk-1 without affecting lymphatic vessels of the chorioallantoic membrane. Dev. Biol. 176, 76–85.

Wolpert, L. (1969). Positional information and the spatial pattern of cellular differentiation. J. Theor. Biol. 25, 1–47.

3.2

A Mechanical Theory of In Vitro Vascular Network Formation

J.D. Murray, D. Manoussaki, S.R. Lubkin, and R. Vernon

Introduction

Spatial Pattern Formation in Development

Understanding the evolution of spatial patterns and the mechanisms which create them are among the most crucial issues in developmental biology. Considerable progress has been made in understanding some of the basic principles that any mechanism must possess to be able to generate spatial patterns. In spite of this we still do not know, with any certainty, definitive details of a single pattern-formation mechanism which is involved in development. Model mechanisms—morphogenetic models—for biological pattern generation can suggest possible scenarios as to how pattern is laid down, and sometimes when [as, for example, in the experimentally confirmed case of stripe patterning on Alligator mississippiensis (Murray et al., 1990)] and how the embryonic form might be created. There has, in the past few years, been an increasing recognition among experimentalists and theoreticians that dramatic progress in biology could come about through a genuine interdisciplinary approach involving experimentalists and theoreticians. The book by Murray (1993) discusses in detail many successful case studies of such an interdisciplinary approach.

Embryogenesis depends on a series of sequential processes which generate specific patterns at each stage in development. The network pattern observed in angiogenesis is such an example, and how they are formed is an important and challenging question. This chapter describes a possible patterning mechanism firmly based on, and motivated by, detailed experiments carried out by Drs. Sage and Vernon and their co-workers in the Health Sciences at the University of Washington. These experiments were also used for comparison between the theory and experiment (see below).

Broadly speaking, there are two prevailing views of pattern generation that have influenced the thinking of embryologists in the past 15 years. One is the longstanding and well-known Turing (1952) chemical pre-pattern approach; the other is the more recent continuum mechanochemical approach developed by G.F. Oster and J.D. Murray and their colleagues (see, for example, Murray et al., 1983; Oster et al., 1983; Murray and Oster, 1984a,b). General descriptions and overview of pattern formation mechanisms are given in the book by Murray (1993).

Turing's (1952) theory of morphogenesis involves hypothetical chemicals—morphogens—which react and diffuse in such a way that if the chemical kinetics and the diffusion coefficients have certain properties, steady-state heterogeneous spatially patterned solutions in chemical concentrations can evolve. Morphogenesis then proceeds by the cells reading and reacting to the chemical pre-pattern and differentiating according to some bauplan, such as Wolpert's (1981) "positional information" concept. This considers the cells to have been pre-programmed to differentiate according to the underlying morphogenetic pre-pattern. This view of morphogenesis is essentially a slave process, since once the chemical pre-pattern is established all else follows. Turing's theory has stimulated a vast amount of research, both mathematical and experimental. Such reaction diffusion models have been widely studied and applied to a variety of biological problems; see, for example, the books by Meinhardt (1983; see also Chapter 3.1 in this book) and Murray (1993). Clear demonstration of the existence and identification of morphogens is still a major problem.

Another class of chemically-based models rely on chemotaxis and the response of cells to gradients in a chemoattractant. Several of these have had considerable success, such as in the patterning sequences exhibited by the slime mold Dictyostelium discoideum (see, for example, articles in the book edited by Othmer et al., 1993) and the complex bacterial patterns displayed by E. coli and Salmonella (Berg and Budrene, 1991; Woodward et al., 1995).

The Murray-Oster mechanochemical models of morphogenesis take a very different approach by directly bringing mechanical forces and known properties of cells and biological tissue into the process of morphogenetic pattern formation. Here pattern formation and morphogenesis go on simultaneously as a single process. The form-shaping movements of the cells and the embryological tissue interact continuously to produce the observed pattern. An important aspect of this approach is that the models are formulated in terms of measurable quantities such as cell densities, elastic forces, cell traction forces, tissue deformation, known chemicals, and so on. This focuses attention on the morphogenetic process itself and is more amenable to experimental investigation. The work described in this chapter is exclusively a mechanical theory.

A major use of any theory is in its quantitative or qualitative comparison with extant experiment, its biological predictions. Even though different theories might be able to create similar patterns, they are mainly distinguished by the different experiments they suggest and their direct connection with specific biological situations. Mechanical models lend themselves to experimental scrutiny more readily than reaction diffusion models. Also, a major point in favor of simultaneous development is that such mechanisms have the potential for self-correction. Embryonic development, which proceeds sequentially, is usually a stable process with the embryo capable of adjusting to many outside disturbances. The process whereby a pre-pattern exists and then morphogenesis takes place is effectively an open loop system. These are potentially unstable processes and make it difficult for the embryo to make the necessary corrective adjustments as development proceeds. It is likely that both types of models—chemical and mechanical—are involved in development, but, until more is known about the morphogens involved, it seems that, at this stage, mechanical models can indicate specific and measurable experimental activity to help elucidate the underlying mechanisms involved in organ morphogenesis.

Mechanical Forces in Vessel Remodeling

Mechanical and fluid mechanical forces play important roles in the overall development of vasculature. As early as 1893, Thoma suggested the importance of mechanical factors in the growth of blood vessels during development. In a review of his and other works on mechanical forces on angiogenesis, Hudlická and Brown (1993) describe Thoma's observations of how vascular sprouting in the growing embryo might occur as a result of a combination of velocity of blood and pressure. Since then a large number of studies has revealed a number of effects that mechanical forces play in the development of vasculature: it may also play a role in the early branching in the developing lung (Lubkin and Murray, 1995).

The study of how mechanical forces influence the development of vasculature has been organized along two main lines. The first has focused on describing the macroscopic aspects of vessel remodeling. The term "vessel remodeling" has been used to describe changes in vessel wall structure as well as changes in the vascular tree, generated by capillary generation and regression. It is known, for example, that changes in blood pressure induce changes in the thickness of various vascular structures. Elevated pressure in pulmonary arteries during hypertension, for example, results in the thickening of the intima and adventitia layers (Fung and Liu, 1991). Blood volume overload together with bradycardia are important for angiogenesis during cardiac hypertrophy (Brown et al., 1994; Tomanek and Torry, 1994). Moreover, changes in the vessel wall stress could be important in remodeling of the vascular plexus (Price and Skalak, 1994). Vascular shear stresses or blood flow patterns have also been suggested as one of the causes of vascular network remodeling via intussusception (Patan et al., 1996).

The second approach has concentrated on elucidating the mechanism by which mechanical forces applied to vascular cells alter gene expression (Ando and Kamiya, 1996) and the molecular effects on the signaling pathway. Shear stress resulting from blood flow has been shown to change cell shape and affect the cells' cytoskeletal organization (Dewey et al., 1981; Wechezak et al., 1985), which, in turn, can influence the cell cycle (Ingber et al., 1995). Increased capillary pressure has been found to make cells sensitive to mitogenic factors (Folkman and Greenspan, 1975) and shear stress will stimulate endothelial cell DNA synthesis (Ando et al., 1990) and the migration and proliferation of endothelial cells (Ando et al., 1987).

Reviews on some of the roles of mechanical forces in vascular development and remodeling are given, for example, in Hudlická (1984), Hudlická and Brown (1993) and Skalak and Price (1996).

Cell–Extracellular Matrix Interactions for Vasculogenesis

The development of in vitro angiogenesis systems provides a controlled means of studying vessel formation (Folkman and Haudenschild, 1980). Such systems have shown the important mechanical role that the extracellular matrix (ECM) plays in angiogenesis. It provides molecules for cellular adhesion onto the matrix; it provides a substrate for cell spreading, a process shown to be important in the control the cell cycle (Ingber et al., 1995) and it provides a scaffold necessary for cell migration and morphogenesis (Markwald et al., 1979; Vernon et al., 1992). Cells cannot only produce or degrade their ECM, but also alter its structure by applying mechanical forces. Through matrix production and degradation, cells can influence the mechanical properties of their ECM. Through mechanical forces, they can reorganize its fibrous components into matrix lines that the cells use as migratory pathways. Scenarios as to how cell-ECM interactions can be orchestrated to form complex spatial patterns in development is the main thrust of the Murray-Oster mechanical theory of biological pattern formation (Murray et al., 1983; Oster et al., 1983; Murray, 1993).

Mechanical interactions between the cells and their ECM have been important, for example, in the development of epicardiac cushion and the heart valves. Markwald et al. (1979) suggested that these structures appear in the developing embryo as the cardiac jelly is organized by endocardiac cells in "tracks" that serve to guide the cells away from the endocardium and toward the myocardium. They observed cells deforming the interstitium and aligning the matrix fibers, thus generating pathways on which the cells would then migrate (Markwald et al., 1979).

In vitro angiogenesis systems have exhibited a similar mechanism through which endothelial cells reorganize into vascular networks (Ingber and Folkman, 1989). The networks that form in vitro resemble the development of vasculature in the early quail embryo (Poole and Coffin, 1987), and the in vitro mechanism was considered as a possible model for the study of the in vivo appearance of the vascular plexus (Vernon et al., 1992; Vernon and Sage, 1995).

Recent studies have also helped to further elucidate the mechanism of vascular network formation in vitro (Vernon et al., 1992). This work has shown that networks are not specific to endothelial cells, since a variety of traction exerting cells, not necessarily endothelial cells, can form networks when cultured on gelled basement membrane (Matrigel). Moreover, cells cultured on different substrates (Matrigel, type I collagen gels) also form networks, provided the matrix is malleable enough. These studies suggest that the mechanism lacks cellular/matrical specificity and that a physical/mechanical mechanism could possibly explain the generality of the network-forming process.

Here we describe a continuum mathematical model mechanism which attempts to capture the interaction between the mechanical forces generated by the cells and their extracellular matrix. Our aim is to demonstrate that a purely mechanical mechanism could be responsible for the observed pattern and how they are actually formed in development. Such a mechanical model does not specify the type of cells and matrix involved but rather only considers the possible mechanical interactions between the various components. We test the model against the patterns observed by particular experiments and we further use the model to quantitatively describe the interaction between the various mechanical properties and the network and their pattern-forming abilities.

A Model of In Vitro Vascular Network Formation

Experiments by Vernon, Sage, and their co-workers show that bovine aortic endothelial cells (BAEC), cells of the murine Leydig cell line TM3, adult human dermal fibroblasts, and human smooth muscle cells, when cultured on Matrigel, are observed to form networks (Vernon et al., 1992). The process for all cell lines is similar: cells adhere onto the matrix and start pulling on it. The pulling results in movement of the matrix and the cells that have adhered to it, eventually forming cell aggregates with significant amounts of matrix accumulated underneath the aggregates. As a result, cellular traction tension lines appear around the clusters. Such tension lines have also been observed as a result of traction forces that fibroblasts exert onto silicone rubber (Harris et al., 1980). In time, the matrix components appear to form fibrous lines between neighboring clusters and along the tension lines. Once the lines form, the cells become actively motile and migrate onto the matrix pathways, thus forming cellular cords. Eventually, the culture plane is tessellated with polygons with sides defined by the cellular cords (Vernon et al., 1992).

Network formation is not particular to Matrigel. BAE cells and TM3 cells form networks on hydrated gels of 0.12–0.16 mg/ml collagen I. Collagen content influences the formation of networks, presumably by altering matrix stiffness and thereby the effect of the cell-exerted traction onto the matrix. Moreover, thickness of the matrix layer also influences the size of the networks formed: thin layers of matrix resulted in small or no network formation. On a ramp of increasing matrix thickness, larger cellular networks form on the thicker areas of matrix.

These results suggest that mechanical interactions are of primary importance for the development of pattern. Mathematical descriptions of the interactions, which we describe below, confirm the crucial role that mechanical forces play in the network pattern formation and provide a tool for assessing which mechanical properties of the ECM control the patterning process, as well as how the parameters may modulate pattern size. It also provides various experimental scenarios to highlight the effect of experimentally variable parameters, such as gel thickness, density of cells, and so on.

Mathematical Model

Our model is based on the Murray-Oster mechanochemical theory, which provides a detailed and quantitative mathematical description of possible cell–matrix interactions (Murray et al., 1983; Oster et al., 1983). The continuum model mechanism we present here describes the cell–extracellular matrix mechanical interactions. The matrix is considered to be a monophasic continuum and accounts for the hypothesized rheological properties of the matrix. The mechanism includes cellular traction exerted on the matrix; cell proliferation and cell death, as well as secretion of chemicals and the cellular response to them can also be included in the mechanochemical theory.

In the simple model here we keep only the mechanical aspects of the extracellular matrix–cell interaction to highlight their role in the genesis of vascular networks. In the next section, we briefly describe the general assumptions behind the mathematical formulation of forces in the mechanistic model, and sequester the specific model equations in the Appendix.

Cell Movement

We denote the average local cell density by the variable n (cells/mm^2). In our model we do not include cell proliferation: on the Matrigel cultures the first cell aggregates and lines of tension appear after 4 hours and the networks are complete within about 24 hours. The time between two subsequent cell mitoses, however, is about 17 hours for endothelial cells. We therefore assume that no significant changes in the cell populations occur which could influence the pattern-forming process. Models including cell and matrix proliferation are currently under investigation.

Local changes in the cell density occur mainly as a result of two different mechanisms for cell movement: (i) cells move passively together with the matrix, as the matrix is moving across the dish (the process we call convection); and (ii) cells locomote (active motion) along areas of aligned matrix fibers. We model these two processes mathematically via a cell density conservation equation:

$$\text{rate of change in cell density} = \text{convective flux} + \text{flux due to active movement} \qquad (1)$$

The specific quantification of these terms is given in the Appendix. We initially only consider the convective movement of cells to test the formation of networks solely by the mechanical, matrix-distorting forces. Subsequently, we model cell locomotion as a random motion (diffusive movement), biased toward areas of matrix alignment.

Forces Within the Extracellular Matrix: Force Balance Equation

The matrix, Matrigel, used in the experiments described is derived from the Engelbreth-Holms-Swarm tumor sarcoma. It contains collagen IV, laminin, and fibronectin. Under cell-exerted traction it deforms and is dragged over large distances in the dish. While it deforms, the collagen fibrils probably polymerize, to eventually yield the cord-like matrix structures that arise during the intermediate stages of the network development. It therefore changes its properties while it deforms. Due to lack of material descriptions for Matrigel or collagen I gels in the strain rate of relevance, we restrict ourselves to describing the initial stages of matrix deformation and before significant changes in the matrix structure have occurred.

The movement of the matrix is, of course, resisted by the attachment of the matrix onto the dish. In vivo planar angiogenesis is, by comparison, influenced by the attachment of the fibrous and cellular components to the adjacent tissue structures.

The key forces are, of course, those generated by the cells. The others are generated as a direct consequence. So, forces that are present in the tissue are: (i) the cell-exerted traction; (ii) the resistance due to the matrix–dish contact; and (iii) the viscoelastic forces of the matrix material which are resisting the deformation. Since the time required for generation of the pattern is relatively very long and the size of the pattern is small in absolute terms, inertia effects are negligible and the forces at any given point are considered to be in equilibrium.

Cell Traction

The cell-exerted traction, F_{cells}, within a certain area depends on the traction per cell (which we denote by τ) as well as on the local cell density. More cells within an area exert a larger total traction force. However, cell size limits the number of cells that can be in a certain area and so the total traction that can be exerted within the area saturates. This, in effect, cell–cell inhibition as regards force generation gives rise to a saturating function for cell-generated traction forces in the model equations.

Traction per cell depends on the related cellular cytoskeletal components and the transmembrane cell-surface molecules (integrins) (Vernon and Sage, 1995) that link the cell's cytoskeleton to the ECM, but we do not explicitly model these dependencies.

Matrix Viscoelastic Response

In response to a force, the matrix is displaced slowly due to its viscous properties. If strains developed in collagen gels are small (less than 10%), then the gels have been found to respond linearly to applied stresses (Barocas et al., 1995). We therefore describe the matrix response as that of a linear viscoelastic body (see Appendix for further discussion).

Matrix Drag Across Dish

As the cells pull the matrix, it moves across the dish. Experiments indicate that some matrix fibrils remain attached to the dish, while the rest are dragged across the lower parts of the matrix. The net effect of the attachment of the fibrils on the dish is a resistance to the movement of the matrix, which, in a planar situation, we treat as a resistive viscous drag.

The equation reflecting the balance of these forces is then

$$\underbrace{F_{cells}}_{\text{cell-exerted traction}} + \underbrace{F_{anchoring}}_{\text{attachment on dish}} + \underbrace{F_{matrix}}_{\text{viscoelastic restoring forces}} = 0 \tag{2}$$

The word Eqs. (1) and (2) constitute the model mechanism for mechanical generation of the network. As mentioned above, the quantification of the various terms is given in the model equations in the Appendix. There are, of course, several parameters in the model and these play crucial roles in whether or not a network is generated, its geometric characteristics, the time for formation, and so on.

Parameter Values

The various parameters describing cell behavior and matrix properties, for example, its stiffness (Young's modulus), the Poisson ratio, and the cellular random motility coefficient, may depend on either or both the stress or strain (and therefore fiber orientation) of the matrix, which means that their value within a culture varies, especially at the later stages of the pattern appearance. This is a property of many parameters of biomaterials, and, as such, makes estimates difficult in the absence of sophisticated experimentation. Some characterization of the material properties can be achieved by employing appropriate creep and relaxation experiments for the matrix (Fung, 1993).

However, studies on Matrigel have not been reported. Studies of the rheological properties of collagen gels have been done by subjecting the gel to oscillations that are generally larger than 0.1 Hz. However, rheological properties of the gels depend on the straining rate, and 0.1 Hz is a straining rate significantly larger than the one occurring in experiments involving cell–matrix interactions that we consider. Few studies have used the appropriate creep, or oscillatory shear testing, for our case dealing with smaller straining rates. Rheological properties of collagen I gels (Barocas et al., 1995) and tissue-derived gels such as the vitreous body (Lee et al., 1993, 1994a) have been described by applying these slower-strain techniques (Barocas et al., 1995; Lee et al., 1993). We therefore use such results where they refer to tissues of approximately similar structure to the matrices used in the experiments by Vernon et al. (1992).

The stiffness of the matrix is represented by the elastic modulus of the material, E (dynes/cm^2), and is determined primarily by the type, amount, and organization of its fibrous components. Collagen is an important contributor to the gel matrix stiffness, but the proteoglycans and soluble molecules that it contains can also influence the elastic modulus in a significant way (Lee et al., 1994b). For example, the anterior porcine vitreous body and the posterior bovine vitreous body were found to have similar collagen contents (0.156 mg/ml, 0.146 mg/ml) but different concentrations of hyaluronic acid and electrolytes; their elastic moduli were 26.93 dyne/cm^2 and 8.01 dyne/cm^2, respectively (Lee et al., 1994a). Clearly, hyaluronic acid and the electrolytes play an important role, but the exact interactions of the various ECM components in regulating its stiffness have not yet been completely explained.

Since the collagen gels used in the vascular network-forming experiments have collagen content 0.12–0.16 mg/ml (Vernon et al., 1992) comparable to that of the vitreous bodies mentioned, we assume that these gels' elastic moduli are within a similar range.

The zero shear viscosity of 2.1 mg/ml collagen I gel has been measured using creep tests to be 7.4×10^6 Poise (Barocas et al., 1995), while that of the posterior bovine vitreous body was estimated to be 2.5×10^2 Poise (Lee et al., 1994a). It is possible that the disparity in values is due to the differences in composition of the two gels (the former has collagen content that is in weight 14 times larger). The effect of the viscosity (μ_1, μ_2, Poise) on the predicted pattern forming process is studied for a range of possible values.

Traction per cell (τ dynes/cell) we consider to be comparable to that of human umbilical vein endothelial cells, which is approximately 6.1×10^4 dynes/cm^2 (Kolodney and Wysolmerski, 1992). The values were calculated for a confluent monolayer of cells. Estimating the cell density from the figures presented at $2.25 \times 10^5 - 4 \times 10^5$ cells/cm^2, we calculate the traction force per cell to be in the region of $\tau \approx 0.15 - 0.27$ dynes/cell.

The Poisson ratio (v cm/cm) is another parameter characterizing a material. It can be theoretically measured by extending a material by a certain amount and measuring its compression on the perpendicular direction. The ratio of relative extension to compression on the perpendicular direction gives the magnitude of the parameter. In practice, such measurements for soft materials such as collagen gels can be very difficult. This is because the material is both difficult to handle mechanically, and the value of v changes in time as the fibrous component may separate from the water component of the gel. Each of the two different phases, solid and aqueous, will have a different Poisson ratio. The Poisson ratio for the fiber network of such gels, which is draining of water, is estimated at $v \approx 0.2$ (see Scherer et al., 1991).

Cell motility for endothelial cells has been calculated using migration assays (Hoying and Williams, 1996). The values reported for endothelial (human microvessel endothelial cells) on gelatin were $9.5 \pm 1.2 \times 10^{-9}$ cm^2/sec in the presence of endothelial cell growth factor and $2.6 \pm 0.6 \times 10^{-9}$ cm^2/sec in its absence, and $19.3 \pm 4.22 \times 10^{-9}$ cm^2/sec on fibronectin (Hoying and Williams, 1996; motility remained high with no endothelial growth factor).

In our study, we are concerned with the initial stages of network pattern emergence where the strains are still small and we can assume that the parameters stay approximately constant.

Results

The model mechanism system of equations (see Appendix) was analyzed using mathematical, analytical, and numerical techniques described elsewhere (Manoussaki et al., 1996). Here we present and discuss the results of the analysis: the resulting relations between the various model parameters (stiffness, cell traction, viscosity, Poisson ratio) and how these are likely to influence the formation of networks.

The first indication of spatial pattern generation consists of small displacements in cells and matrix. We asked whether small variations within the cell distribution could cause certain regions to comply to cellular forces more than others and thereby influence the displacement of the matrix. Mathematical analysis showed that small variations in cell density could trigger matrix displacements, even in the absence of active cell movement, if the following relation between the model (and real biological) parameters is true:

$$\frac{\tau}{E} > \frac{1}{1-v^2} \cdot \frac{1}{n_0} \cdot \frac{(1+\alpha\, n_0^2)^2}{1-\alpha n_0^2} \tag{3}$$

where E is the matrix stiffness, τ is the traction per cell, v is the Poisson ratio, n_0 is the initial cell seeding density, and α is a parameter determining the decrease in total cell traction upon confluence and is chosen so that $\alpha N^2 \approx 0.1 - 0.4$ where N is the cell density at confluence. Since the seeding density is an order of magnitude less than N, the last term of the inequality (3) is close to 1, and the inequality takes the approximate form

$$n_0 \frac{\tau}{E} > \frac{1}{1-v^2} \tag{4}$$

Of course, a restriction on n_0 is that the initial seeding density is less than the density at confluence $n_0 < N$. Relation (4) points at a simple rule that might make patterns possible: patterns are more likely to form if the cell traction τ is sufficiently large, or if the matrix stiffness E is sufficiently low or if the initial seeding density is relatively large (the relation, however, does not guarantee the formation of pattern). The results are in broad agreement with experimental findings: when cell traction was inhibited or matrix stiffness greatly increased, no patterns formed (Vernon et al., 1992). On the other hand, increases in the cell plating densities facilitates the formation of pattern. If not many cells are seeded, then they have difficulty in deforming the matrix and starting the patterning process.

The above inequalities predict when pattern could form, but do not predict the type of pattern formed. The pattern described by the equations can be determined by solving the governing equations numerically. Numerical simulations show that indeed formation of networks is possible through such a purely mechanical process (Fig. 1).

The results observed are the following. The numerical simulations clearly showed network formation and subsequent remodeling: some polygons grew larger and others closed in, in a sphincter-like manner. There was a direct correspondence between areas of large cell density and areas of high matrix density, just as we would expect.

The choice of parameters influenced the formation of pattern and rate at which the network pattern formed. If traction was too low, the plating density too low, or the stiffness too high, no patterns formed. If, on the other hand, stiffness was too low or traction too high, cells and the matrix formed clusters but no matrix cords were observed.

Viscosity of the matrix, as to be expected intuitively, influenced the time it took for the pattern to appear, provided traction and stiffness were within the appropriate range, but did not determine whether pattern would form or not.

Importantly, we found that random motility of cells was not necessary for the formation of pattern. Networks would still form, provided the seeding density was sufficiently large.

To test the model further, we simulated the experiment of cells grown on a ramp of increasing matrix thickness. On areas of greater thickness, the polygons that formed as a numer-

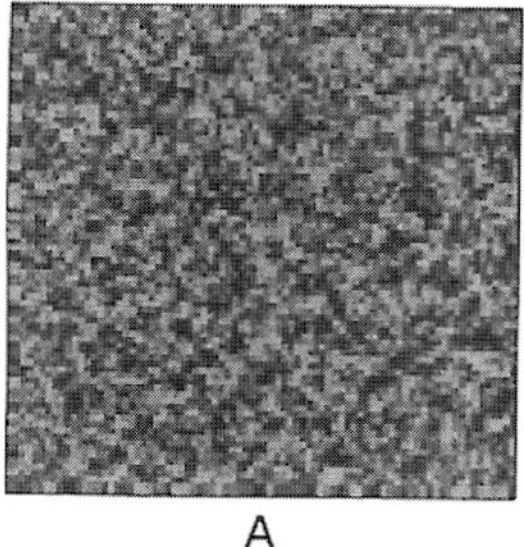
A

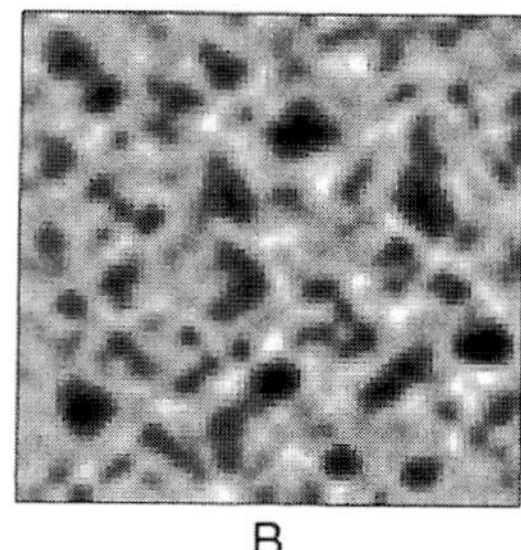
B

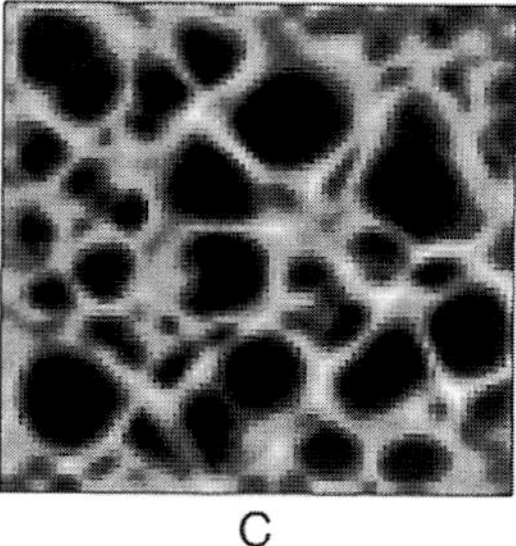
C

Fig. 1 Numerical results of the model equations. The figure shows the cell density distribution across the dish. White areas denote areas of large cell densities and dark areas denote regions of low cell densities. Initially the cells are approximately uniformly distributed throughout the dish, with some small random variations in the density from one point to the next. With time, some cells cluster and other areas become less populated. Small sized polygons appear, some of which grow in size while others close in, in a sphincter-like manner. From left to right, figures represent pattern at 7 hr (A), 20 hr (B), and 24 hr (C). The size of one square corresponds to 800 μm. The parameter values used in the simulations were: $\tau = 0.06$ dynes/cell, $E = 20$ dynes/cm^2, $\mu_1 = 1.5 \times 10^7$ Poise, $\mu_2 = 0.9 \times 10^7$ Poise, $D = 0.46 \times 10^{-11}$ cm^2/sec, $v = 0.2$, $s = 10^{10}$ dynes·sec/cm^3, $n_0 = 10^5$ cells/cm^2, $\alpha = 4 \times 10^{-13}$. The distribution of matrix is similar to that of the cell densities (not shown).

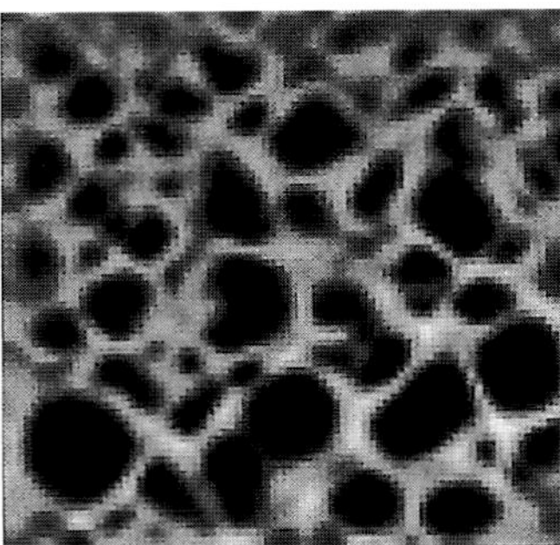

Fig. 2 Numerical simulations of cellular networks formed on a ramp of matrix. On the top we had initially a matrix that was a tenth of the thickness of the matrix on the lower side of the square. Smaller networks form in the presence of a thinner matrix (top side). Parameter values are the same as in Fig. 1.

ical solution to the equations were clearly larger than those on areas of thinner matrix. We thus felt a certain confidence that the equations were able to capture the actual mechanical interaction between cells and matrix (Fig. 2). Although the model presented is valid for small strains, we present simulations with large strains and for long times.

Discussion

Mathematical analysis and simulation of the model mechanism give insight as to which elements are crucial to the development of pattern, help assess the relative importance of such elements, and predict how changes in these factors affect pattern growth and shape. In this way, the model can be useful in assessing the biological conditions under which a pattern may or may not form.

We have presented a simple, purely mechanical model for network pattern formation in vitro, in which cellular traction forces play a crucial role in forming the planar pattern of cellular aggregates, cords, and matrix deformations observed in vitro. In the simulations we chose to consider some important biological parameters as being constant throughout the process. For example, we assumed that the traction a cell exerts remains constant with time and with matrix properties. We know, however, that cell traction can vary with time, and is influenced by the cells' ability to attach to the matrix. We also know that the matrix properties will change as cells remodel the matrix, by reorganizing the fibers and also by secreting proteases that degrade it. Cells may also secrete other molecules that could have an effect on the mechanical properties of the matrix.

In spite of the simplifying assumptions in the model, the results indicate several key aspects. In the absence of traction, as in the in vitro system, no pattern forms, supporting our hypothesis that the pattern forms as a result of cellular traction forces. The evolution of the pattern from the model proceeds in the same manner as the in vitro pattern, with irregular polygons both increasing in size and decreasing in number, with the smallest polygons pinching off and vanishing.

As in the in vitro system, pattern forms in our system only when the ratio of traction forces and seeding cell density to gel stiffness is above a certain threshold. Our model results confirm that on thicker gel, larger polygons will form. It also indicates that isotropic cellular traction is sufficient for pattern formation. We further found that if cellular traction is sufficiently high or the plating densities large, then traction forces alone can lead to formation of networks. Biased cellular migration may be a component of pattern formation in vivo or in vitro, but we have shown it is not a necessary feature.

We also found that matrix thickness is an important factor influencing the pattern. Cells spread on a matrix whose thickness increased from one end of the dish to the opposite, and formed polygons whose size increased with increasing thickness of the gel. The numerical simulation results are in agreement with the experimental results of Vernon et al. (1992).

The model mechanism is the first mathematical description of cell–matrix interactions for the formation of vascular networks in which all of the component variables are, in principle, measurable. The theory can be extended to account for the effect of cell-secreted molecules on the matrix material properties, cell proliferation, and cell secretion, thus providing an analytical tool for assessment of the relevant importance of these factors during angiogenesis. It can also be used to study the effect on the pattern if there is spatial variability in the parameters. We believe that the success of the current model in mimicking experimental observations gives some reason for optimism that it can be used as a guide to predict the outcome of other experiments and, optimally, the consequences of experimentally disturbing the mechanism which obtains in vivo.

Acknowledgments

The authors would like to thank Mr. E.A. Hinkle for his skillful assistance in the preparation of this manuscript. This work was supported in part by grants DMS-9500766 from the National Science Foundation (SRL, JDM, DM), DMS-9306108 from the National Science Foundation (SRL), and 2 P41 RR01243-12 and 2P41 RR01243-17 from the National Institutes of Health (JDM).

Appendix

Mathematical Model

Here we quantify in mathematical terms the word Eqs. (1) and (2) in the model mechanism described above. Fuller details are given in Manoussaki et al. (1996).

Local changes in the cellular density are determined by a combination of two movements:

(i) A convective flux $J_{convection} = vn$ where v(x,y,t) denotes the matrix velocity at the point (x,y) in space, at time t. Here, $v = u_t$, where u is the displacement of the matrix.

(ii) A flux due to cellular locomotion. The process is modeled via an anisotropic strain-dependent random motion tensor. We thus consider that cell movement can be approximated by a random walk that is biased along areas of matrix alignment.

The resulting conservation equation for the cell density is:

$$\underbrace{\partial n/\partial t}_{\text{rate of change in cell density}} = \underbrace{-\nabla\cdot(vn)}_{\text{convection}} + \underbrace{\nabla\cdot\nabla\cdot(D(\varepsilon)n)}_{\text{strain-biased random motion}} \qquad (5)$$

where v denotes the velocity with which the cells move as they ride on the matrix; it is the same as the matrix velocity. In this equation, therefore, D(ε) is the random motion tensor, dependent on the matrix strain ε, whose particular form depends on the specific assumption about how cells perform their random movements. Its specific form for small strain was derived under the assumption that the movement bias increases in expansion along one direction and/or compression in the perpendicular direction (Cook, 1995):

$$D(\varepsilon) = D_0 \begin{pmatrix} 1 + \dfrac{\varepsilon_{xx} - \varepsilon_{yy}}{2} & \dfrac{\varepsilon_{xy} + \varepsilon_{yx}}{2} \\ \dfrac{\varepsilon_{xy} + \varepsilon_{yx}}{2} & 1 - \dfrac{\varepsilon_{xx} - \varepsilon_{yy}}{2} \end{pmatrix} \quad (6)$$

Here ε_{xx}, ε_{yy}, ε_{xy}, ε_{yx} represent the components of the strain tensor, ε, and D_0 is the motility coefficient when no strain is present. In the initial analysis we did not consider strain-dependent active movement so as to test whether networks are possible under just cell-exerted traction.

Forces Within the Extracellular Matrix

By considering only small strains, we can model the matrix as a viscoelastic material whose properties remain constant during the initial stages of the patterning process.

Since the width of the matrix sample is much larger than the thickness of the matrix, we approximate the matrix as a two-dimensional material. Moreover, we assume that the traction exerted by the cells remains confined in the plane parallel to the dish (or, approximately, the matrix surface). In engineering terms, we could say that we consider the material to be under a plane stress assumption. We assume that the presence of cells does not affect the deformations of the matrix (that is, it is not viewed as a two-phase medium) and therefore the strain field through their own volume. The movement of the matrix is considerably resisted by the attachment of the matrix to the dish. In vivo planar angiogenesis is, by comparison, influenced by the potential attachment and continuation of the fibrous and cellular components of the adjacent tissue.

The forces that are present in the tissue are: (i) the cell-exerted traction; (ii) the resistance due to the matrix-dish contact; and (iii) the viscoelastic forces of the matrix material which are resisting the deformation caused by the cells. Since the time required for generation of the pattern is very long and the size of the pattern is small in absolute terms, inertia effects are negligible and the forces at any given point are considered to be in equilibrium so we have (from Eq. 2 above):

$$\underbrace{F_{cells}}_{\text{cell-exerted traction}} + \underbrace{F_{anchoring}}_{\text{attachment on dish}} + \underbrace{F_{matrix}}_{\text{viscoelastic restoring forces}} = 0 \quad (7)$$

We consider the cell-exerted forces and the matrix response as arising from the corresponding stresses within the matrix medium. We set $F_{cells} = \nabla \cdot \sigma_{cells}$, with σ_{cells} the stress tensor in the matrix due to the cells and $F_{matrix} = \nabla \cdot \sigma_{matrix}$ with σ_{matrix} the viscoelastic stress tensor in the matrix material. $F_{anchoring}$ is an external (body) force resisting the matrix displacement.

Cell Traction. The cell-exerted traction F_{cells} depends on the local cell density as described earlier. We write the cell-exerted stress on the matrix due to the cells as

$$\sigma_{cells} = \tau \frac{n}{1 + \alpha n^2} \quad (8)$$

The parameter τ (dynes/cell) represents the stress that one cell exerts on the matrix at low cell densities and α is the parameter discussed earlier. For such densities with which we are concerned, the traction force is approximately equal to $\sigma_{cells} = \tau n$. We assume a decrease in cell traction for unrealistically large cell densities in order to prevent such densities.

Matrix Viscoelastic Response. For the viscoelastic stresses arising in the matrix σ_{matrix} we assume a Voigt model (see, for example, Fung, 1994): in response to a force, the matrix will displace slowly due to its viscous properties. The displacement of the matrix is denoted by u(x,y,t) and also by $(u_1(x,y,t),u_2(x,y,t))$, since displacement is a two-dimensional vector quantity. We assume that if the strain

$$\varepsilon = \frac{1}{2}(\nabla \cdot u + \nabla \cdot u^T)$$

arising from the matrix displacements is small, then the material will tend to return to its original position. Under these assumptions, the material stress σ_{matrix} is written as a sum of its elastic and viscous components:

$$\sigma_{matrix} = \sigma_{viscous} + \sigma_{elastic} \tag{9}$$

The viscous component is proportional to the rate of change of strain ε in the matrix:

$$\sigma_{viscous} = \mu_1 \varepsilon_t + \mu_2 \theta_t I \tag{10}$$

The parameters μ_1,μ_2 (dynes/sec) are the bulk and shear viscosities of the matrix and $\theta = \nabla \cdot u$ is the dilation of the matrix along the plane of the dish.

Under the plane-stress assumption, the elastic stress component is written as:

$$\sigma_{elastic} = E'(\varepsilon + v'\theta I) \tag{11}$$

where $E' = E/(1 + v)$ and $v' = v/(1 - v)$. E is the Young's modulus, and measures the stiffness of the material; v (mm/mm) is the Poisson ratio which measures how much a strip of gel will contract in one direction when it is stretched by a unit length on the transverse direction.

Matrix Drag Across Dish. As the cells pull the matrix, it moves across the dish. Experiments indicate that some matrix fibrils remain attached to the dish, while the rest are dragged across the lower parts of the matrix. In a two-dimensional model, the resistance to the movement of the matrix that arises from the fibril attachment to the dish is modeled as a viscous drag:

$$F_{anchoring} = -s \frac{u_t}{\rho} \tag{12}$$

where $\rho(x,y,t)$ (mm) represents the thickness of the matrix and s is a measure of the strength of the resistance contributed by the attachment. We thereby assume that the resistance is proportional to the velocity $v = u_t$ of displacement of the matrix.

Matrix Thickness

Since $\sigma_{zz} = 0$, we get a relation between the strain components in the three directions, ε_{xx}, ε_{yy}, ε_{zz}. From Hooke's law in three dimensions we have

$$\sigma_{zz} = \frac{E}{1+v}\,[(1-v)\varepsilon_{zz} + v(\varepsilon_{xx} + \varepsilon_{yy})] = 0$$

from which we get

$$\varepsilon_{zz} = \frac{-v}{1-v}\,(\varepsilon_{xx} + \varepsilon_{yy}) \tag{13}$$

When there is a strain ε_{zz} in the z-direction, the thickness is calculated using $\rho(x,y,t) = \rho_0(x,y)(1 + \varepsilon_{zz})$ from which we derive the relation giving the thickness ρ as a function of space and time:

$$\rho(x,y,t) = \rho_0(x,y)(1 + \varepsilon_{zz}) = \rho_0\left[1 - \frac{v}{1-v}\,(\varepsilon_{xx} + \varepsilon_{yy})\right]$$

$$\Rightarrow \rho(x,y,t) = \rho_0\left(1 - \frac{v}{1-v}\,\theta\right) \tag{14}$$

where $\theta = \varepsilon_{xx} + \varepsilon_{yy}$ is the dilation.

Boundary Conditions for the Equations

Experiments have shown that the matrix does not move very much along the edge of the dish. This we incorporate into the model simulation by assuming zero displacement as boundary condition: hence, for a square dish (square rather than circular simply for ease of numerical programming and reduction of computer time) of dimensions [ab] × [ab], the condition becomes:

$$u(x,y = a,t) = u(x,y = b,t) = u(x = a,y,t) = u(x = b,y,t) = 0 \tag{15}$$

where u(x,y,t) denotes displacement at position x = (x,y) and at time t. The cells are restricted to remain within the dish. This gives rise to the no-flux boundary condition for the cells across the boundary:

$$J_{\text{cells}} = vn - \nabla \cdot (D(\varepsilon)n) = 0 \tag{16}$$

Equations (5) and (7) constitute the model mechanism with the various terms given by the other Eqs. (6) and (8)–(14). The equations were solved with boundary conditions (15) and (16) with initial conditions in cell density and matrix distribution in keeping with the experimental set-up.

References

Ando J. and Kamiya A. Flow-dependent regulation of gene expression in vascular endothelial cells. Japanese Heart Journal, 37(1): 19–32, 1996.

Ando J. Komatsuda J. Ishikawa, C. and Kamiya, A. Fluid shear stress enhanced DNA synthesis in cultured endothelial cells during repair of mechanical denudation. Biorheology, 27(5): 675–684, 1990.

Ando J., Nomura H., and Kamiya A. The effect of fluid shear stress on the migration and proliferation of cultured endothelial cells. Microvascular Research, 33(1):62–70, 1987.

Barocas V.H., Moon A.G., and Tranquillo R.T. The fibroblast-populated collagen microsphere assay of cell traction force—Part 2: Measurement of the cell traction parameter. Journal of Biomechanical Engineering, 117:161–170, 1995.

Berg H., and Budrene E. Complex patterns formed by motile cells in E. coli. Nature, 349:630–633, 1991.

Brown M.D., Davies M.K., and Hudlická O. Effect of long-term bradycardia on heart microvascular supply and performance. Cellular and Molecular Biology Research, 40(2): 137–142, 1994.

Cook J. Mathematical Models for Dermal Wound Healing: Wound Contraction and Scar Formation. Ph.D. thesis, University of Washington, 1995.

Dewey C.F. Jr., Bussolari S.R., Gimbrone M.A., and Davis P.F. Jr. The dynamic response of vascular endothelial cells to fluid shear stress. Journal of Biomechanical Engineering, 103: 17–84, 1981.

Folkman J. and Greenspan H.P. Influence of geometry on control of cell growth. Biochimica et Biophysica Acta, 417(3–4):211–213, 1975.

Folkman J. and Haudenschild C. Angiogenesis in vitro. Nature, 288:551–556, 1980.

Fung Y.C. Biomechanics. Mechanical Properties of Living Tissues, 2nd edition. Springer-Verlag, New York, 1993.

Fung Y.C. and Liu S.Q. Changes of zero-stress state of rat pulmonary arteries in hypotoxic hypertension. Journal of Applied Physiology, 70(6):2455–2470, 1991.

Harris A.K., Wild P., and Stopak D. Silicone rubber substrata: A new wrinkle in the study of cell locomotion. Science, 208:177–179, 1980.

Hoying J.B. and Williams S.K. Measurement of endothelial cell migration using an improved linear migration assay. Microcirculation, 3(2), 167–174, 1996.

Hudlická O. Development of microcirculation: Capillary growth and adaptation. In American Handbook of Physiology, volume IV of sect. 2: The Cardiovascular System, chapter 5, pages 165–205. American Physiological Society, Bethesda, MD, 1984.

Hudlická O. and Brown M.D. Physical forces and angiogenesis. In Mechanoreception by the Vascular Wall (G.M. Rubanyi, editor), chapter 10, pages 197–241. Futura Publishing Company, Inc., Mount Kisco, NY, 1993.

Ingber D. and Folkman J. Mechanochemical switching between growth and differentiation during fibroblast growth factor-stimulated angiogenesis in vitro: Role of extracellular matrix. Journal of Cell Biology, 109:317–330, 1989.

Ingber D.E., Prusty D., Sun Z., Betensky H., and Wang N. Cell shape, cytoskeletal mechanics, and cell cycle control in angiogenesis. Journal of Biomechanics, 28(12):1471–1484, 1995.

Kolodney M.S. and Wysolmerski R.B. Isometric contraction by fibroblasts and endothelial cells in tissue culture: a quantitative study. Journal of Cell Biology, 117(1):73–82, 1992.

Lee B., Mitchell L., and Buchsbaum G. Rheology of the vitreous body. Part 1: Viscoelasticity of human vitreous. Biorheology, 29:521–533, 1993.

Lee B., Mitchell L., and Buchsbaum G. Rheology of the vitreous body. Part 2: Viscoelasticity of bovine and porcine vitreous. Biorheology, 31:327–338, 1994a.

Lee B., Mitchell L,. and Buchsbaum. G. Rheology of the vitreous body. Part 3: Concentration of electrolytes, collagen and hyaluronic acid. Biorheology, 31:339–351, 1994b.

Manoussaki D., Lubkin S.R., Vernon R.B., and Murray J.D. A mechanical model for the formation of vascular networks in vitro. Acta Biotheoretica, 44:271–282, 1996.

Markwald R.R., Fitzharris T.P., Bolender D.L., and Bernanke D.H. Structural analysis of cell:

matrix association during the morphogenesis of atrioventricular cushion tissue. Developmental Biology, 69(2):634–654, 1979.

Meinhardt H. Models of Biological Pattern Formation. Academic Press, London, 1983.

Murray J.D. Mathematical Biology, 2nd corrected edition. Springer-Verlag, Heidelberg, 1993.

Murray J.D., Oster G.F., and Harris A.K. A mechanical model for mesenchymal morphogenesis. Journal of Mathematical Biology, 17:125–129, 1983.

Murray J.D. and Oster G.F. Cell traction models for generating pattern and form in morphogenesis. Journal of Mathematical Applications in Medicine & Biology, 1, 51–75, 1984a.

Murray J.D. and Oster G.F. Generation of biological pattern and form. IMA Journal of Mathematical Biology, 19:265–279, 1984b.

Murray J.D., Deeming D.C., and Ferguson M.W.J. Size dependent pigmentation pattern formation in embryos of Alligator mississippiensis: time of initiation of pattern generation mechanism. Proc. Roy. Soc. (Lond.), B239, 279–293, 1990.

Oster G.F., Murray J.D., and Harris A.K. Mechanical aspects of mesenchymal morphogenesis. Journal of Embryology and Experimental Morphology, 78, 83–125, 1983.

Othmer H.G. Maini P.K. and Murray J.D. (editors). Experimental and Theoretical Advances in Biological Pattern Formation, NATO ASI Series A: Life Sciences, volume 259. Plenum Press, New York, 1993 (388 pages) [Proceedings of a NATO Advanced Research Workshop of that name in Oxford, 1992.]

Patan S. Munn L.L., and Jain R.K. Intussusceptive microvascular growth in a human colon adenocarcinoma xenograft: a novel mechanism of tumor angiogenesis. Microvascular Research, 51(2):260–272, 1996.

Poole T.J. and Coffin J.D. Developmental angiogenesis: quail embryonic vasculature. Scanning Microscopy, 2(1):443–448, 1987.

Price R.J. and Skalak T.C. Circumferential wall stress as a mechanism for arteriolar rarefaction and proliferation in a network model. Microvascular Research, 47(2):188–202, 1994.

Scherer G.W., Hdach H., and Phalippou J. Thermal expansion of gels: a novel method for measuring permeability. Journal of Non-Crystalline Solids, 130:157–170, 1991.

Skalak T.C. and Price R.J. The role of mechanical stresses in microvascular remodeling. Microcirculation, 3(2): 143–165, 1996.

Thoma R. Untersuchungen über die Histogenese und Histomechanik des Gefassystems. Enkeverlag, Stuttgart, 1893.

Tomanek R.J. and Torry R.J. Growth of the coronary vasculature in hypertrophy: Mechanisms and model dependence. Cellular and Molecular Biology Research, 40(2):129–136, 1994.

Turing A.M. The chemical basis for morphogenesis. Phil. Trans. Royal Society (London), B237: 37–72, 1952.

Vernon R.B., Angello J.C., Iruela-Arispe M.L., Lane T.F., and Sage E.H. Reorganization of basement membrane matrices by cellular traction promotes the formation of cellular networks in vitro. Laboratory Investigation, 66:536–547, 1992.

Vernon R.B., Lara S.L., Drake C.J., Iruela-Arispe M.L., Angello J.C., Little C.D., Wight T.N., and Sage E.H. Organized type I collagen influences of endothelial patterns during "spontaneous angiogenesis in vitro": Planar cultures as models of vascular development. In Vitro Vascular and Developmental Biology, 31:120–131, 1995.

Vernon R.B. and Sage E.H. Between molecules and morphology. Extracellular matrix and creation of vascular form. American Journal of Pathology, 147(4):873–883, 1995.

Wechezak A.R., Viggers R.F., and Sauvage L.R. Fibronectin and F-actin redistribution in cultured endothelial cells exposed to shear stress. Laboratory Investigation, 53(6):639–647, 1985.

Wolpert L. Positional information and pattern formation. Phil. Trans. Royal Society (London), B295, 441–450, 1981.

3.3

Blood Vessel Growth: Mathematical Analysis and Computer Simulation, Fractality, and Optimality

Haymo Kurz, Konrad Sandau, Jörg Wilting, and Bodo Christ

Introduction

The structural complexity of the circulatory system exceeds the available genetic information. In the developmental process, therefore, self-organization on epigenetic levels can be postulated, which exploits information that is being generated during embryogenesis. We used mathematical tools to analyze patterns and complexity, and designed a computer model to predict geometrical and biophysical properties of bifurcating vessel systems. In particular, some boundary conditions during development, and the problem of optimality are addressed. We propose that the complexity of blood vessel formation in vivo and in sapio may be adequately described with a combination of various classical geometrical and physical concepts, supplemented by concepts of fractal geometry.

Efficiency and Complexity

The evolutionary concept "survival of the fittest" implies that those individuals have the greatest chance to reproduce which obtain a maximum share of available resources with a minimum effort: they are more efficient systems. However, there are different and possibly divergent aspects to efficiency that may be loosely linked to the concepts of information, mass, and energy. With respect to warm-blooded vertebrates with their genomic complexity, large body mass and high metabolic rate, one might wonder what the boundary conditions are that govern the development of their circulatory systems in order to: (1) achieve a maximum structural and regulatory complexity with minimum genetic information; (2) build a maximum vessel length and capillary density with minimum tissue mass; (3) perform a maximum transport with minimum heart power? We may safely assume that no global optimum solution, independent of species, organ, or developmental age exists. Rather, we expect that a variety of compromises have evolved, which define locally optimal vascular patterns. For a first approach toward addressing these questions, the bird's egg, in particular the extraembryonic vascular bed of the chorioallantoic membrane (CAM), is the system of choice because the egg is an (essentially) closed system, and the CAM is comparatively simple in structure and function. Moreover, the CAM has been used frequently to study embryonic respiratory physiology and to assess growth factor effects (Auerbach et al., 1991; Wilting et al., 1993, 1996; Kurz et al., 1995). We therefore focused our studies on this organ with respect to both quantitative analysis, computer simulation and considerations of optimality.

Bifurcations: Fractal or Nonfractal Blood Vessel Growth?

Quantitative analysis of blood vessel geometry has a longstanding tradition, which in some aspects appears as a series of variations over a common theme: the bifurcation. Since the early work of Wilhelm Roux (1878, 1895; cf. Kurz et al., 1997), diameter relations and branching angles at (mostly arterial) bifurcations were studied by anatomists, physiologists, mathematicians, and theoretical biologists (cf. Thompson, 1917). The assumption of self-similarity, i.e., that the geometric relations around bifurcations are independent of their absolute size, was implicit in most older papers, and has become explicit in many more recent ones since the advent of fractal geometry. For instance, the bifurcation angle per se does not contain information on length and diameter, Roux's (1878) quotient of branch diameters eliminates their actual size, and the calculation of a bifurcation exponent, Δ, from branch radii before (R_0) and after (R_1, R_2) a bifurcation does the same (Eq. 1).

$$R_0^{\Delta} = R_1^{\Delta} + R_2^{\Delta} \tag{1}$$

The significance of Δ for describing arterial bifurcations was first recognized by Thoma (1901). While he had no general solution for this defining equation, he suggested $2.0 < \Delta < 4.0$, a very reasonable assumption in real systems (Sandau and Kurz, 1994; Kurz and Sandau, 1997). The pivotal role of Δ for understanding the physiology of the circulatory system was underlined by several attempts to answer the question of "optimum design" of blood vessels (Hess, 1913; Murray, 1926; Suwa et al., 1963; Uylings, 1977; Woldenberg and Horsfield, 1986; LaBarbera, 1990; Pries et al., 1990, 1992; Pollanen, 1992; Rossitti and Löfgren, 1993; Frame and Sarelius, 1995). Murray's (1926) minimum work condition $\Delta = 3$ perhaps was the most elegant and famous solution, but which is thought to "better describe the fluid transport systems of sponges than the circulatory system of mammals" (LaBarbera, 1990). Mandelbrot (1982) applied the concept of fractal dimension, D, to blood vessel patterns, assuming size-independent self-similarity. He suggested that bifurcation exponent Δ and fractal dimension D could have the same value. All these approaches were useful, because one could neglect the influence of animal size and developmental stage on bifurcation geometry. However, this concept has been challenged by others (Panico and Sterling, 1995; Kurz and Sandau, 1997): When we look at the details of developmental mechanisms and of anatomy (e.g., finite-sized capillaries), the highly idealized nature of the mathematical concepts involved becomes obvious. The different nature of Δ (cf. Eq. 1) and D (defined by a limit assuming infinitely small structures) is implicit in their mathematical definitions. More detailed remarks on the issue of fractal vs. nonfractal growth and form can be found in Tautu (1994), Murray (1995), Sandau and Kurz (1997), and Kurz and Sandau (1997).

During development, an abundant capillary meshwork is formed initially that gradually reorganizes into the definite vessel pattern (for details see Wilting et al., 1995; Wilting et al., this volume). Interestingly, both intussusceptive mesh formation (e.g., in the CAM; Patan et al., 1993) and sprouting from stem vessels (e.g., in the central nervous system; Kurz et al., 1996) lead to capillary loops. In the adult, anastomoses also appear frequently, as the well-known arterial arcades and circles, or the interconnections between veins. Therefore, the more general term "network" better describes blood vessel topology, than the commonly-used term "tree." To make things more complicated, real vessel systems are generally arranged in three dimensions, but may also be arranged within thin tissue sheets. All of the above have important consequences: (1) the assumption of a single self-similar branching generator, which is implicit to most fractal growth models, does not correspond to multiple real growth mechanisms; (2) self-similarity is not present at the level of the capillary bed, and may be absent from parts of the larger vessels; (3) lower and upper limits exist for blood vessel size, and the geometry of bifurcations may vary, in relation to their size; (4) blood flow through a bifurcation is not always as clearly defined as it may seem. However, we do observe regions with beautifully bifurcating vessels that are arranged on quasi two-dimensional surfaces: e.g., the pia mater, the mammalian retina, and the CAM (Mainster, 1990; Matsuo et al., 1990; Kurz et al., 1994; Masters, 1994; Kirchner et al., 1996). In such vascular beds, the assumptions of one typical growth process, and of a self-similar arterial tree may be realized in some regions. We are able therefore to analyze such patterns with methods derived from fractal geometry, although their actual growth mechanism is not "fractal." On the other hand, we may be able to develop a growth model that does not make "fractal" assumptions like an infinite branching generator, but nevertheless produces patterns that mimic their natural counterparts.

Quantitative Analysis of Blood Vessel Patterns

Tissue vascularity has been characterized using traditional quantitative parameters like vessel volume, length, or surface per tissue volume. Stereology (Weibel, 1979; Sandau, 1995) has provided tools that facilitate unbiased estimates of these parameters even in highly anisotropic, e.g., muscle tissues. These classical measures provide essential information on physiological and pathological relations between vessels and their surrounding tissues, but they do not fully characterize vascular patterns. In particular, characteristics like tortuosity, vessel endpoint density and distribution, or bifurcation geometry contain important information as well, but are not easily determined in most organs. Even more difficult to quantify are size, number, and functional status of blood vessel cells (endothelial, EC, and smooth muscle cells, SMC), and their interactions with other cells in intact tissues. For the latter tasks, tools of the emerging single-cell-stereology (Sandau and Hahn, 1994; Cruz-Orive, 1996) appear most promising, whereas the former may be treated with tools of point field statistics and fractal geometry (Stoyan and Stoyan, 1994). Sandau (1996a) has reviewed the integration of classical and advanced quantitative techniques with image analysis. In the following text, we briefly describe the use of measurements and stereology, of point field statistics and fractal geometry to characterize normal CAM vessel development and the effects of vascular endothelial growth factor (VEGF) on the CAM.

Quantitative CAM Assays for Angiogenesis

The chicken embryo CAM, which is the major respiratory organ from day 7 to hatching at day 21, shows two distinct growth periods: before incubation day 11, it expands to cover the egg shell from inside, but after day 11, it can expand no further. One can predict that endothelial proliferation and blood vessel formation run at maximum levels before day 11, when oxygen supply is critical for the rapidly growing embryo, but are reduced considerably after day 11. This is exactly what was observed by Ausprunk et al. (1974); nonetheless this fact did not prevent researchers during the early years of CAM assay studies to apply their putative angiogenic growth factors before day 11. The result of the assay may then be of limited value, because the growing CAM is easy to stimulate and normal angiogenic activity is in place. Consequently, a considerable amount of spoke-wheel patterns were published, but as the studies became more elaborate, only few of a host of so-called angiogenic factors proved to be biologically significant. The most promising candidate was vascular endothelial growth factor (VEGF; cf. Wilting et al., 1996, 1997; Wilting et al., this volume). Therefore, we evaluated the application of VEGF to the day 13 to 15 CAM using vessel length density, EC proliferation density and distribution, and "fractal dimension" or complexity.

A control (carrier alone) and a $VEGF_{121}$-treated day 15 CAM after 2 days of treatment are shown in Fig. 1A and B. Additional control and VEGF-treated specimens received BrdU before fixation to enable detection of DNA-synthesizing EC nuclei. Vessel length density, proliferation pattern, and complexity were assessed with computer-assisted microscopy: on digitized images of the CAM, an automatic intersection count was used to measure length density, and a semi-automatic counting grid was applied for detection of proliferating EC nuclei (Kurz et al., 1995). Initially, the classical box counting method (BCM) was used to estimate "fractal dimension" (Kurz et al., 1994; Wilting et al., 1996). We found that the two first approaches were fairly stable under variable tissue and image processing. For the measurement of "fractal dimension," however, an extremely high degree of standardization had to be obeyed because rotation or translation of the pattern, various threshold settings and image noise strongly affected the BCM result. A more refined method for the measurement of complexity was therefore developed (Sandau, 1996b). This method proved to be more stable against adverse effects and is referred to as the extended counting method, XCM (Sandau and Kurz, 1996).

Evaluation of various CAM assays and of control CAMs with length density, L_A, proliferation intensity, N_A, and complexity dimension, D_{XCM}, is summarized in Table 1. The data show

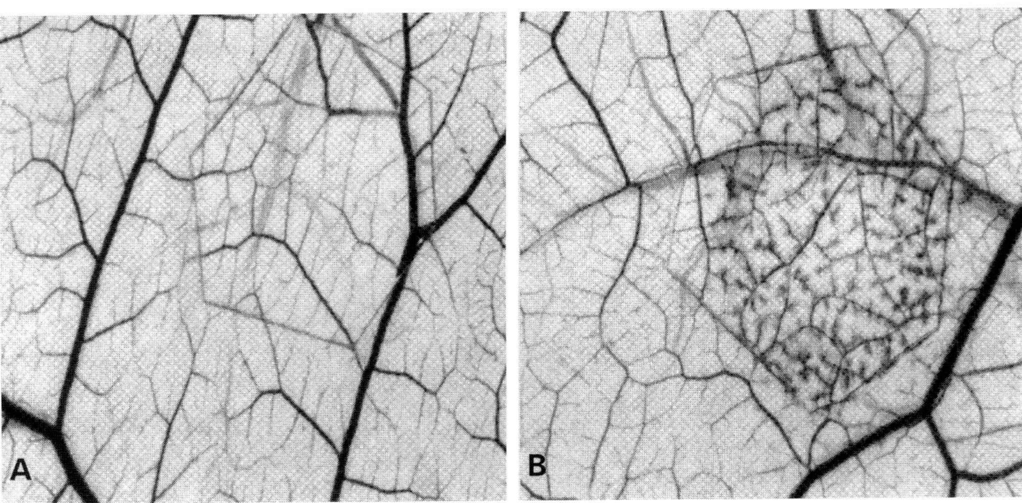

Fig. 1 (A) Control CAM with unchanged vascular pattern under the polygonal carrier. (B) $VEGF_{121}$-treated CAM with increased complexity of the capillary and precapillary vascular pattern. Image analysis in combination with XCM enables automatic detection of the angiogenic effect.

Table I. Comparison of day 10 and 15 control and day 15 VEGF-treated CAMs. For measurements of vessel length density, L_A, and endothelial proliferation intensity, N_A, see Kurz et al. (1995), and of vascular bifurcation complexity, D_{XCM}, see Sandau and Kurz (1997). Values are means ± standard deviation, n indicates number of specimens.

Parameter	Unit	10d control	15d control	15d VEGF
L_A	mm^{-1}	1.9 ± 0.1 ($n = 6$)	3.4 ± 0.2 ($n = 6$)	4.4 ± 0.3 ($n = 6$)
N_A	mm^{-2}	672 ± 30 ($n = 3$)	286 ± 24 ($n = 3$)	1150 ± 35 ($n = 3$)
D_{XCM}		1.51 ± 0.023 ($n = 10$)	1.56 ± 0.029 ($n = 23$)	1.64 ± 0.025 ($n = 25$)

that L_A is augmented by 80%, and N_A diminished by 50% in the normal 15 day CAM as compared to the 10 day CAM. This is accompanied by a slightly increased complexity, which indicates that the mature bifurcation pattern is almost fully established at day 10. Treatment with VEGF leads to enhanced EC proliferation, to an additional 30% rise in L_A, and to an altered bifurcation pattern as indicated by elevated D_{XCM}. These studies also showed that the distribution of proliferating EC in the VEGF-treated 15-day CAM was similar to that of the untreated 6- or 8-day CAM (Kurz et al., 1995).

Spatial Statistics of CAM Development

A further parameter for understanding tissue perfusion in general, and CAM respiratory efficiency in particular, is vascular endpoint density (cf. Strick et al., 1991). However, this parameter is difficult to determine precisely, because in fixed CAMs, few endpoints are readily detected, and because in sections, spatial relationships over large distances are difficult to maintain. Determining endpoints is rendered nearly meaningless in VEGF-treated CAMs because a complicated, three-dimensional blood sinus is formed, where definition of an endpoint becomes virtually arbitrary. We therefore studied a related property, the distribution of arteriovenous endpoint distances in normal CAMs on days 10 and 16. Clearly, a low arteriovenous distance means a shorter capillary length through which oxygenation can take place. We rarely observed arteriovenous distances shorter than 100 μm in the day 10 CAM, but found that about 33% of these distances were below 100 μm in the day 16 CAM (Kurz et al., 1994). Combined with the approximately threefold higher blood pressure in the older CAM (VanMierop and Bertuch, 1967), this means greatly reduced mean capillary passage time and may hence explain the functional shunt that has been postulated from respiratory data (Tazawa, 1980). VEGF treatment leads to a further enhanced vascularity, but it appears likely that the additional vessels do not add to CAM function, because they cannot contribute much to oxygenation (due to their position deep in the chorionic mesenchyme), but rather provide additional shunt vessels that bypass the superficial capillaries.

As mentioned above, the bifurcation exponent Δ is fundamental to the geometric design of branching systems. We therefore measured the diameters of arteries (>40 μm) around bifurcations in the day 10 and day 15 CAM and calculated mean values of $\Delta = 2.5$ and $\Delta = 2.6$. This is close to the value $\Delta = 2.7$ calculated by Suwa et al. (1963) for adult humans and corresponds to $\Delta = 2.5$ to 2.8 estimated by Thoma (1901) for human placental and chick yolk sac arteries. We have shown recently (Kurz and Sandau, 1996) that the value of Δ may be decisive for arterial design, and possibly for the control of arterial growth.

Computer Simulation of Blood Vessel Growth

A prime objective in developing our model (Sandau and Kurz, 1994) was to simulate the arterial branching pattern of the CAM. Growth processes in a meshwork-like capillary layer, which may be idealized to hexagons, drive the elongation and ramification of the larger vessels. In the computer model, the arterial tree grows "on" a hexagonal grid, which defines the details of the structure. The "generator" is idealized as an elementary bifurcation with three branches of equal length. Starting in one seed element, each subsequent element has a predecessor, whose diameter increases in relation to blood flow as required to support its successors. The biophysical properties were derived from Poiseuilles' law, the continuity condition and Δ. They were implemented in computer algorithms such that the stochastic rules for adding successors were influenced by, e.g., local pressure and flow (stochastic growth function). All simulations reported here included a twofold increase in viscosity during development that corresponded to the increase in haematocrit. A typical simulation is shown in Fig. 2A, which illustrates the stochastic nature of the growth process.

In terms of geometry, our approach is distinct from models based on a priori self-similarity as described by, e.g., Mandelbrot (1982), Bittner (1991), or Spatz (1991). In these previous models, branch lengths tend to zero. Our model, which considers finite-sized elementary structures, is more akin to the invasion percolation model described by Stark (1991); his model, like most "fractal" pattern generators, however, remained purely geometric, because it neglected (blood) flow.

In terms of physics, we defined a self-organizing branching vascular bed, without anastomosing vessels, and under increasing perfusion pressure, increasing blood viscosity, and increas-

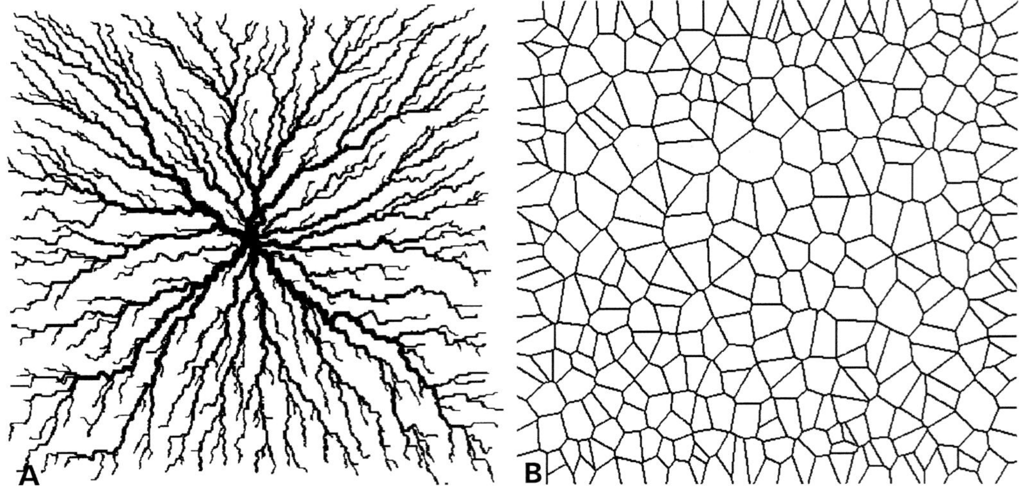

Fig. 2 (A) Complete simulation for $\Delta = 2.4$. The pattern illustrates the self-avoidance of the tree, with small vessels branching off even close to the source point. (B) The endpoints of the tree induce a Voronoi tessellation of the "tissue." The polygonal cells are formed by lines which are farthest away from the nearest arterial endpoints, thus indicating regions for a possible veinous drainage. Note that endpoints can be found close to the source point and that extremely large or small cells are rare.

ing flow. During the computer simulation, the algorithm adds "capillary" elements to preexisting elements with free ends in the network, and subsequently transforms them into "arteries." The probability to add a new element relates to local pressure, flow, velocity, and direction. Hence, shear and wall stresses, which all depend on the previously generated network, are implicit in the stochastic growth rules. This model can describe significant mechanical stimuli for ECs, and for SMCs, respectively; it thus provides a simulation of arterial development rather than of primordial angiogenesis.

The most important physical simplifications stipulated that at all branch tips the pressure difference and the radius were constant elementary units. Consequently, the flow at each ultimate branch tip was also constant. All mature simulated systems differ negligibly in endpoint number and deliver approximately the same volume per unit time to the covered area. Consequently, the graph of a simulation (Fig. 2A) gives an indication as to how wide—for any given Δ value—the arterial lumen should be in order to supply a given tissue mass.

Geometric Properties

Complex branching patterns such as those shown in Fig. 2A emerged from a perfectly regular elementary unit, under linear physical constraints, and with limited instructions. The patterns were remarkably insensitive to variation in the stochastic growth function. We also characterized the patterns with respect to branching complexity, branch length distribution, and endpoint distribution. Whereas a limited range of complexity could be expected in view of the common hexagonal generator, it was surprising to see that XCM measured a nearly constant complexity of the simulated patterns, independent of Δ. The value of 1.69 ± 0.01 was somewhat higher than the value in the day 15 CAM (cf. Table 1). This could indicate that our model produces a slightly more erratic vessel pattern, or a higher bifurcation density than its untreated biological paradigm. However, values of up to 1.7 were observed in VEGF-treated CAMs, and could indicate a natural upper boundary for arterial bifurcations.

In the simulated system, vessel branch lengths can be easily recorded. They show a gamma distribution, similar to that observed in the branching patterns of rivers (Stark, 1991), and similar to the lognormal distribution observed in brain vasculature (Mironov et al., 1994). In a few CAMs, we performed branch length measurements and also found a gamma distribution (Kurz and Sandau, 1997). However, true branch lengths are difficult to measure in a CAM preparation, and we therefore consider our data as preliminary.

As mentioned above, we previously applied basic ideas of point field statistics (Stoyan and Stoyan, 1994) to assess distribution of EC proliferation in CAM capillaries (Kurz et al., 1995), and to analyze vascular patterns in the neural tube (Kurz et al., 1995). We here consider the CAM capillary layer being partitioned by vascular endpoints that induce variably-sized zones of influence, and which thus produce tessellations of the tissue (Fig. 2B). Three-dimensional examples of such tessellations are liver lobules or lung segments that appear as polygons of various shapes and sizes in tissue sections. In this study we used the Voronoi tessellation as a first approximation to model the zones of influence of each endpoint. A cell of a Voronoi diagram is the set of all points in the plane, which are closer to the generating point than to any other point within the point field. We then calculated the amount of tissue that is supplied by any one endpoint. Our preliminary results, which are not presented in detail, show size variations of Voronoi cells, which explain the well-known heterogeneity of perfusion (VanBeek et al., 1989; Caruthers and Harris, 1994) without the necessity for fractal geometry. However, they yield more regular patterns than the purely random Voronoi tessellation induced by a Poisson point process (Sandau and Kurz, 1995; Kurz and Sandau, 1997). The results suggest that applying point field statistics for the evaluation of computer models and of real growth patterns should be used more often, because not only regular, but also random patterns can be classified and analyzed with respect to morphogenetic events.

Biophysical Properties

In any perfused vascular system, static and dynamic interactions of the blood with the vessel wall have to be considered. It is important to distinguish between the major static entity, blood pressure, and the major dynamic entity, blood flow. Note that pressure is present in any arterial vessel connected to the heart, whereas flow may or may not be present. On interaction of these physical entities with the components of the vessel wall, two different stresses are produced: blood flow induces shear stress, which is exclusively experienced by the EC (Davies, 1995), whereas pressure induces tangential wall stress, a tensile stress which is primarily experienced by the SMC of the tunica media (Owens, 1995). From the functional characteristics of the embryonic circulation (Clark, 1991), we calculated wall stress in the developing chicken aorta to be much lower than in the adult, whereas shear stress approached the adult value very early (Kurz and Sandau, 1997).

Flow, Fluid Shear Stress, and Passage Time

Flow is defined as displaced blood volume per unit time, which is the same as vessel cross-sectional area times mean blood velocity. These simple relations can be used to calculate shear stresses at the endothelial surface (if blood viscosity is known) and passage times for cells or signals to travel through the vascular system. An immediate consequence of the fundamental Eq. (1) and Poiseuille's law is that the mean velocity at any point in the system is low and constant for $\Delta = 2.0$, but is linearly related to vessel radius for $\Delta = 3.0$. In any particular simulation, where the number, j, of succeeding "capillaries" supplied by any "artery," and Δ are known, and for a suitable constant c_1, we calculate fluid shear stress, τ, by

$$\tau(j) = c_1 \cdot j^{1-\frac{3}{\Delta}} \tag{2}$$

This equation indicates constant τ in all branches of the system for $\Delta = 3$, which was implicit in Murray's (1926) famous approach, and was recognized in recent publications (LaBarbera, 1990; Rossitti, and Löfgren, 1993; Pries et al., 1995). The equation also predicts the distribution of τ for other Δ values, with one perhaps surprising consequence: τ has to increase toward the smallest branches for $\Delta < 3.0$. However, blood viscosity is reduced in narrow tubes, which could compensate for this effect such that shear stress may even decrease toward smaller vessels, especially if Δ also changes (Frame and Sarelius, 1995). Moreover, the moderate dependence of τ from radius and pressure in the main vessels of different systems is confirmed by real values. Perhaps biophysical models may help to settle incongruities in the literature concerning τ values, which are considered low and almost constant by some authors (LaBarbera, 1990), and high and variable by others (Pries et al., 1995).

From the mean velocity in any vessel segment, we calculate the passage time, t, (of an erythrocyte or a chemical signal) in the arterial system from source to each endpoint as

$$t\,(j) = c_2 \cdot \Sigma\, j^{\frac{2}{\Delta}-1} \tag{3}$$

where the sum is taken over all predecessor vessel segments of the considered endpoint (c_2: scaling factor). For $\Delta = 2.0$, passage time does not depend on the size of the vessels, but only on the distance from source to endpoint, because mean velocity is constant in all branches. Hence, the spatial distribution of t in a vascular bed depends on the bifurcation exponent, which is illustrated in Fig. 3A and B. It is obvious that for $\Delta > 2.0$ (i.e., in all real systems) a "chaotic" pattern evolves on a perfectly regular lattice, and from perfectly linear physical properties. One could speculate, whether such spatiotemporal distributions contribute to heterogeneity of perfusion, or quasiturbulent mixing (Sernetz et al., 1985) in a laminar flow system. More important with respect to efficiency is the absolute value of t: we have shown that the mean passage time varies by about a factor of two for a variation of Δ between 2.4 and 2.8, or 2.8 and 3.2 (Kurz and Sandau, 1997). A short recirculation period means more frequent oxygen loading and unloading cycles of erythrocytes, and hence a higher transport efficiency for a given amount of hemoglobin.

Pressure, Wall Tension, and Wall Tissue Mass

Pressure loss, dp, in each arterial segment is related to the number of succeeding vessel segments, j, and to Δ, (for an appropriate constant c_3) by

$$dp\,(j) = c_3 \cdot j^{1-\frac{4}{\Delta}} \tag{4}$$

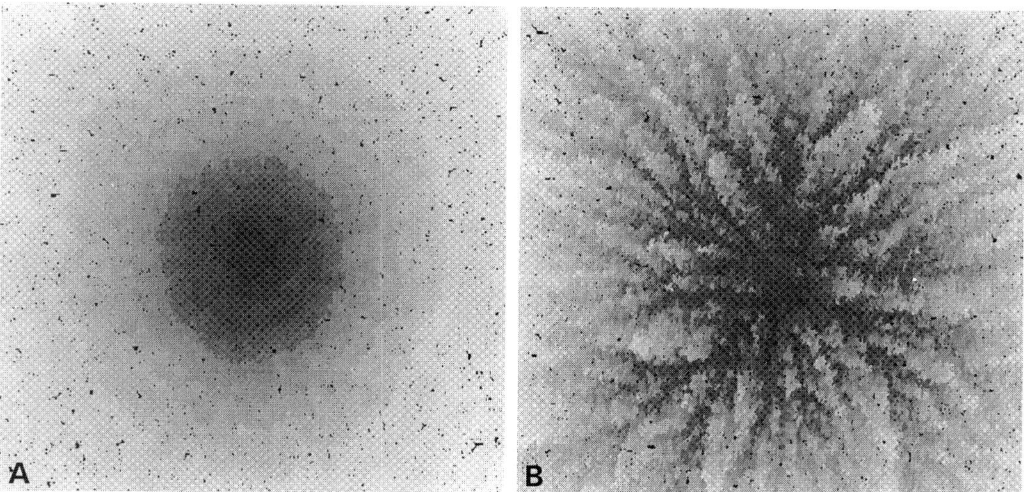

Fig. 3 Spatial distribution of passage time, t, for simulations with (A) $\Delta = 2.0$, and (B) $\Delta = 3.2$. Dark regions represent short, and bright regions represent long passage times from the central source to the endpoints. This illustrates the intrinsic chaotic nature of bifurcating systems. Gray scales are adapted to the total time range in each system: median t was 264 in (A), but only 18 units in (B). Therefore, transport efficiency of, e.g., erythrocytes in a system with $\Delta = 2.0$ is nearly 15 times that of one with $\Delta = 3.2$.

This means that dp is constant throughout the system for $\Delta = 4$, whereas for lower Δ values, dp is minimized in the periphery of the system. From Eq. (5) we can obtain the local pressure, p, in every segment of known radius, R, in our simulations. We therefore can calculate tangential wall tension, T, from Laplace's law as $T = p\,R$ for any simulated vessel segment, and for any Δ value; interestingly, there is no straightforward way to express p, and hence T, as a function of R or j in analogy to Eq. (5); nevertheless, T can be calculated numerically in our simulation.

It is clear that wall tension varies greatly in any conceivable fluid transport system that is driven from a single source point; this is because the highest pressures are found in the largest vessels (if hydrostatic pressure is ignored). To cope with the variation in tensile stress, either stress resistance of the wall tissue, or vascular wall thickness have to increase in proportion to T. Several lines of evidence support the view that the specific strength of SMC (in terms of their contractile proteins and of extracellular matrix components) is the same for all warm-blooded animals, and that, e.g., actin expression is related to mechanical stress (Owens, 1995). Although detailed knowledge about development of cells and matrix in the arterial media is emerging only recently (Yablonka-Reuveni, 1995; Hungerford et al., 1996), the varying time course of SMC development in veins and arteries (Li et al., 1996; Moessler et al., 1996), and our own data on actin expression (not shown) indicate that this protein is produced in proportion to T. In the model, therefore, we assume that tensile stress in the wall, $\sigma = T/h$ is kept constant by adapting wall thickness, h, which is equivalent to $h \sim T$. Finally, the total wall tissue mass, M_W, in the simulated system was obtained from the local knowledge of pressure and radii in all vessel segments. M_W should be included in an analysis of efficiency, because tissue mass definitely resembles an amount of material, of energy and of information for building and maintaining it.

Optimality

As initially stated, we cannot expect to formulate a global optimum strategy for all conceivable vascular beds. In this study we focus on the avian CAM vessels, where one optimality constraint appears obvious: conservation of a minimum of tissue in the extraembryonic membranes that will be abandoned upon hatching. In the posthatching competition for food, the larger chicks win while their smaller siblings are often eliminated. Selection therefore favors offspring that incorporate as much of the egg's contents into their bodies as possible.

We calculated vascular tissue mass M_W for Δ values between 2.0 and 4.0. A minimum for $\Delta = 2.7$ was found in our simulations, which assumed constant vessel wall stress resistance (Kurz and Sandau, 1997). What happens, if Δ deviates from this minimum tissue optimum too much? In systems with excessively wide arteries ($\Delta < 2.4$), the embryo has to pay for an (energy saving) low perfusion pressure not only with up to twice the wall tissue, but also with greatly enhanced blood volume and prolonged erythrocyte recirculation time. In systems with excessively narrow arteries ($\Delta > 3.0$), the embryo has to pay for a reduced blood volume and a shortened erythrocyte recirculation time with a greatly enhanced blood pressure, and hence a manyfold increase in wall tissue and cardial energy consumption.

In the CAM, the minimum tissue condition seems to be closely obeyed, as our aforementioned measurements indicated. The slightly lower values could correspond to the initially lower perfusion pressure, which favors wider arteries. It remains to be seen, what the molecular basis for controlling the minimum tissue condition is. Nevertheless, we would suggest several lines of evidence that a mean value of $\Delta = 2.7$ governs the design of arterial systems in most warm-blooded animals:

For example, the minimum-tissue optimum corresponds to $\Delta = 2.7$ as measured by Suwa et al., (1963) in various organs of adult humans. This is different from Murray's (1926) hydraulic minimum-work optimum at $\Delta = 3.0$, which implies a—not too realistic—constant fluid shear stress (Rossitti and Löfgren, 1993). No data exist for the embryo or fetus proper, but in early studies on the arteries of the human placenta and of the chick yolk sac, Thoma (1901) supposed Δ to be between 2.5 and 3.0. The minimum tissue value $\Delta = 2.7$ fits embryonic development, because it corresponds to a constant tensile wall-stress condition accompanied by continually increasing length, pressure, and viscosity. This is in contrast to the assumptions of constant length, pressure, and viscosity, which led to Murray's (1926) $\Delta = 3.0$. In addition, his constant cost-of-blood approach fails with respect to embryonic development, where an initially "inexpensive," watery blood containing few erythrocytes and plasma proteins gradually becomes more "expensive."

Recently, further evidence emerged regarding the nonconstancy of shear stresses (LaBarbera, 1990; Pries et al., 1995; Frame and Sarelius, 1995) which suggests that Δ must be below 3.0. We have also shown that a diameter exponent $\Delta = 2.7$ is in excellent agreement with allometry (Schmidt-Nielsen, 1989): The observed aortic lumen diameters in animals of varying size—"from mouse to elephant"—comply with predictions based upon capillary diameter and number of bifurcations (Kurz and Sandau, 1997). This implies that essentially the same set of genetic instructions could be used for defining the design of arteries in all warm-blooded animals, irrespective of their size. As a further consequence, the well-known scaling law of metabolism (Kleiber, 1932; Spatz, 1991) could be related to the constant-stress, minimum-tissue condition in the arterial wall. Moreover, our approach suggests that expression of (regulatory) genes in cells of the arterial wall may correspond to the distribution of mechanical stresses during development. It remains to be seen, to what extent the information for vascular pattern regulation lies within the tissue-autonomous genetic program, and to what extent it is derived from local mechanical factors. Efficient use of information would favor the use of epigenetic clues.

Conclusions and Perspectives

By analyzing and simulating blood vessel growth, we have shown that a variety of mathematical concepts are needed to adeaquately address various aspects of the circulatory system. Most of these concepts, however, are not related to fractal geometry. In our estimation, real blood vessel growth parallels only few of the basic assumptions underlying fractals. In particular, we would like to warn of a concept like fractal dimension in the context of developmental biology; for example, a certain D value does not prove or disprove a particular growth mechanism (Murray, 1995; Kurz and Sandau, 1997), and measuring D may be hampered by severe restrictions in practice (Soille and Rivest, 1996; Sandau and Kurz, 1997). Furthermore, a classical geometric parameter like Δ obviously provides a deeper understanding of geometry and biophysics. We nevertheless suggest that fractal-related methods, like XCM, should be used to characterize texture parameters for comparing the complexity of patterns. We are skeptical against the naive use of BCM and related methods (Kurz et al., 1994; Kirchner et al., 1996). True fractals may not be everywhere as suggested by Barnsley (1993), and much of biological design presumably may be explained without fractal geometry (Weibel, 1991). Blood vessel patterns perhaps are best characterized by a combination of simple deterministic, stochastic, and chaotic properties.

From our quantitative studies on the CAM assay, we would like to emphasize two aspects. First, enhanced vascularity of an organ may not be beneficial to that organ, because small arteriovenous distances or excessively wide and tortuous capillaries can impair oxygen and metabolite exchange with the tissue. Second, screening for anti-angiogenic factors (Auerbach and Auerbach, 1994) may be achieved by quantitative evaluation of the CAM assay. The CAM also seems to be of particular value for studying formation of the arterial wall, and for analyzing the development of regulatory interactions between endothelial and smooth muscle cells.

Mathematical and physical models have accompanied developmental biology since the seminal work of Wilhelm Roux (1878, 1895). More than 100 years ago, Roux recognized fundamental concepts of vascular morphogenesis, although he did not have the means to solve the mathematical and biological problems (Kurz et al., 1997). A broader discussion of mathematical models in vascular morphogenesis (cf. Kurz and Sandau, 1997), and in general biology, is beyond the scope of this article; we therefore refer to a WWW site (http://www.gdb.org/Dan/mathbio/intro.html) that provides general information, including references, on the interface between mathematics and biology. Moreover, we would like to recall that Thompson (1917) emphasized the notion that the design of biological structures is governed by physical boundary conditions, in particular by optimality criteria, and perhaps one should be aware that no genetic or regulatory measure can overcome physical constraints.

We have demonstrated that an optimum value exists for the bifurcation exponent in the avian extraembryonic circulation. Measurements of bifurcation geometry in adult systems (Suwa et al., 1963; LaBarbera, 1990) and our considerations on allometry (Kurz and Sandau, 1997) appear to indicate that the same optimality criterion determines the design of the complete arterial system. We now can speculate whether this minimum mass condition influenced the evolution of developmental mechanisms such that a minimum of genetic information is needed for realizing a vascular network. With respect to the efficient use of energy, it is clear that maintenance of both genetic information and tissue mass contribute to energetic costs for the organism. But their share probably is low compared to the cost for maintaining blood pressure. We cannot expect to fully address the energetic aspect with our simulations, because pressure is needed not only for circulation within the vascular system but also for filtration through the vessel wall, e.g., in the kidney. In the future, refined models will be needed that include properties like filtration, elastic walls, turbulent flow, etc. for a more complete analysis of energetics. We wish to emphasize that great care should be taken in building mathematical models to include available biological information, and that both geometrical and physical properties should be predicted by the simulation.

References

Auerbach W., Auerach R. (1994): Angiogenesis inhibition: a review. Pharmac Ther 63:265–311.

Auerbach R., Auerbach W., Polakowski I. (1991): Assays for angiogenesis: a review. Pharmac Ther 51:1–11.

Ausprunk D.H., Knighton D.R., Folkman J. (1974): Differentiation of vascular endothelium in the chick chorioallantois: a structural and autoradiographic study. Dev Biol 38:237–248.

Barnsley M.F. (1993): Fractals Everywhere. Boston: Academic Press.

Bittner H.R. (1991): Modelling of fractal vessel systems. In: Fractals in the Fundamental and Applied Sciences, Peitgen H.O., ed. Amsterdam: Elsevier, 59–71.

Caruthers S.D., Harris T.R. (1994): Effects of pulmonary blood flow on the fractal nature of flow heterogeneity in sheep lungs. J Appl Physiol 77:1474–1479.

Clark E.B. (1991): Functional characteristics of the embryonic circulation. In: The Development of the Vascular System. Basel: Karger.

Cruz-Orive L.M. (1997): Stereology of single objects. J Microsc 186:93–107.

Davies P.F. (1995): Flow-mediated endothelial mechanotransduction. Physiol Rev 75:519–560.

Frame M.D.S., Sarelius I.H. (1995): Energy optimization and bifurcation angles in the microcirculation. Microvasc Res 50:301–310.

Hess W.R. (1913): Das Prinzip des kleinsten Kraftverbrauchs im Dienste hämodynamischer Forschung. Leipzig: Veit & Co.

Hungerford J.E., Owens G.K., Argraves W.S, Little C.D. (1996): Development of the aortic vessel wall as defined by vascular smooth muscle and extracellular matrix markers. Dev Biol 178:375–392.

Kirchner L.M., Schmidt S.P., Gruber B.S. (1996): Quantitation of angiogenesis in the chick chorioallantoic membrane model using fractal analysis. Microvasc Res 51:2–17.

Kleiber M. (1932): Body size and metabolism. Hilgardia 6:315–349.

Kurz H., Sandau K. (1997): Modelling of blood vessel development—bifurcation pattern and hemodynamics, optimality and allometry. Comments Theor Biol 4/4:261–291.

Kurz H., Wilting J., Christ B. (1994): Multivariate characterization of blood vessel morphogenesis in the avian chorioallantoic membrane (CAM): cell proliferation, length density and fractal dimension. In: Fractals in Biology and Medicine, Losa G.E., Nonnenmacher T.H., Weibel E., eds., Boston: Birkhäuser.

Kurz H., Ambrosy S., Wilting J., Marmé D., Christ B. (1995): Proliferation pattern of capillary endothelial cells in chorioallantoic membrane development indicates local growth control, which is counteracted by vascular endothelial growth factor application. Dev Dyn 203:174–186.

Kurz H., Gärtner T., Eggli P.S., Christ B. (1996): First blood vessels in the avian neural tube are formed by a combination of dorsal angioblast immigration and ventral sprouting of endothelial cells. Dev Biol 173:133–147.

Kurz H., Sandau K., Christ B. (1997): On the bifurcations of blood vessels—Wilhelm Roux's doctoral thesis (Jena 1878)—a seminal work for biophysical modelling in developmental biology. Annals Anat 179:33–36.

LaBarbera M. (1990): Principles of design of fluid transport systems in zoology. Science 249:992–1000.

Li L., Miano J.M., Mercer B., Olson E.N. (1996): Expression of the SM22a promotor in transgenic mice provides evidence for distinct transcriptional regulatory programs in vascular and visceral smooth muscle cells. J Cell Biol 132:849–859.

Mandelbrot B.B. (1982): The Fractal Geometry of Nature. New York: Freeman.

Mainster M.A. (1990): The fractal properties of retinal vessels: embryological and clinical implications. Eye 4:235–241.

Masters B.R.(1994): Fractal analysis of normal human retinal blood vessels. Fractals 2:103–110.

Matsuo T., Okeda R., Takahashi M., Funata M. (1990): Characterization of bifurcating structures of blood vessels using fractal dimensions. Forma 5:19–27.

Mironov V.A., Hritz M.A., LaManna J.C., Hudetz A.G., Harik S.I. Architectural alterations in rat cerebral microvessels after hypobaric hypoxia. Brain Res 660:73–80.

Murray J.D. (1995): Use and abuse of fractal theory in neuroscience. J Comp Neurol 361:369–370.

Murray C.D. (1926): The physiological principle of minimum work: I. The vascular system and the cost of blood volume. Proc Natl Acad Sci USA 12:207–304.

Moessler H., Mericskay M., Li Z., Nagl S., Paulin D., Small J.V. (1996): The SM22 promoter directs tissue-specific expression in arterial but not in venous or visceral smooth muscle cells in transgenic mice. Development 122:2415–2425.

Owens G.K. (1995): Regulation of differentiation of vascular smooth muscle cells. Physiol Rev 75:487–517.

Panico J., Sterling P. (1995): Retinal neurons and vessels are not fractal but space filling. J Comp Neurol 361:479–490.

Patan S., Haenni B., Burri P.H. (1993): Evidence of intussusceptive capillary growth in the chicken chorio-allantoic membrane (CAM). Anat Embryol 187:121–130.

Pollanen M.S. (1992): Dimensional optimization at different levels of the arterial hierarchy, J Theor Biol 159:267–270.

Pries A.R., Secomb T.W., Gaehtgens P., Gross J.F. (1990): Blood flow in microvascular networks: experiments and simulation. Circ Res 67:826–834.

Pries A.R., Secomb T.W., Gaehtgens P. (1995): Design principles of vascular beds. Circ Res 77:1017–1023.

Rossitti S., Löfgren J. (1993): Vascular dimensions of the cerebral arteries follow the principle of minimum work. Stroke 24:371–377.

Roux W. (1878): Über die Verzweigungen der Blutgefässe. Jena: Dissertation.

Roux W. (1895): Gesammelte Abhandlungen über Entwicklungsmechanik der Organismen. Leipzig: Engelmann.

Sandau K. (1995): Spatial fibre and surface processes—stereological estimations and applications. In: From Data to Knowledge: Theoretical and Practical Aspects of Classification, Data Analysis and Knowledge Organisation, Gaul W, Pfeifer D, eds. Berlin: Springer.

Sandau K. (1996a): Quantitative microscopy and stereology with tools of image analysis. EJ Pathol 2.2:962–02.

Sandau K. (1996b): A note on fractal sets and the measurement of fractal dimension. Physica A 233:1–18.

Sandau K., Hahn U. (1994): Some remarks on the accuracy of surface area estimation using the spatial grid. J Microsc 173:67–72.

Sandau K., Kurz H. (1994): Modelling of vascular growth processes: a stochastic biophysical approach to embryonic angiogenesis. J Microsc 175:205–213.

Sandau K., Kurz H. (1995): Some properties about the endpoints of a 2–D arterial tree simulated by a stochastic growth model. 8th Int Workshop Stoch Geom Stereol Image Anal, Sandbjerg.

Sandau K., Kurz H. (1997): Measuring fractal dimensions and complexity—an alternative approach with an application. J Microsc 186:164–176.

Schmidt-Nielsen K. (1989): Scaling: Why Is Animal Size So Important? Cambridge: Cambridge University Press.

Sernetz M., Gelléri B., Hofmann J. (1985): The organism as bioreactor. Interpretation of the reduction law of metabolism in terms of heterogeneous catalysis and fractal structure. J Theor Biol 117:209–230.

Soille P., Rivest J.F. (1996): On the validity of fractal dimension measurements in image analysis. J Visual Comm Image Represent 7:217–229.

Spatz H.C. (1991): Circulation, metabolic rate, and body size in mammals. J Comp Physiol B 161:231–236.

Stark C.P. (1991): An invasion percolation model of draining network evolution. Nature 352:423–425.

Stoyan D., Stoyan H. (1994): Fractals, Random Shapes and Point Fields. Chichester: Wiley & Sons.

Strick D.M., Waycaster R.L., Montani J.P., Gay W.J., Adair T.H. (1991): Morphometric measurements of chorioallantoic membrane vascularity: effects of hypoxia and hyperoxia. Am J Physiol 260:H1385–H1389.

Suwa N., Takahashi T., Fukasawa H., Sasaki Y. (1963): Estimation of intravascular blood pressure gradient by mathematical analysis of arterial casts. Tohoku J Exp Med 79:168–198.

Tautu P. (1994): Fractal and non-fractal growth of biological cell systems. In: Fractals in Biology and Medicine, Losa G.E., Nonnenmacher T.H., Weibel E., eds. Boston: Birkhäuser.

Tazawa H. (1980): Oxygen and CO_2 exchange and acid-base regulation in the avian embryo. Am Zool 20:395–404.

Thoma R. (1901): Über den Verzweigungsmodus der Arterien. Archiv Entwicklungsmechanik 12:352–413.

Thompson D.W. (1917): On Growth and Form. Cambridge University Press.

Uylings H.B.M. (1977): Optimization of diameters and bifurcation angles in lung and vascular tree structures. Bull Math Biol 39:509–520.

Van Beek J.H.G.M., Roger S.A., Bassingthwaighte J.B. (1989): Regional myocardial flow heterogeneity explained with fractal networks. Am J Physiol 257:H1670–H1680.

Van Mierop L.R.S., Bertuch C.J. (1967): Development of arterial blood pressure in the chick embryo. Am J Physiol 212(1):43–48.

Weibel E.R. (1979): Stereological Methods. New York: Academic Press.

Weibel E.R. (1991): Fractal geometry: a design principle for living organisms. Am J Physiol 162:L361–L369.

Wilting J., Christ B., Bokeloh M., Weich H.A. (1993): In vivo effects of vascular endothelial growth factor on the chicken chorioallantoic membrane. Cell Tissue Res 274:163–172.

Wilting J., Brand-Saberi B., Kurz H., Christ B. (1995): Development of the embryonic vascular system. Mol Cell Biol Res 41/4:219–232.

Wilting J., Birkenhäger R., Eichmann A., Kurz H., Martiny-Baron G., Marmé D., McCarthy J.E.G., Christ B., Weich H.A. (1996): $VEGF_{121}$ induces proliferation of vascular endothelial cells and expression of flk-1 without affecting lymphatic vessels of the chorioallantoic membrane. Dev Biol 176:76–85.

Woldenberg M.J., Horsfield K. (1986): Relation of branching angles to optimality for four cost principles. J Theor Biol 122:187–204.

Yablonka-Reuveni Z., Schwartz S.M., Christ B. (1995): Development of chicken aortic smooth muscle: expression of cytoskeletal and basement membrane proteins defines two distinct cell phenotypes emerging from a common lineage. Cell Mol Biol Res 41:241–249.

3.4

Mathematical Modeling of Tumor-Induced Angiogenesis

Mark A.J. Chaplain and Michelle E. Orme

Introduction

Cancer is a multistep process. Once a mutation of a single, normal cell has taken place, the transition from the transformed cell to a tumor and then to metastasis is a complex process. In the case of a solid tumor (e.g., carcinoma) a crucial step in its development is the process of angiogenesis—the formation of a capillary network arising from pre-existing vasculature (as opposed to vasculogenesis; Vernon et al., 1992). Angiogenesis itself is a complex phenomenon, a dynamical system, involving the successful and timely interaction of many variables. Although on the one hand this complexity can make the task of the mathematical modeler difficult, on the other, it simultaneously provides the biomathematician with a full palette of rich colors to work with. In this chapter, we present several mathematical models describing the important aspects of tumor-related angiogenesis.

Solid tumors are known to progress through two distinct phases of growth—the avascular phase and the vascular phase (Folkman, 1975, 1976, 1985). The initial avascular growth phase can be studied in the laboratory by culturing cancer cells in the form of three-dimensional multicell spheroids (Mueller-Klieser, 1987; Sutherland, 1988; Durand, 1990). It is well known that these spheroids, whether grown from established tumor cell lines or actual in vivo tumor specimens, possess growth kinetics which are very similar to in vivo tumors. Typically, these avascular nodules grow to a few millimeters in diameter. Cells toward the center, being deprived of vital nutrients, die and give rise to a necrotic core. Proliferating cells can be found in the outer three to five cell layers. Lying between these two regions is a layer of quiescent cells, a proportion of which can be recruited into the outer layer of proliferating cells. Much experimental data has been gathered on the internal architecture of spheroids, and studies regarding the distribution of vital nutrients (e.g., oxygen) and metabolites within the spheroids have been conducted (Durand, 1990; Sutherland et al., 1971; Wibe et al., 1981; Sutherland and Durand, 1984; Groebe and Mueller-Klieser, 1991; Bourrat-Floeck et al., 1991; Bredel-Geissler et al., 1992). Indeed, recent research into the phenomenon of hypoxia (Brown and Gaccia, 1994) has meant that such studies are enjoying a new lease of life. The spherical symmetry of these multicellular aggregates and relative abundance of data has also meant that avascular tumor growth has been widely studied theoretically through the use of mathematical models (Adam, 1987; Britton and Chaplain, 1993; Burton, 1966; Byrne and Chaplain, 1995c, 1996; Casciari et al., 1992; Chaplain and Britton, 1993; Chaplain et al., 1994; Degner and Sutherland, 1988; Greenspan, 1972, 1974, 1976; Landry et al., 1982; McElwain and Ponzo, 1977; Maggelakis and Adam, 1990; Shymko and Glass, 1976).

The transition from the dormant avascular state to the aggressive vascular state, with the tumor now possessing the ability to invade surrounding tissue and metastasize to distant parts of the body, depends upon its ability to coerce new blood vessels from the surrounding tissue to sprout toward and then subsequently penetrate the tumor, thus providing it with an adequate blood supply and microcirculation. In order to accomplish this neovascularization, it is now a well-established fact that tumors secrete a number of diffusible chemical substances into the surrounding tissues and extracellular matrix. Much work has been carried out into the nature of such substances and their effect on endothelial cells since initial research began in the early 1970s with Folkman, culminating in the extraction and purification of several angiogenic factors, the cloning of their genes from libraries of complementary DNA (cDNA) and the determination of their amino acid sequences (Folkman and Klagsbrun, 1987). Throughout the remainder of this chapter we shall refer to these substances generically as tumor angiogenesis factors or TAF.

The first events of angiogenesis are rearrangements and migration of endothelial cells (EC) situated in nearby vessels (Paku and Paweletz, 1991). The main function of EC is in the lining of the different types of vessels such as venules and veins, arterioles and arteries, small lymphatic vessels, and the thoracic duct. They form a single layer of flattened and extended cells and the intercellular contacts are very tight. Large intercellular spaces are not visible and easy penetration of the established layer of cells is impossible. Special processes must take place for the intra- and extravasation of different cellular elements of blood or lymphatic fluids and tumor cells. Even intravascular tumor cells have to induce the formation of gaps in the single layer of EC in order to leave the respective vessels and hence one can consider these cells to be the principal characters in the drama of angiogenesis (Sholley et al., 1977; Paweletz and Knierim, 1989).

In response to the angiogenic stimulus, EC in the neighboring normal capillaries which do not possess a muscular sheath are activated to stimulate proteases and collagenases. The endothelial cells destroy their own basal lamina and start to migrate into the extracellular matrix. Small capillary sprouts are formed by accumulation of EC which are recruited from the parent vessel. The sprouts grow in length by migration of the endothelial cells. Experimental results have demonstrated that EC are continually redistributed among sprouts, moving from one sprout to another. This permits the significant outgrowth of a network of sprouts even when cell proliferation is prevented (Sholley et al., 1984). At some distance from the tip of the sprout the EC divide and proliferate to contribute to the number of migrating EC. The mitotic figures are only observed once the sprout is already growing out and cell division is largely confined to a region just behind the sprout tip. Solid strands of EC are formed in the extracellular matrix. Lumina develop within these strands and mitosis continues. Initially, the sprouts arising from the parent vessel grow outwards toward the tumor in a more or less parallel way to each other. They tend to incline toward each other at a definite distance from the origin when neighboring sprouts run into one another and fuse to form loops or anastomoses. Both tip-tip and tip-branch anastomosis occur and the first signs of circulation can be recognized. From the primary loops, new buds and sprouts emerge and the process continues until the tumor is eventually penetrated.

Once vascularized the tumours grow rapidly as exophytic masses. In certain types of cancer, e.g., carcinoma arising within an organ, this process typically consists of columns of cells projecting from the central mass of cells and extending into the surrounding tissue area. The local spread of these carcinoma often assumes an irregular jagged shape (Chaplain, 1995a,b, 1996). By the time a tumor has grown to a size whereby it can be detected by clinical means, there is a strong likelihood that it has already reached the vascular growth phase.

In this chapter, we present a review of some recent mathematical models which attempt to describe the complex process of tumor-induced angiogenesis (Byrne and Chapain, 1995a,b; Chaplain and Orme, 1996; Chaplain and Sleeman, 1990; Chaplain and Stuart, 1991; 1993; Chaplain et al., 1995; Orme and Chaplain, 1996).

Tumor-Induced Angiogenesis: Capillary Sprout Formation and Growth

In this section and throughout the rest of the chapter, wherever angiogenesis is mentioned, we choose to focus on the model experimental system wherein a fragment of tumor is implanted into the cornea of a test animal and an angiogenic response is elicited from blood vessels situated in the nearby limbus region of the cornea (Gimbrone et al., 1974; Muthukkaruppan et al., 1982). We initially model the process using partial differential equations which describe the rate of change of a given variable at a particular point in space $x = (x,y,z)$ and at a particular time t.

TAF having concentration $c(x,t)$ is secreted by the solid tumor and diffuses into the surrounding tissue. Upon reaching neighboring EC situated in, for example, the limbal vessels, the TAF stimulates the release of enzymes by the EC which degrade their basement membrane. As described in the introduction, after degradation of the basement membrane has taken place, the initial response of the EC is to begin to migrate toward the source of angiogenic stimulus. Capillary sprouts are formed and cells subsequently begin to proliferate at a later stage. Once the capillary sprouts have formed, mitosis is largely confined to a region a short distance behind the sprout tips (Ausprunk and Folkman, 1977; Sholley et al., 1984; Paweletz and Knierim, 1989; Stokes and Lauffenburger, 1991). Ausprunk and Folkman (1977) hypothesized that the reason for this proliferation was that these cells or vessels at the sprout tips were acting as sinks for the TAF. Following Chaplain and Stuart (1991, 1993), we thus incorporate a sink term for the TAF in addition to a natural decay term for the TAF. We assume that the local rate of uptake of TAF by the EC is governed by Michaelis-Menten kinetics with appropriate constants Q and K_m (Chaplain and Stuart, 1991, 1993) and that it also depends on the cell density, i.e., the greater the density of EC, the more TAF will be removed by the cells acting as sinks (Ausprunk and Folkman, 1977). In general, a function $g(n)$ can therefore be chosen to be some strictly increasing function to account for this. For simplicity the actual function used in the model is given by $g(n) = n/n_0$, a simple linear function, where n is the endothelial cell density and n_0 is some reference cell density used to scale the equations (this is described in more detail in the following pages). We also assume that the decay of TAF with time is governed by first-order kinetics with constant decay rate d. With the assumption of linear (Fickian) diffusion with constant diffusion coefficient D_c, this leads to the following equation for the TAF in the external tissue

$$\frac{\partial c}{\partial t} = D_c \nabla^2 c - \frac{Qcn}{(K_m + c)n_0} - dc \tag{2.1}$$

The initial condition is

$$c(x,0) = c_0(x) \tag{2.2}$$

where $c_0(x)$ is a prescribed function chosen to describe qualitatively the profile of TAF in the external tissue when it reaches the limbal vessels. The TAF is assumed to have a constant value

c_0 on the boundary of the tumor and to have decayed to zero at the limbus giving the boundary conditions as

$$c(0,t) = c_0, \quad c(L,t) = 0 \tag{2.3}$$

We now derive our equation for the EC in a similar way. The cascade of events making up the complex process of angiogenesis is essentially driven by the EC. We will thus follow the route of the EC from their origin in their parent vessel (e.g., the limbus), their crossing of the extracellular matrix and other material in the surrounding host tissue, to their destination within the tumor.

The first events of angiogenesis are rearrangements and migration of EC rather than induction of cell division (Paweletz and Knierim, 1989). In response to the angiogenic stimulus, endothelial cells in the neighboring normal capillaries which do not possess a muscular sheath are activated to stimulate proteases and collagenases. The EC destroy their own basal lamina and start to migrate into the extracellular matrix. Small capillary sprouts are formed by accumulation of EC which are recruited from the parent vessel. The sprouts grow in length by migration of the EC (Cliff, 1963; Schoefl, 1963; Warren, 1966; Reidy and Schwartz, 1981; Sholley et al., 1984). Experimental evidence (Sholley et al., 1984) has demonstrated that EC are continually redistributed among sprouts, moving from one sprout to another. This permits the significant outgrowth of a network of sprouts even when cell proliferation is prevented (Sholley et al., 1984). At some distance from the tip of the sprout the EC divide and proliferate to contribute to the number of migrating endothelial cells. The mitotic figures are only observed once the sprout is already growing out and cell division is largely confined to a region just behind the sprout tip. Solid strands of EC are formed in the extracellular matrix. Lumina develop within these strands and mitosis continues. At a certain distance from the parent vessel, the sprouts tend to incline toward each other. Neighboring sprouts run into one another and fuse to form loops or anastomoses. Both tip-tip and tip-branch anastomosis occur and the first signs of circulation can be recognized. From the primary loops, new buds and sprouts emerge and the process continues in a self-similar manner until the tumor is eventually penetrated.

We now attempt to account for the above sequence of events using a population balance equation for the EC and to interpret the processes described above by analysing the endothelial cell density profile. The main events we model are the migration and the proliferation of the EC (the processes of anastomosis and budding will be accounted for implicitly in the model). We note that the migration and replication of EC are not linked together. Different types of stimuli are necessary for these two processes and we take this important fact into account in our model. We begin then with a general conservation equation for the endothelial cell density n(x,t) which is of the form

$$\frac{\partial n}{\partial t} + \nabla . J = F(n)G(c) - H(n) \tag{2.4}$$

where J is the cell flux, F(n) and H(n) are functions representing a normalised growth term and a loss term respectively for the EC. We assume that mitosis is governed by logistic type growth and that cell loss (due to anastomosisis, for example) is a first-order process. Further we assume that the endothelial cell proliferation is controlled in some way by the TAF (Paweletz and Knierim, 1989; Paku and Paweletz, 1991) and this is reflected by the inclusion of the function G(c) which is assumed to be nondecreasing. As stated previously, the initial response of EC to the angiogenic stimulus is one of migration (Paweletz and Knierim, 1989; Paku and Paweletz, 1991). Proliferation is a crucial but secondary response. In order to account for this through the function G(c) we assume that there is a threshold concentration level of TAF below which proliferation does not occur. Thus in the present model we chose G(c) to be of the form

$$G(c) = \begin{cases} 0, & c \leq c^* \\ \dfrac{c - c^*}{c_0}, & c^* < c \end{cases} \tag{2.5}$$

where $c^* \leq c_0$.

There is substantial evidence that the response of the endothelial cells to the presence of the TAF is a chemotactic one (Ausprunk and Folkman, 1977; Stokes and Lauffenburger, 1991) and so we assume that the flux J of EC consists of two parts, one representing random motion and the other chemotactic motion of the cells. Thus

$$J = J_{\text{random}} + J_{\text{chemotaxis}} \tag{2.6}$$

For the EC, we assume nonlinear random motion so that the above equation becomes

$$J = -D_n \left(\frac{n}{n_0} \right)^{\sigma} \nabla n + n\chi_0 \nabla c \tag{2.7}$$

Simple Fickian diffusion is recovered on taking $\sigma = 0$. With all of the above assumptions we thus have the following population diffusion-chemotaxis equation for the EC

$$\frac{\partial n}{\partial t} = D_n \nabla \cdot \left(\left(\frac{n}{n_0} \right)^{\sigma} \nabla n \right) - \chi_0 \nabla \cdot (n \nabla c) + \overbrace{rn\left(1 - \frac{n}{n_0}\right) G(c)}^{F(n)} - \overbrace{k_p n}^{H(n)} \tag{2.8}$$

We assume that initially the endothelial cell density at the limbus is a constant, n_0, and zero elsewhere and that throughout the subsequent motion, the cell density remains constant at the limbus.

As stated previously, the main aim of the model is to monitor the progress of the EC (in particular those at the sprout tips) as they cross the ECM and eventually reach the tumor. Once they reach the tumor and penetrate it, interactions with the tumor cells become important (Paweletz and Knierim, 1989) and the assumptions of the present model no longer hold. The modeling of this stage of the process is considered elsewhere (Liotta et al., 1977; Orme and Chaplain, 1996).

Following Chaplain and Stuart (1991, 1993), we normalize the equations using as reference variables for the TAF concentration, cell density, space and time $c_0, n_0, L, L^2/D_c$, respectively (note that the diffusion coefficient has dimensions $(\text{length})^2(\text{time})^{-1}$). We initially choose a one-dimensional geometry. The equations now become

$$\frac{\partial c}{\partial t} = \frac{\partial^2 c}{\partial x^2} - \frac{\alpha n c}{\gamma + c} - \lambda c \tag{2.9}$$

$$\frac{\partial n}{\partial t} = D_1 \frac{\partial}{\partial x}\left(n^{\sigma} \frac{\partial n}{\partial x} \right) - \kappa \frac{\partial}{\partial x}\left(n \frac{\partial c}{\partial x} \right) + \mu n(1 - n)G(c) - \beta n \tag{2.10}$$

where

$$G(c) = \begin{cases} 0, & c \le c^* \\ c - c^*, & c^* < c \end{cases} \tag{2.11}$$

$$\alpha = \frac{L^2 Q}{D_c c_0}, \gamma = \frac{K_m}{c_0}, \lambda = \frac{L^2 d}{D_c}$$

$$D_1 = \frac{D_n}{D_c}, \kappa = \frac{c_0 \chi_0}{D_c}, \mu = \frac{L^2 r}{D_c} \tag{2.12}$$

$$\beta = \frac{L^2 k_p}{D_c}$$

The above equations are then solved numerically using parameter values taken, as far as possible, from experimental data (see Chaplain and Stuart, 1993 for details). The initial condition for TAF was taken to be $c(x,0) = c_0(x) = 1 - x^2$ which is of the correct qualitative shape for the TAF profile in the external tissue, i.e., a constant value of 1 at the tumor edge decaying away to 0 at the limbus. A zero-flux boundary condition for n was imposed at $x = 0$.

Figures 1 and 2 show the profiles of the TAF concentration and the endothelial cell density respectively in the external host tissue at various times. As can be seen from Figure 2, shortly after $t = 0.7$ the model loses its validity since the EC reach the tumor and interactions between tumor cells and EC become important. The time taken for the EC to first reach the tumor corresponds to a real time of approximately 11 days which is within the experimentally observed timescale (Balding and McElwain, 1985). By varying the parameters μ and κ the time taken for the EC (and hence the capillary sprouts) to reach the tumor can also be varied.

Figure 3 shows the endothelial cell density profile when the chemotactic response parameter $\kappa = 0$. As can clearly be seen, there is very little cell response indicating that chemotaxis, and therefore cell migration, plays a major initiating role in the angiogenic process.

Figure 4 shows the endothelial cell density profile when the cell proliferation parameter $\mu = 0$. Once again the effect of this is significant. The figure shows that cell outgrowth, and hence sprout outgrowth, has virtually ceased after $t = 0.4$ which corresponds to a real time of 3.5 days. This agrees both with experimental evidence (Sholley et al., 1984) and with the model of Stokes and Lauffenburger (1991). The previous two figures demonstrate that neither cell migration (via chemotaxis) nor cell proliferation alone is sufficient for a completion of angiogenesis, agreeing with experimental observations. In order for a completion of angiogenesis both events must be included in the model as observed experimentally.

Comparing Figure 2 with Figure 4 demonstrates a good qualitative agreement with the available experimental evidence. Both simulations give the same profile of cell density for times $0 \le t \le 0.4$. This shows that the initial response of the endothelial cells is essentially one of migration with proliferation of the cells occurring at a later time. The form of the function G(c) ensures that there is always a region within the sprouts where there is zero cell proliferation and also that once cells have started to proliferate the proliferation is mainly confined to a region a short distance behind the sprout tips. Thus the model distinguishes, as far as is possible within its limitations, between proliferating cells near the sprout tip and nonproliferating cells within the rest of

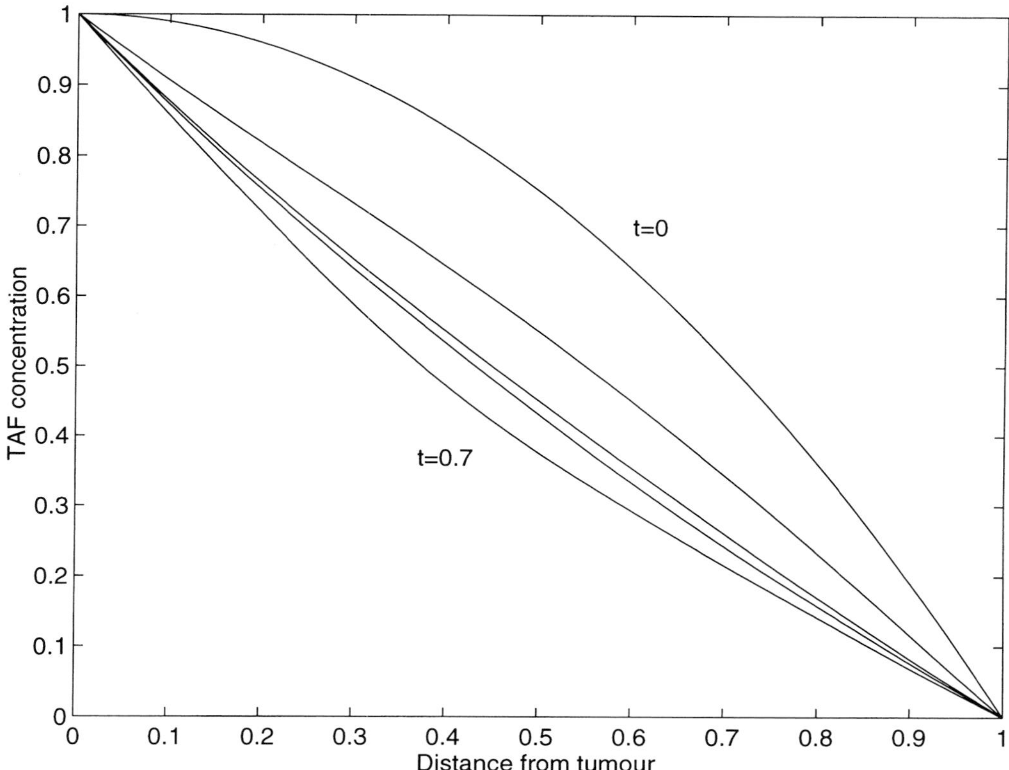

Fig. 1 Profile of the TAF concentration in the external host tissue at times t = 0,0.1,0.3,0.5,0.7 showing changing gradient profile. Parameter values: $\alpha = 10$, $\gamma = 1$, $\lambda = 1$, $D = 0.001$, $\sigma = 2$, $\kappa = 0.7$, $\mu = 100$, $\beta = 4$, $c^* = 0.2$.

the sprout. Also once the cells have started to proliferate the endothelial cell density is greatest (locally) a short distance behind the sprout tips which is what is observed experimentally (Paweletz and Knierim, 1989; Sholley et al., 1984; Ausprunk and Folkman, 1977). Similar results were obtained for different functions G(c) which were of the same qualitative form as (2.11).

For all numerical simulations carried out we took $c^* = 0.2$. Qualitatively similar results were obtained for various other values of c^* between 0.1 and 0.4. We note that other numerical simulations were carried out with $D = 0$ and $\alpha = 0$. In each case, the profiles of TAF concentration and endothelial cell density remained almost the same as when these parameters were nonzero. In the former case this would seem to confirm the results of the models of Stokes and Lauffenburger (1991) where diffusion was also seen to have a negligible effect on the results.

Finally, we note that the numerical simulations of the previous section indicate that the concentration profile for c rapidly reaches a steady-state profile. This is not entirely unexpected since the TAF diffuses much faster than the EC. Thus a reasonable approximation to the TAF concentration profile may be obtained by solving

$$\frac{\partial^2 c}{\partial x^2} - \lambda c = 0 \tag{2.13}$$

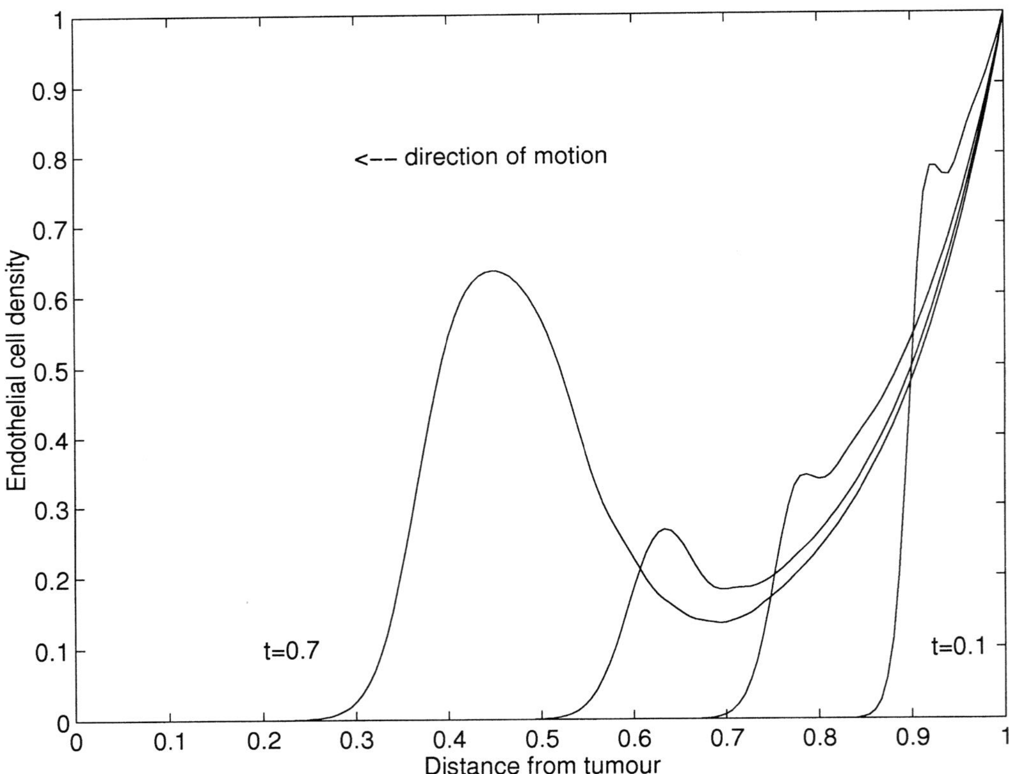

Fig. 2 Profile of the endothelial cell density in the external host tissue at times t = 0,0.1,0.3,0.5,0.7. Shortly after t = 0.7 the endothelial cells reach the tumor and the assumptions of the model no longer hold. Parameter values as per Figure 1.

subject to c = 1 at tumor boundary, c = 0 at all other boundaries. We first solve the above equation in a one-dimensional domain, normalized to be [0,1], yielding c as a function of x only. We assume the tumor implant is situated at x = 0 and the EC at x = 1. The solution to (2.13) (in one-dimensional space) is given by

$$c(x) = \sinh[(1 - x)\sqrt{\lambda}]/\sinh \sqrt{\lambda}$$

It can be shown that a good approximation to the TAF concentration profile with $\lambda = 1$ can be obtained on taking $c = 1 - x$. The single equation to be solved is thus

$$\frac{\partial n}{\partial t} = D_1 \frac{\partial}{\partial x}\left(n^{\sigma} \frac{\partial n}{\partial x}\right) + \kappa \frac{\partial n}{\partial x} + \mu(1 - x)n(1 - n) - \beta n \tag{2.14}$$

with initial conditions

$$n(x,0) = \begin{cases} 1, & x = 1 \\ 0, & \text{elsewhere} \end{cases} \tag{2.15}$$

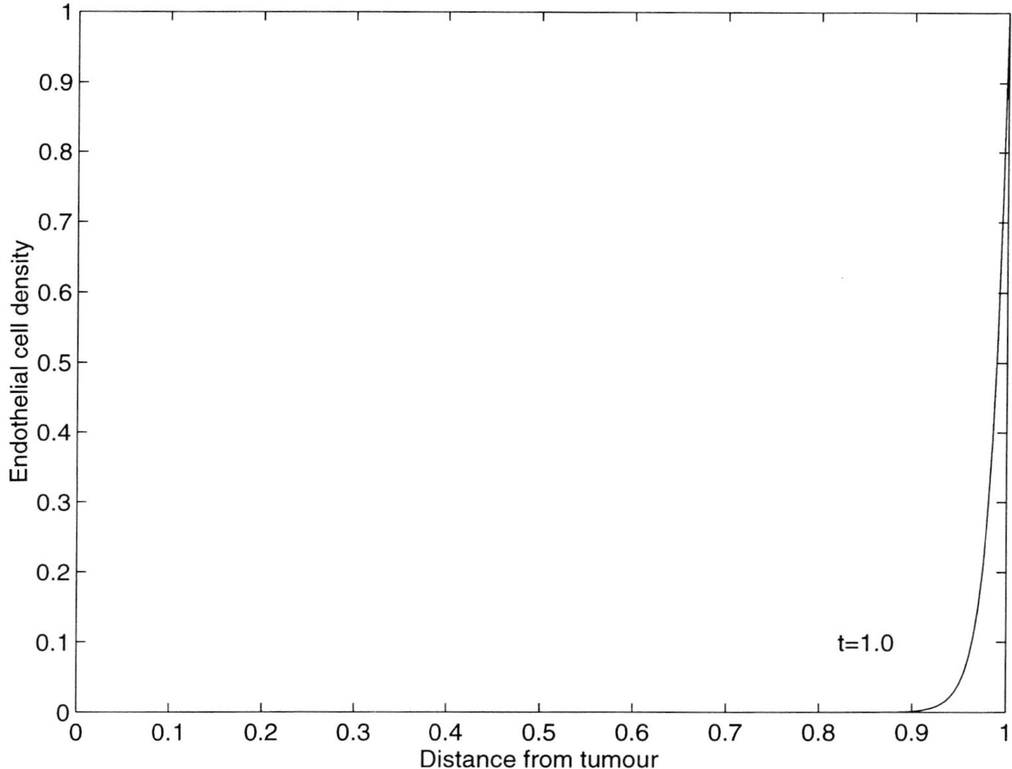

Fig. 3 Profile of the endothelial cell density in the external host tissue at t = 1.0 when the chemotaxis parameter κ is set to zero. The figure shows very little cell, and hence sprout, growth. This illustrates that cell migration is vital to the complete process of angiogenesis. Parameter values: $\alpha = 10$, $\gamma = 1$, $\lambda = 1$, $D = 0.001$, $\sigma = 2$, $\kappa = 0.0$, $\mu = 90$, $\beta = 4$, $c^* = 0.2$.

and boundary conditions $n(1,t) = 1$, and zero-flux for n at $x = 0$. Numerical solutions of this are given in Figure 5 which demonstrates that all the salient features of the full model are captured in this caricature model.

Two-Dimensional Models of Tumor Angiogenesis and Anti-Angiogenesis Strategies

The one-dimensional models of the previous section are in good qualitative agreement with observed experimental data with the following characteristics—the formation of a high density of EC behind the leading tip giving rise to the brush-border effect, dependence upon proliferation of EC for successful completion of vascularization of the tumor, and migration dependent upon chemotaxis. However, experiments are carried out in very thin regions (< 50 μm) of the cornea or ear chamber of test animals (Gimbrone et al. 1974; Muthukkaruppan et al., 1982;

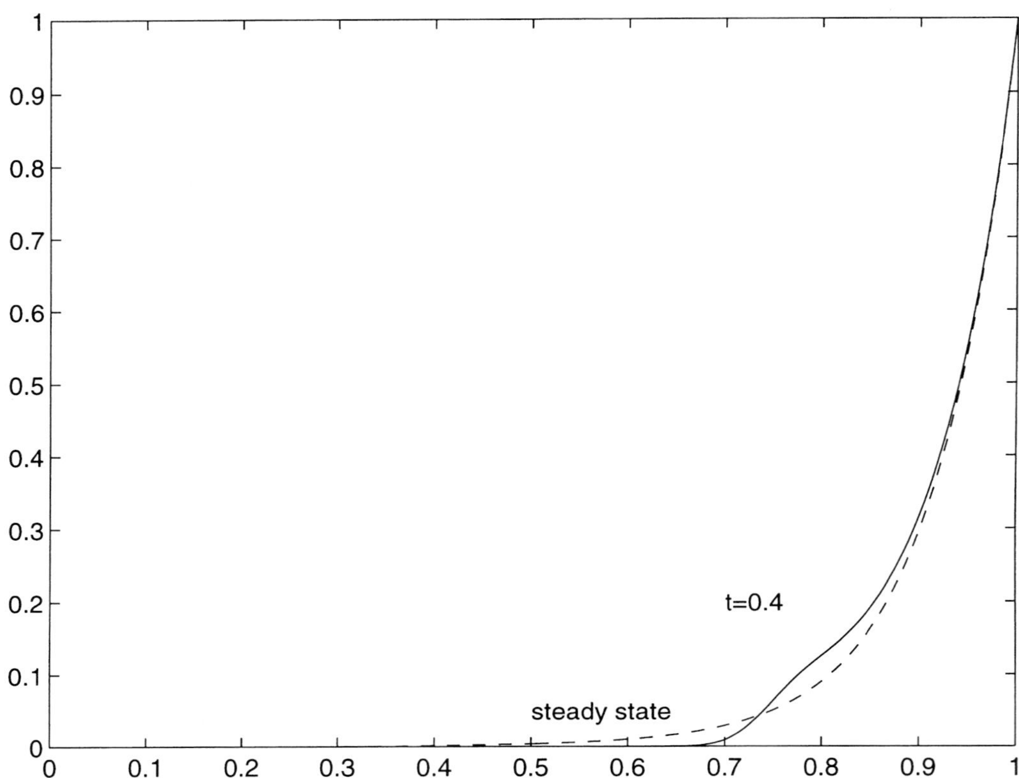

Fig. 4 Profile of the endothelial cell density in the external host tissue at time $t = 0.4$ (full line) and at steady state (hatched line) when the cell proliferation parameter μ is set to zero. All outgrowth has stopped by steady state. The figure clearly shows that at $t = 0.4$ sprout outgrowth has virtually ceased, which is in agreement with the experimental evidence. This illustrates that cell proliferation is essential for the completion of angiogenesis. Parameter values: $\alpha = 10, \gamma = 1, \lambda = 1, D = 0.001, \sigma = 2, \kappa = 0.5, \mu = 0.0, \beta = 6, c^* = 0.2$.

Stokes and Lauffenburger, 1991) and so a two-dimensional domain is a far more realistic approximation of both in vitro and in vivo capillary sprout growth. In this section we build upon the work of the previous section and develop a two-dimensional model of capillary network formation. Using normalized variables, we work on the unit square $[0,1] \times [0,1]$ and assume that the tumor implant occupies the region $(x - 0.5)^2 + y^2 \le 0.1$. We assume that the TAF diffuses in a radially symmetric manner and it can easily be seen that the corresponding solution (in polar coordinates with radial symmetry) is given in terms of Bessel functions but once again we find a very good approximation is given by

$$c(x,y) = c(r) = \begin{cases} 1, & r \le 1 \\ (1-r)^2/0.81, & r > 1 \end{cases} \tag{3.16}$$

where $r^2 = (x - 0.5)^2 + y^2$. This profile is illustrated in Figure 6.

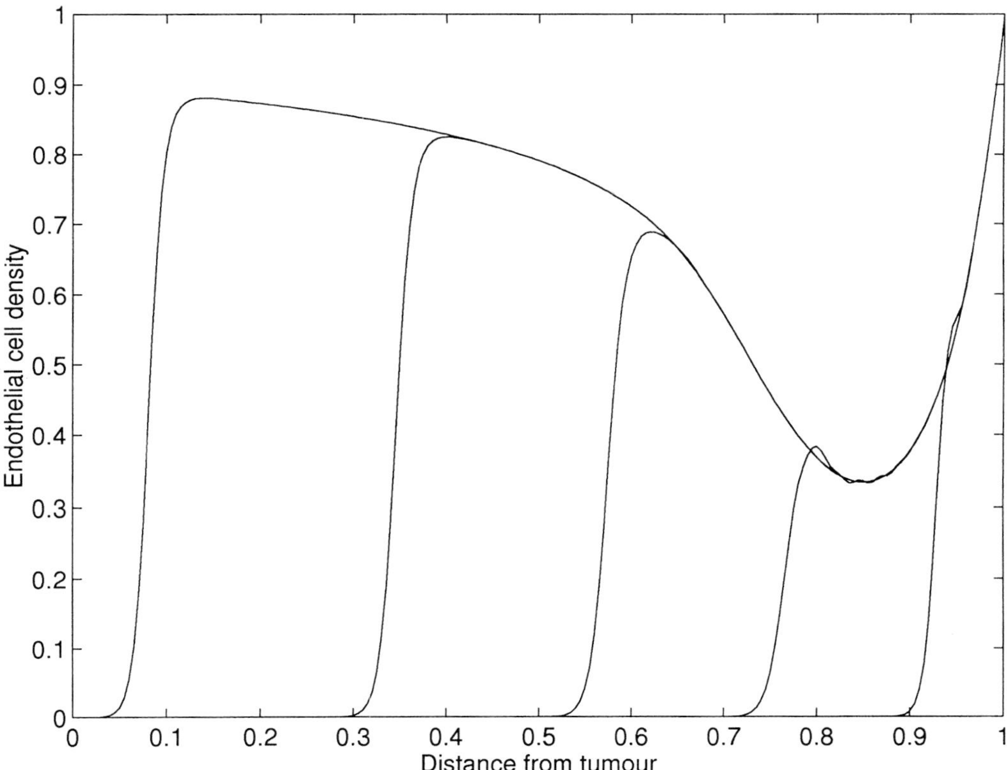

Fig. 5 Profile of endothelial cell density at times t = 0.1,0.3,0.5,0.7,0.9 calculated from caricature Eq. (2.14). The profiles are very similar to those on Figure 2. Parameter values as per Figure 1.

The single equation we now have to solve is given by

$$\frac{\partial n}{\partial t} = D_1 \frac{\partial}{\partial x}\left(n^{\sigma}\frac{\partial n}{\partial x}\right) + D_1 \frac{\partial}{\partial y}\left(n^{\sigma}\frac{\partial n}{\partial y}\right) - \kappa \frac{\partial}{\partial x}\left(n\frac{\partial c}{\partial x}\right) - \kappa \frac{\partial}{\partial y}\left(n\frac{\partial c}{\partial y}\right)$$

$$+ \mu c(x,y)n(1-n) - \beta n - s(x,y)n \tag{3.17}$$

In the one-dimensional model of tumor angiogenesis, the first order loss term $-\beta n$ was considered to be a loss due to anastomosis or the formation of secondary capillary buds. In two or more dimensions, however, these processes are made explicit in the simulations, and hence the inclusion of such a term in the above equation represents a loss due to cell death. Since EC have a long half life (Paweletz and Knierim, 1989), we assume that any death occurs due to external intervention, i.e., the introduction of a cytotoxic drug. The term s(x,y) will be used to model explicitly the formation of the brush-border once the initial capillary buds have fused together via anastomosis.

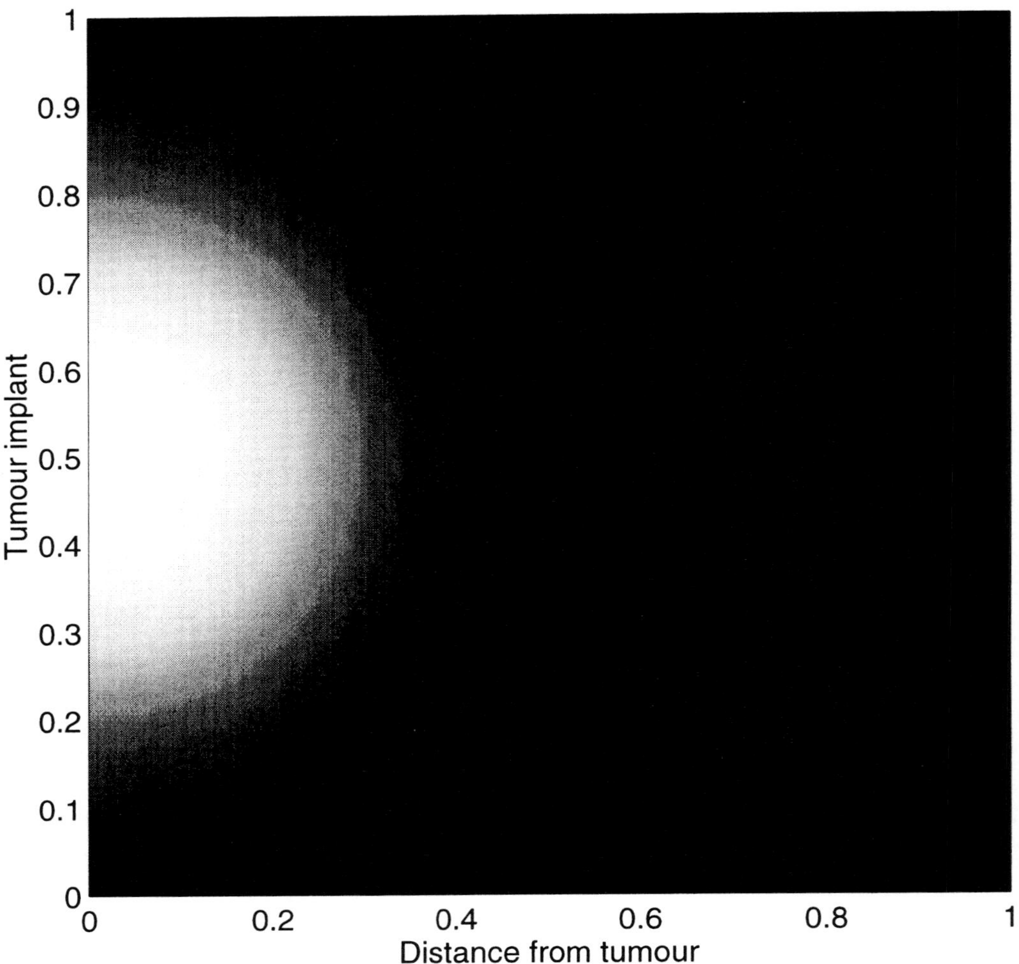

Fig. 6 Profile of the steady state TAF concentration profile used in the two-dimensional simulations of (3.17). White = high concentration, black = low.

We assume that the initial phase of capillary sprout growth has taken place, i.e., the cells have secreted enzymes which degrade their basement membrane permitting the formation of small buds (Paweletz and Knierim, 1989; Paku and Paweletz, 1991). Indeed, Muthukkaruppan et al. (1982) do not consider tumor-induced angiogenesis to be initiated until 2 days after implant and only designate a stage I of angiogenesis after 4 days recognizing that there may well be some limbal response to the actual preparation of the implant site. Paweletz and Knierim (1989) report two different initial mechanisms from which angiogenesis develops:

- Loops of vessels already present in the parental microcirculation are elongated toward the angiogenic stimulus (TAF) by intercalating EC. They always penetrate the tumor implant as loops. Blood flow is possible within these elongating loops, but is realized only after completion of neovascularization. This kind of growth of new vessels apparently occurs after mild angiogenic stimulation and slow development of vascularization.

- During strong angiogenic induction and rapid growth, mostly sprouts which have to fuse at their tips to develop loops are formed.

We present initial conditions for each of the above scenarios. In the first case, we assume a loop (or loops) have already formed from pre-existing vessels (e.g., in the limbus) and assume that the endothelial cell density within this loop is initially a constant $n_0 = 1$ and zero elsewhere giving initial conditions

$$n(x,0) = \begin{cases} n_0 = 1, & x \text{ within loop} \\ 0, & x \text{ elsewhere} \end{cases} \tag{3.18}$$

In the second case, we assume that there are several capillary buds which have formed and we focus attention exclusively upon those EC which are close to the sprout tips since only these cells proliferate. Here the endothelial cell density is initially a constant $n_0 = 1$ and zero elsewhere, giving initial conditions

$$n(x,0) = \begin{cases} n_0 = 1, & x \text{ near sprout tip} \\ 0, & x \text{ elsewhere} \end{cases} \tag{3.19}$$

Since we are attempting to focus attention on the EC near to the sprout tips (since these are the cells which proliferate) and also since all the EC are confined to within the capillary sprouts, we impose a zero-flux boundary condition on the EC density on all boundaries.

Figures 7, 8, and 9 show the evolution of two loops at times $t = 0, 0.3, 0.5$ as they migrate toward the tumor implant with initial conditions as per Figure 7. In this case we chose $s(x,y,) = 0$ since anastomosis and side-branching do not appear as frequently in this situation (Paweletz and Knierim, 1989). As can be seen from the figures the loops continue to develop and progress as loops. Further figures at later times show that they penetrate the tumour as loops which is what is observed experimentally (Paweletz and Knierim, 1989).

Figures 10, 11, and 12 show the EC density profile within the capillary sprouts as they migrate toward the tumor with initial conditions as per Figure 10. In this case we chose

$$s(x,y) = \begin{cases} 10\cos^6(25\pi x), & y < 0.65 \\ 0, & y \geq 0.65 \end{cases} \tag{3.20}$$

representing explicitly the effect of the brush-border once the initial sprouts/loops have all anastomosed (Muthukkaruppan et al., 1982).

Model Improvements and Extensions

While the results of the previous two-dimensional model represent a big step forward in capturing many of the features of angiogenesis explicitly (movement of capillary sprouts toward each other, anastomosis, formation of the brush-border) there is still scope for further refinement and development of the model. Although faithfully reproducing many of the experimental observations (Gimbrone et al., 1974; Muthukkaruppan et al., 1982), criticisms of the model are (i) the coming-together of the sprouts is due, in large measure, to the imposition of the radially-symmetric TAF concentration gradient, (ii) secondary branching or tip-splitting is not observed. We therefore seek to address these questions in a final development of the model. In doing this, we also attempt to (a) devise an experiment against which our model results can be tested, and (b)

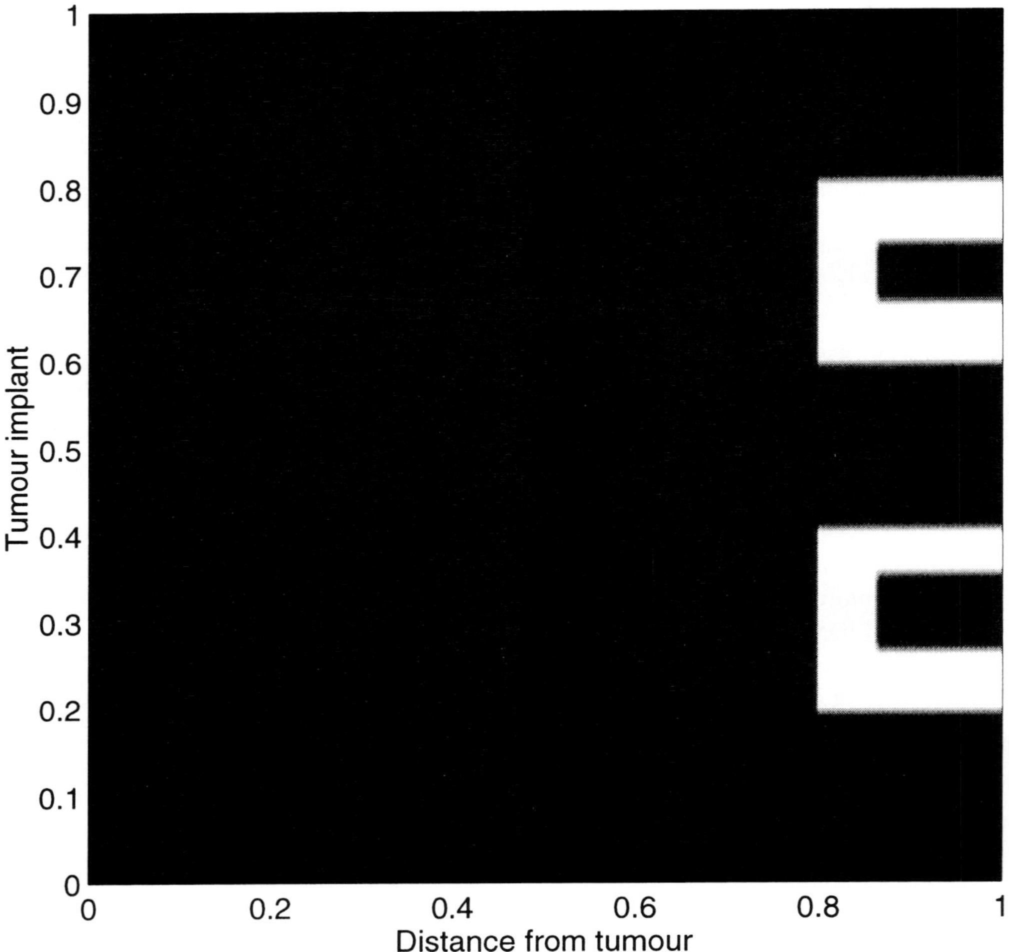

Fig. 7 Profile of the initial conditions used for the endothelial cell density simulating the formation of two loops. White = high density, black = low.

devise anti-angiogenesis strategies for use in cancer treatment. The reason for the latter suggestion is that there is a very strong link between the vascularization of a tumor and the spread of the disease, both locally and to distant sites. The direct supply of nutrients into the tumor results in a rapid increase in growth (Ellis and Fidler, 1995; Gimbrone et al., 1974; Muthukkarruppan et al., 1982). Solid tumor growth is dependent on angiogenesis and any significant increase in tumor size must be preceded by an increase in the vasculature (Norton, 1995; Paweletz and Knierim, 1989). Furthermore, vascularization increases the possibility of tumor cells entering the blood stream (Blood & Zetter, 1990; Ellis and Fidler, 1995; McCulloch et al., 1995) which may consequently lead to metastases. It has been suggested that some measure of the intensity of the tumor vasculature could be used as a prognostic factor (Ellis and Fidler, 1995; Folkman, 1995; McCulloch et al., 1995; Norton, 1995). For example, Frank et al. (1995) have developed a technique for

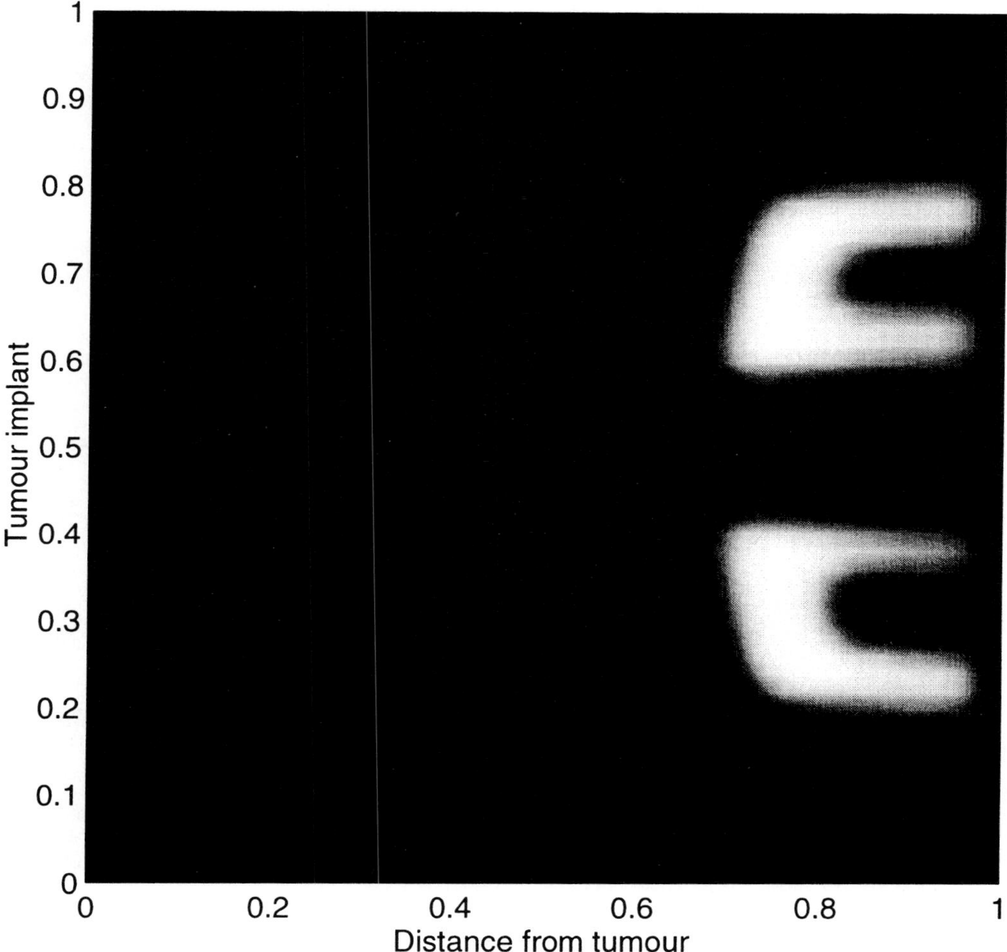

Fig. 8 Profile of the endothelial cell density at time t = 0.3 showing evolution of the two loops. Parameter values: $D = 0.001$, $\sigma = 2$, $\kappa = 0.6$, $\mu = 75$, $\beta = 0$.

grading angiogenesis in order to determine whether increased angiogenesis correlates to higher recurrence and reduced survival in patients suffering from cancer of the colon. It seems clear that anti-angiogenesis strategies could be used to augment existing treatment modalities (Folkman, 1985; Harris et al., 1996). Mathematical models such as the one given here can help us to understand the mechanisms behind angiogenesis and to identify, and perhaps optimize, the different ways by which the angiogenic process can be interrupted.

While our previous models have focused on the role of chemotaxis as the main cause of the EC migration, this is, of course, a simplification of what actually happens. It is known that different TAFs provoke different responses in the EC (Folkman and Klagsbrun, 1985). For example,

- Some TAFs act as a chemoattractant, whereby the EC move up the chemical gradient toward the tumor (chemotaxis).

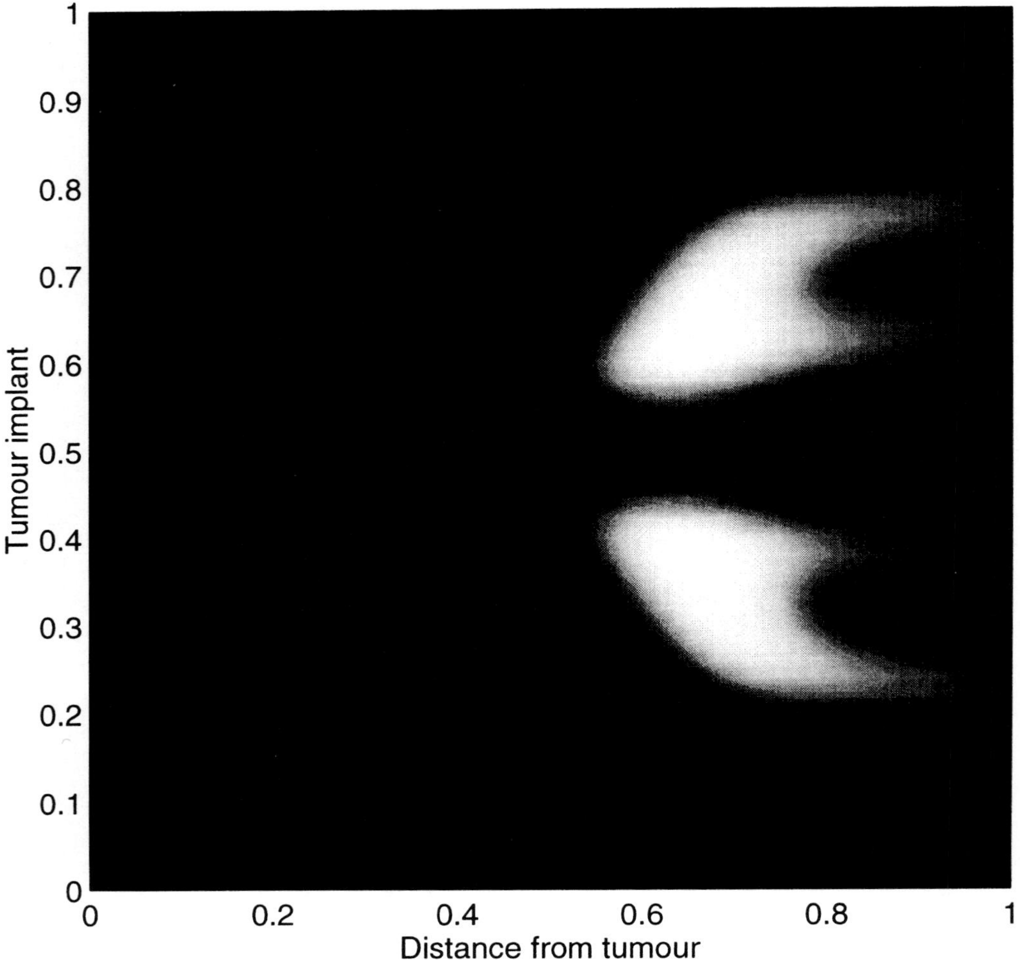

Fig. 9 Profile of the endothelial cell density at time t = 0.5 showing further evolution of the two loops. Parameter values: $D = 0.001$, $\sigma = 2$, $\kappa = 0.6$, $\mu = 75$, $\beta = 0$.

- Other TAFs induce the EC into secreting adhesive substances, e.g., fibronectin, collagens (Alessandri et al., 1986; Terranova et al., 1985; Ungari et al., 1985) and this creates an adhesive gradient which the EC move up—haptotaxis (Carter, 1965).

We now seek to modify our model in light of the latter observation. By doing so we hope to capture some of the observations on a finer scale (e.g., tip-splitting). We therefore incorporate into our model another equation (in addition to those for TAF concentration and EC density) representing a third chemical as outlined below.

Once again we denote by n(x,t) the density of the EC, c(x,t) the concentration of TAF, and p(x,t) the density of an adhesive chemical, such as fibronectin, at position x and time t, and then the general conservation equations are

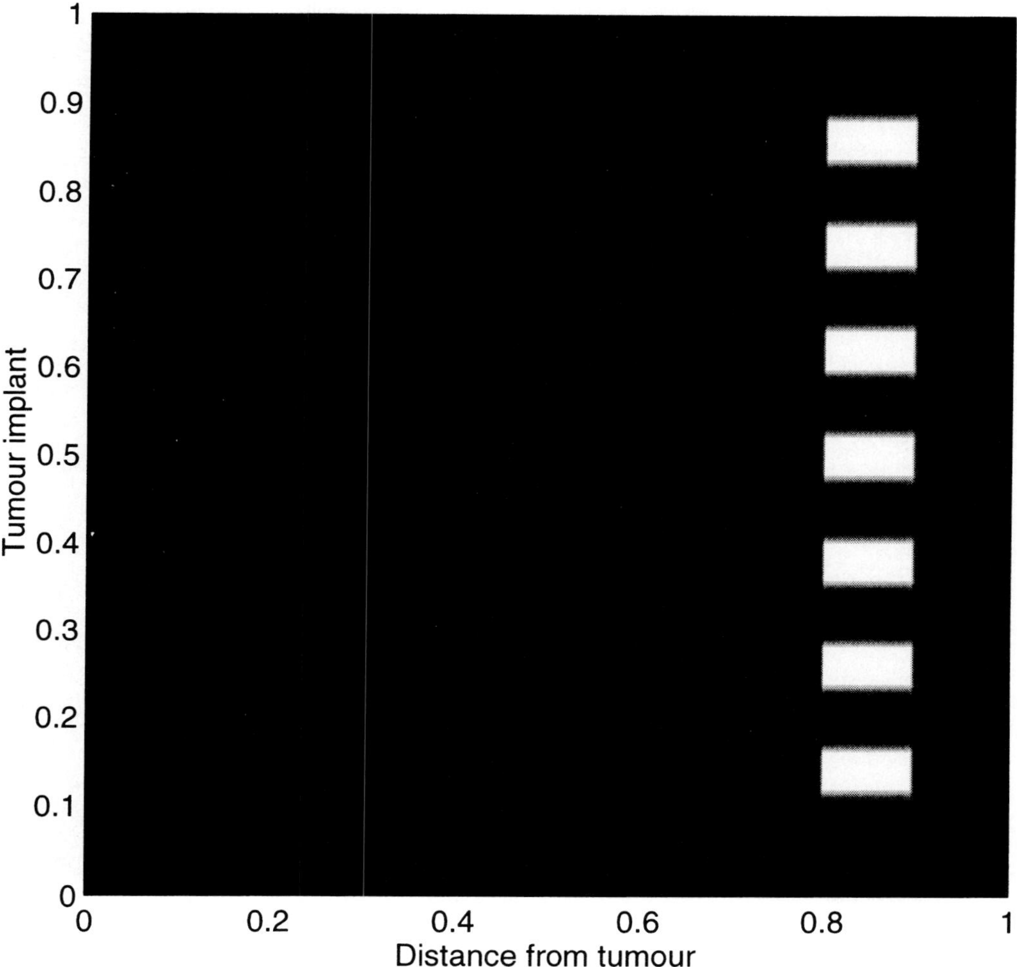

Fig. 10 Profile of the initial conditions used for the endothelial cell density simulating the formation of several capillary sprouts. White = high density, black = low.

$$\frac{\partial n}{\partial t} = -\nabla.J + P_1(n,p,c) \tag{4.21}$$

$$\frac{\partial p}{\partial t} = D_2\nabla^2 p + S_1(n,p,c) + P_2(n,p,c) \tag{4.22}$$

$$\frac{\partial c}{\partial t} = D_3\nabla^2 c + S_2(n,p,c) + P_3(n,p,c) \tag{4.23}$$

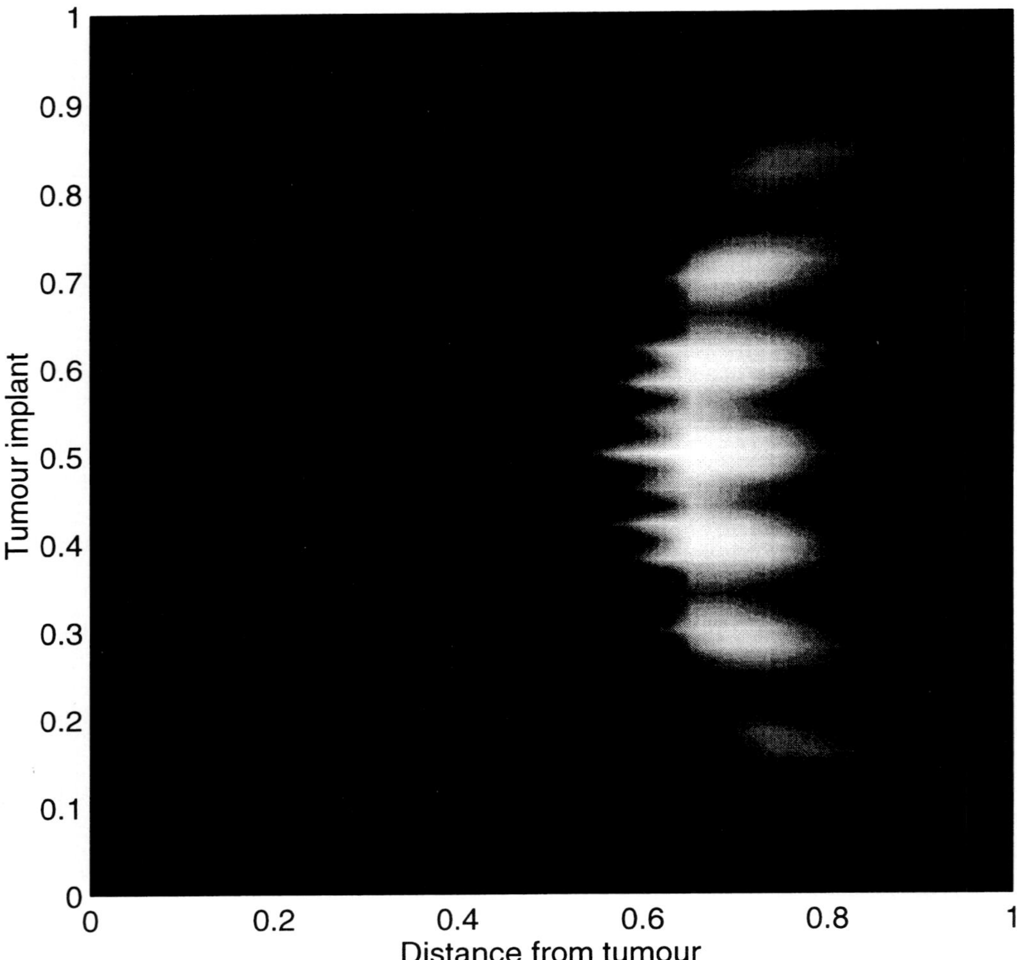

Fig. 11 Profile of the endothelial cell density after the capillary network has been established and anastomosis has taken place corresponding to a real time of 5–6 days after initial sprouts have formed as per Figure 3.9. Parameter values: $D = 0.001$, $\sigma = 2$, $\kappa = 0.6$, $\mu = 75$, $\beta = 0$.

where J is the flux of the EC, D_i, $(i = 2,3)$ are the (constant) diffusion coefficients, P_i, $(i = 1,2,3)$ are net production/loss terms and S_i, $i = 1,2$ are sink terms modeling the uptake of the chemotactic/haptotactic chemical by the EC. These terms will be made explicit below.

First, we assume that the flux of the EC is governed by random motion, chemotaxis and haptotaxis, such that we have

$$J = -D_1 \nabla n + \chi n \nabla p + \kappa n \nabla c$$

where D_1 is the cell random motility coefficient, χ is the haptotaxis coefficient and κ is the chemotaxis coefficient. For simplicity, we assume that D_1, χ, κ are constant. Once again we assume that

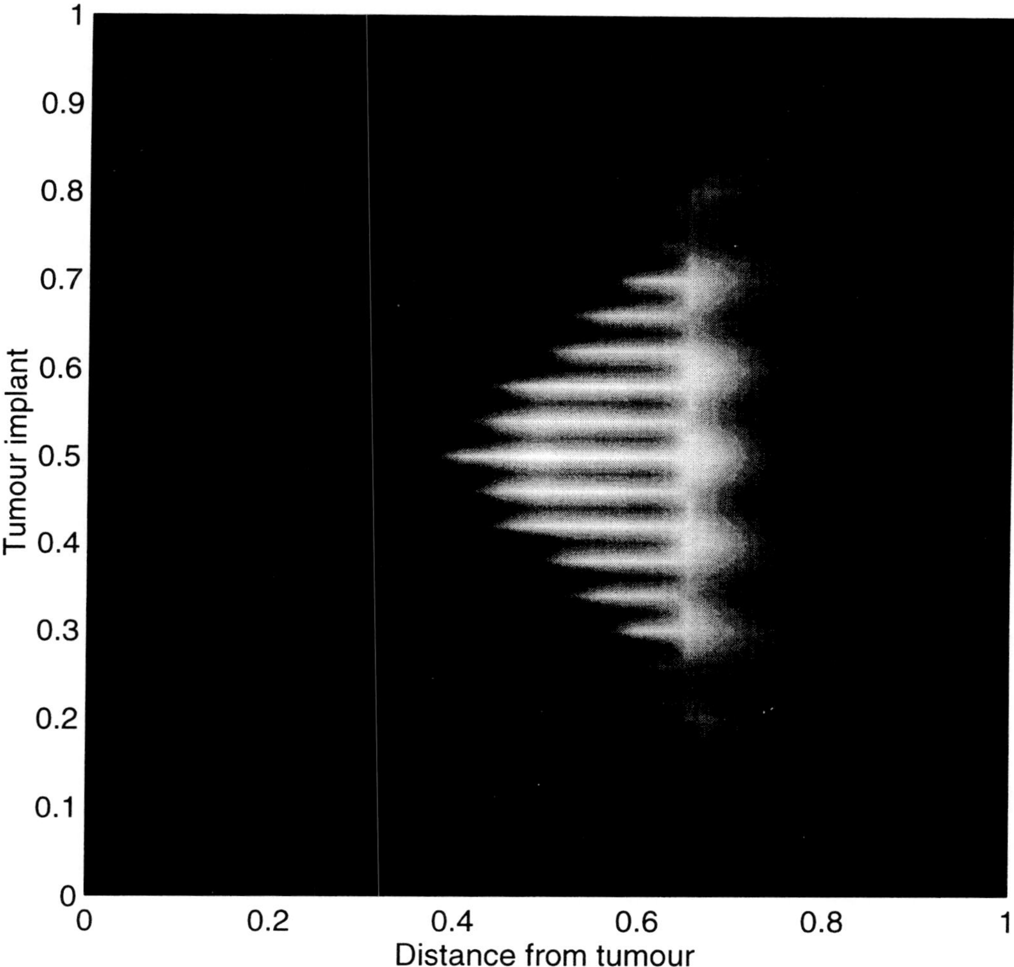

Fig. 12 Profile of the endothelial cell density after the capillary network has been established and anastomosis has taken place corresponding to a real time of 7–8 days after initial sprouts have formed as per Figure 3. Brush-border effect is clearly visible at the vascular front. Parameter values: $D = 0.001$, $\sigma = 2$, $\kappa = 0.6$, $\mu = 75$, $\beta = 0$.

the proliferation of the EC is governed by logistic type growth and that any cell loss is linear. Hence we assume that P_1 takes the form

$$P_1(n,p,c) = \mu n\left(1 - \frac{n}{n_0}\right) - \beta n,$$

where μ is the proliferation rate of the cells, n_0 is the maximum sustainable cell density and β is the rate of cell loss due to external intervention, i.e., the introduction of a cytotoxic drug. In this model, we suppose that the initial release of TAF induces the EC into secreting an adhesive (haptotactic) chemical p, which saturates as p increases. If B is the threshold level of the haptotactic chemical above which the production of p by the EC is switched on and if α is the maximum production rate per cell, then

$$P_2 = \frac{\alpha np}{B + p} - \text{decay}$$

Here, the chemical production term shows the response of the EC to changes in the adhesive chemical density and the term saturates as p increases, as required. We assume that any uptake of the haptotactic and chemotactic chemical by the EC to be of the form

$$S_1 = -s_1 np,\ S_2 = -s_2 nc$$

where s_1 and s_2 are the rate of uptake of the haptotactic and chemotactic chemical, per cell. The decay of the two chemicals is assumed to be linear. Hence, the full model is

$$\frac{\partial n}{\partial t} = D_1 \nabla^2 n - \chi \nabla.(n \nabla p) - \kappa \nabla.(n \nabla c) + \mu n \left(1 - \frac{n}{n_0}\right) - \beta n \tag{4.24}$$

$$\frac{\partial p}{\partial t} = D_2 \nabla^2 p + \frac{\alpha np}{B + p} - s_1 np - \lambda_1 p \tag{4.25}$$

$$\frac{\partial c}{\partial t} = D_3 \nabla^2 c - s_2 nc - \lambda_2 c \tag{4.26}$$

We assume a two-dimensional geometry such that the model equations hold on the square domain $D = [0,L] \times [0,L]$. We assume that the tumour is located along the x-axis and that the parent capillary vessel lies along the line $y = 1$ so that L is the perpendicular distance from the tumor to the parent vessel. By using the above two-dimensional geometry, the model is, in theory, experimentally reproducible. For example, this model could represent an in vitro experiment, whereby tumor cells are placed in a line along one edge of a square petri dish with EC placed along the opposite edge. Alternatively, we could focus upon haptotaxis by suspending TAF in gel, so that the gradient of TAF is constant.

In order to normalize the equations, we define the following reference variables. Let n_0 be a reference endothelial cell density, such as the carrying capacity of the system, p_0 be a typical concentration of the adhesive chemical during angiogenesis, and c_0 be the initial TAF concentration at the tumor boundary. Hence, we nondimensionalize by making the following substitutions:

$$\tilde{n} = \frac{n}{n_0},\ \tilde{p} = \frac{p}{p_0},\ \tilde{c} = \frac{c}{c_0},\ \tilde{t} = \frac{t}{\tau},\ \tilde{x} = \frac{x}{L},\ \tilde{y} = \frac{y}{L}$$

$$\tilde{\chi} = \frac{p_0 \tau \chi}{L^2},\ \tilde{\kappa} = \frac{c_0 \tau \kappa}{L^2},\ \tilde{\mu} = \mu\tau,\ \tilde{\beta} = \beta\tau,\ \tilde{D}_i = \frac{D_i \tau}{L^2},\ i = 1,2,3 \tag{4.27}$$

$$\tilde{\alpha} = \frac{\alpha n_0 \tau}{p_0},\ \tau = \frac{1}{s_1 n_0},\ \tilde{s}_2 = \frac{s_2}{s_1},\ \tilde{B} = \frac{B}{p_0},\ \tilde{\lambda}_i = \lambda_i \tau,\ i = 1,2$$

Dropping the tildes for notational convenience, the full model equations are

$$\frac{\partial n}{\partial t} = D_1\left(\frac{\partial^2 n}{\partial x^2} + \frac{\partial^2 n}{\partial y^2}\right) - \chi\left(\frac{\partial n}{\partial x}\frac{\partial p}{\partial x} + \frac{\partial n}{\partial y}\frac{\partial p}{\partial y} + n\left(\frac{\partial^2 p}{\partial x^2} + \frac{\partial^2 p}{\partial y^2}\right)\right)$$

$$-\kappa\left(\frac{\partial n}{\partial x}\frac{\partial c}{\partial x} + \frac{\partial n}{\partial y}\frac{\partial c}{\partial y} + n\left(\frac{\partial^2 c}{\partial x^2} + \frac{\partial^2 c}{\partial y^2}\right)\right) + \mu n(1-n) - \beta n \qquad (4.28)$$

$$\frac{\partial p}{\partial t} = D_2\left(\frac{\partial^2 p}{\partial x^2} + \frac{\partial^2 p}{\partial y^2}\right) + \frac{\alpha np}{B+p} - np - \lambda_1 p \qquad (4.29)$$

$$\frac{\partial c}{\partial t} = D_3\left(\frac{\partial^2 c}{\partial x^2} + \frac{\partial^2 c}{\partial y^2}\right) - s_2 nc - \lambda_1 c \qquad (4.30)$$

Initial and Boundary Conditions

The initial conditions are as follows:

- If $y \geq 0.9$ and $0.11 \leq x \leq 0.17$, or $0.35, \leq x \leq 0.41$, or $0.59 \leq x \leq 0.65$, or $0.83 \leq x \leq 0.89$, then $n(x,y,0) = 1$. Otherwise, $n(x,y,0) = 0$.
- $p(x,y,0) = n(x,y,0)/2$.
- $c(x,y,0) = 1 - y$.

Thus, we assume that there are initially four capillary sprouts and equivalently, four foci of fibronectin. For all the numerical simulations, we took zero flux boundary conditions, except for the EC at the boundary $y = 1$, for which we assumed that the capillary sprouts were fixed to that parent vessel, i.e., if $0.11 \leq x \leq 0.17$, or $0.35 \leq x \leq 0.41$, or $0.59 \leq x \leq 0.65$, or $0.83 \leq x \leq 0.89$, then $n(x,y, = 1, t) = 1$. Otherwise, $n(x,y = 1,t) = 0$.

Model Simplification Which Emphasizes the Role of Haptotaxis

In order to focus attention upon the role of haptotaxis in angiogenesis, we simplify the equations by once again assuming that the TAF concentration profile does not vary drastically over time. As in the previous sections, it is reasonable to assume that the TAF profile is in some kind of steady state, since the TAF diffuses much faster than the EC and is also secreted by the tumor cells before the EC secrete the haptotactic chemical. Henceforth, we assume that the TAF has reached its steady state and we approximate the TAF profile by $c(x,y) = 1 - y$. Thus, in the next section, we solve numerically the simplified model,

$$\frac{\partial n}{\partial t} = D_1\left(\frac{\partial^2 n}{\partial x^2} + \frac{\partial^2 n}{\partial y^2}\right) - \chi\left(\frac{\partial n}{\partial x}\frac{\partial p}{\partial x} + \frac{\partial n}{\partial y}\frac{\partial p}{\partial y} + n\left(\frac{\partial^2 p}{\partial x^2} + \frac{\partial^2 p}{\partial y^2}\right)\right)$$

$$+\kappa\frac{\partial n}{\partial y} + \mu n(1-n) - \beta n \tag{4.31}$$

$$\frac{\partial p}{\partial t} = D_2\left(\frac{\partial^2 p}{\partial x^2} + \frac{\partial^2 p}{\partial y^2}\right) + \frac{\alpha np}{B+p} - np - \lambda_1 p \tag{4.32}$$

Numerical Simulations of Simplified Model

The system of Eq. (4.31)–(4.32) was solved numerically. In our first numerical simulation, we solve the system (4.31)–(4.32) with parameters $\kappa = 0.65$, $\mu = 5$, $\beta = 0$, $\chi = 0.5$, $\lambda_1 = 0.5$, $\alpha = 5$, $B = 0.001$, $D_1 = 0.0025$, $D_2 = 0.5$. Figures 13 and 14 show the resultant growth of the cap-

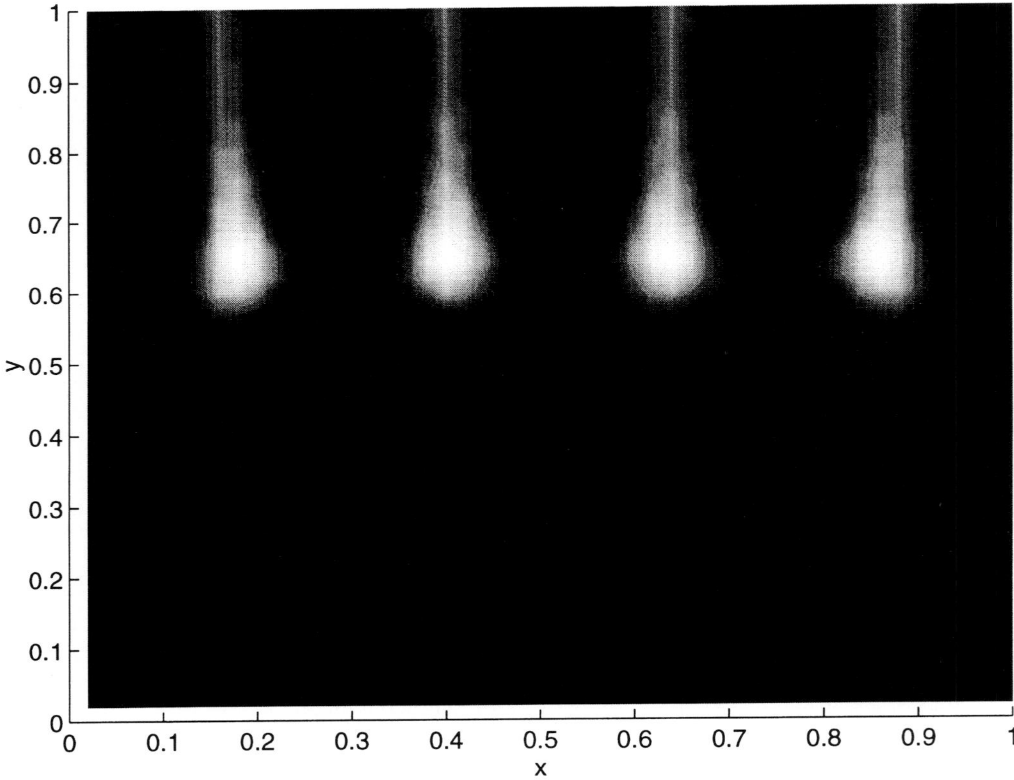

Fig. 13 Profile of the endothelial cell density at time t = 0.8. We can see the beginning of T-shaped branch formation at the tip of each capillary sprout. Fixed TAF profile $c = 1 - y$ and parameter values $\kappa = 0.6$, $\mu = 5$, $\beta = 0$, $\chi = 0.5$, $\lambda_1 = 0.5$, $\alpha = 5$, $B = 0.001$, $D_1 = 0.0025$, $D_2 = 0.5$.

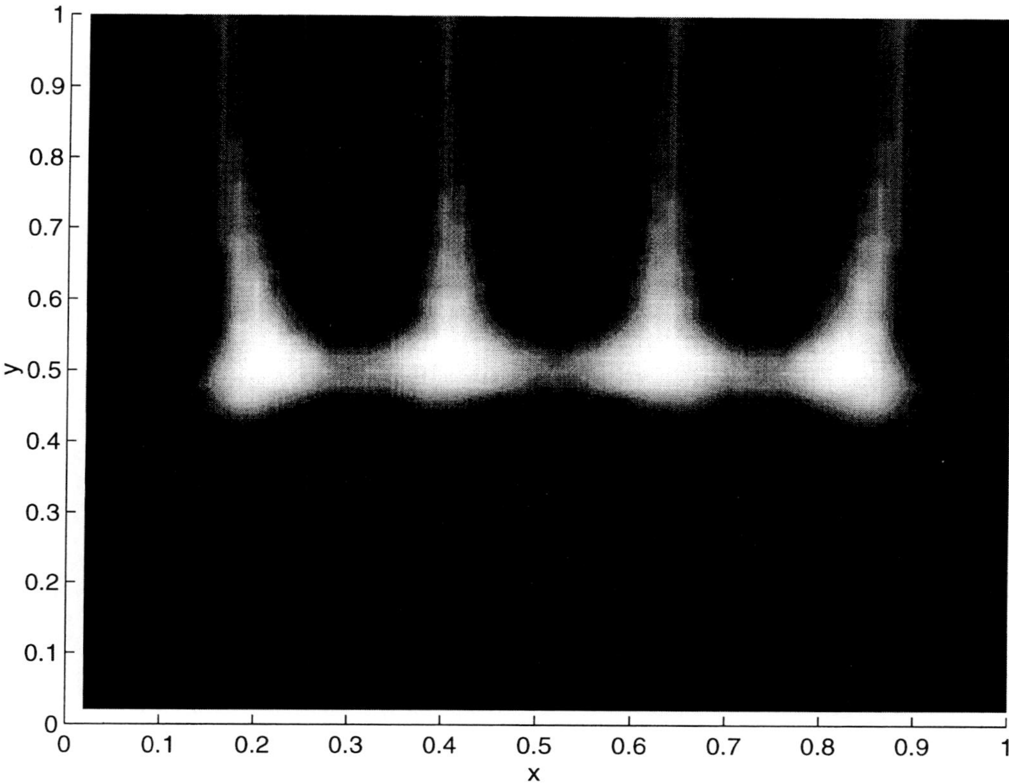

Fig. 14 This figure shows the distribution of the EC n at time t = 1.2. We can see the formation of anastomoses via branch tip to branch tip fusion. Furthermore, the capillaries have a well-defined structure which correlates to a good flow of blood. Fixed TAF profile $c = 1 - y$ and parameter values $\kappa = 0.6$, $\mu = 5$, $\beta = 0$, $\chi = 0.5$, $\lambda_1 = 0.5$, $\alpha = 5$, $B = 0.001$, $D_1 = 0.0025$, $D_2 = 0.5$.

illary vessels through the host tissue. In Figure 13, we can see the beginning of secondary branch formation at the tip of each capillary sprout, and these branches subsequently merge to form anastomoses Figure 14 via branch-tip to branch-tip fusion (Konerding et al., 1992). Furthermore, the capillaries have a well-defined structure which is necessary for blood to flow through the vessels. Note that there is a higher density of tumor cells at the front of the capillary vessel which is the brush border effect observed by (Muthukkaruppan et al., 1982). Figure 15 shows the corresponding distribution of the fibronectin concentration at time t = 1.2.

Anti-Angiogenesis Strategies

In the numerical simulations, we consider four different ways by which the angiogenic process can be disrupted. These four approaches represent viable anti-angiogenesis strategies, which can be used in conjunction with more established treatment modalities (Folkman, 1995;

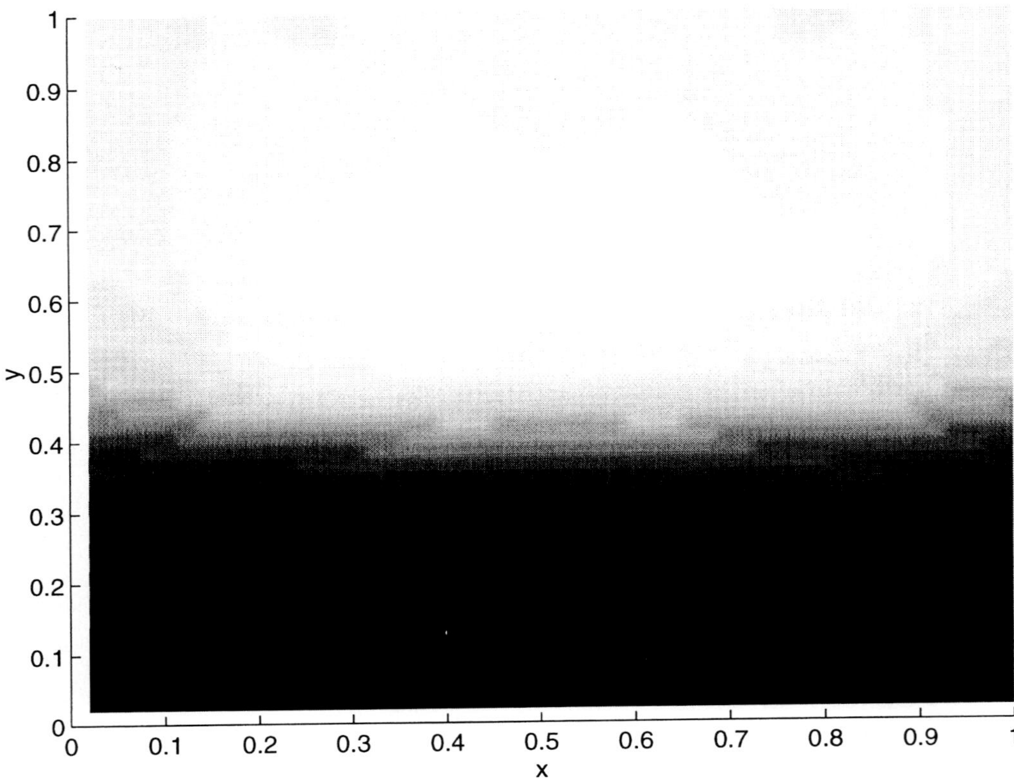

Fig. 15 Profile of the distribution of fibronectin p after a time t = 1.2. Fixed TAF profile c = 1 − y and parameter values $\kappa = 0.6$, $\mu = 5$, $\beta = 0$, $\chi = 0.5$, $\lambda_1 = 0.5$, $\alpha = 5$, B = 0.001, $D_1 = 0.0025$, $D_2 = 0.5$.

Norton, 1995). Furthermore, we explicitly target the endothelium in the neovasculature so that, in theory, the damage to normal tissue is minimized (Brooks et al., 1994; Folkman, 1995).

Cytotoxic Targeting of EC

First, we consider the use of cytotoxic therapy which preferentially kills EC. Such a strategy is most beneficial if preexisting blood vessels can be left unaffected. For example, Brooks et al. (1994) demonstrated that antagonists of integrin $\alpha v \beta_3$ disrupted tumor angiogenesis by selectively inducing apoptosis in EC during the proliferative phase of the cell cycle. Since only the EC in the neovasculature undergo mitosis on the time scale under consideration, the adjacent vessels are left intact. We model this by setting $\beta = 50$. As expected, such a strategy results in the complete regression of the capillary sprouts, leaving no capillaries present in the domain (see Fig. 16).

Inhibition of Cell Mitosis

It has been demonstrated that proliferation of EC is vital for the successful completion of angiogenesis (Ausprunk and Folkman, 1977; Paweletz and Knierim, 1989). Recently, chemical

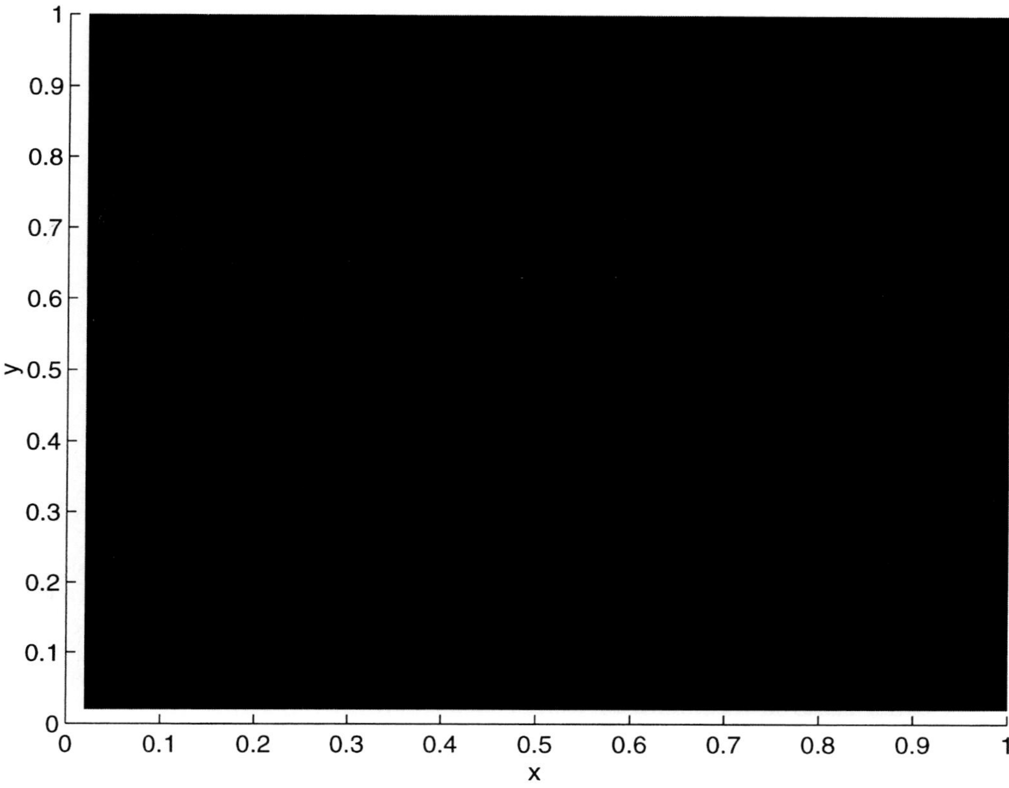

Fig. 16 Profile of the EC density n at time t = 0.1 modeling an anti-angiogenesis strategy whereby the angiogenic process is impeded by the use of endothelial cell-specific drugs, i.e., cytotoxic agents which preferentially kill EC. We see that the capillary sprouts die away completely. Fixed TAF profile $c = 1 - y$ and parameter values $\kappa = 0.6$, $\mu = 5$, $\beta = 50$, $\chi = 0.5$, $\lambda_1 = 0.5$, $\alpha = 5$, $B = 0.001$, $D_1 = 0.0025$, $D_2 = 0.5$.

agents, such as angiostatin (Folkman, 1995; O'Reilly et al., 1994) have been isolated, which specifically inhibit EC proliferation and thus inhibit angiogenesis. Since, the cell doubling time of EC in the absence of TAF is long (months; Paweletz and Knierim, 1989) in comparison with the half life of angiostatin (2.5 days; Folkman, 1995), only the newly formed vasculature would be affected. We model this by setting the cell proliferation rate μ to zero. In the numerical simulation, the capillary sprouts stopped growing after a time t = 0.1 (Fig. 17). O'Reilly et al., 1994 found that angiostatin inhibited angiogenesis 48 hours after implantation, which gives us a value of $\tau = 480$, which is within our estimated range.

Prevention of Cell Migration: (1) Anti-Chemotaxis

Cell migration has been identified as a key event in tumor angiogenesis. Cell migration can be disrupted by interfering with the cells' ability to detect local chemical gradients. For example, EC are known to react chemotactically to hepatocyte growth factor (HGF) (Bussolino et al.,

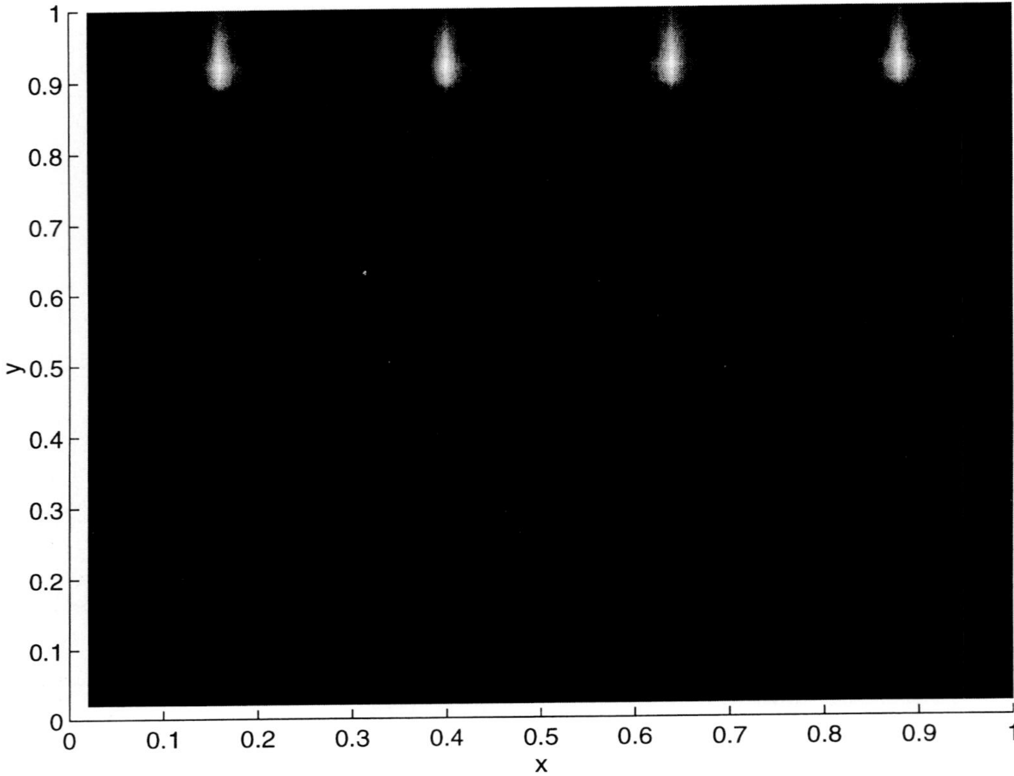

Fig. 17 Profile of the EC density n at time t = 0.1 modeling another potential anti-angiogenesis strategy, namely the prevention of EC mitosis, e.g., by irradiation or by use of an inhibitor. In this case, the growth of the capillary sprouts is arrested. Fixed TAF profile $c = 1 - y$ and parameter values $\kappa = 0.6$, $\mu = 0$, $\beta = 0$, $\chi = 0.5$, $\lambda_1 = 0.5$, $\alpha = 5$, $B = 0.001$, $D_1 = 0.0025$, $D_2 = 0.5$.

1992). It is possible to cultivate antibodies against the HGF receptor (Bussolino et al., 1992) and hence prevent chemotaxis. We model this by setting our chemotaxis coefficient κ to zero to obtain the numerical solution as shown in Figure 18. This shows that in the absence of a detectable TAF gradient, angiogenesis fails.

Prevention of Cell Migration: (2) Anti-Haptotaxis

It is known that fibronectin increases cell–cell and cell–matrix adhesiveness. Yamada and Olden (1978) showed that EC have a specific receptor for fibronectin. By blocking the fibronectin receptors, we prevent the EC from reacting haptotactically to fibronectin. We model this by setting the haptotaxis coefficient χ to zero. At first, this method does not seem to have impeded the growth of the capillary sprouts. At time t = 0.6 there is evidence of anastomoses (Fig. 19), though the loops do not appear to have been formed by the fusion of branch tips. Furthermore, the cap-

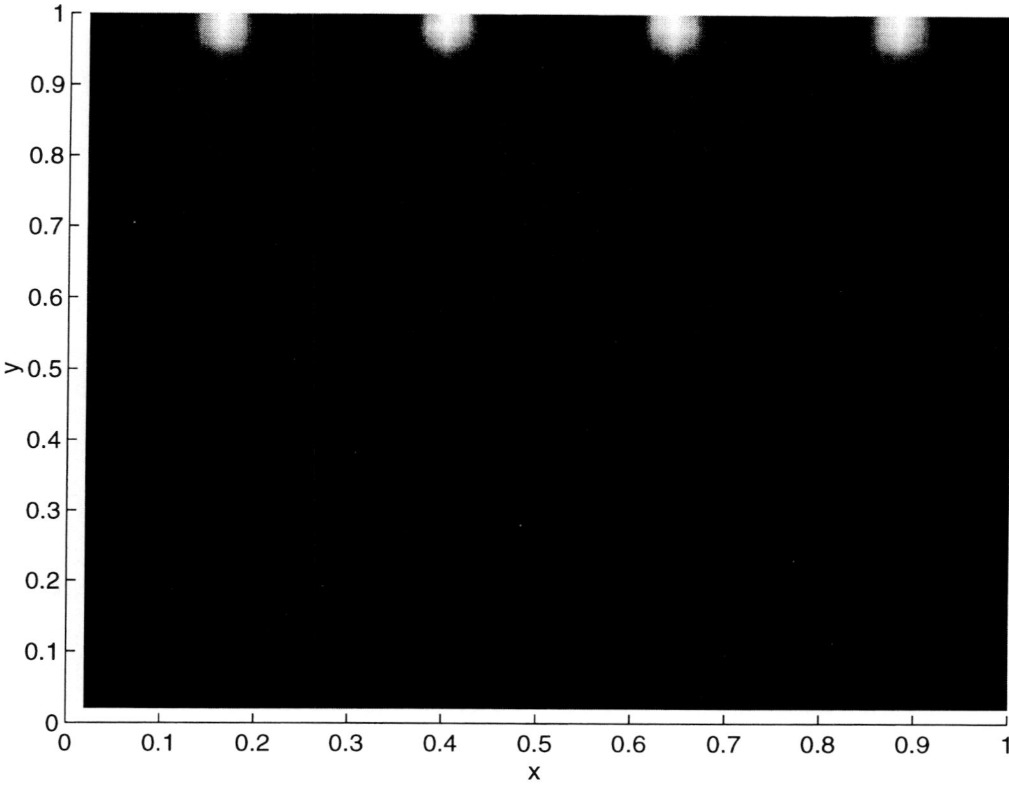

Fig. 18 Profile of EC density n at time t = 0.2 modeling an anti-angiogenesis strategy whereby the EC are unable to react to the chemotactic stimulus, e.g., by disrupting the appropriate receptors on the surface of the cell. The capillary vessels regress and angiogenesis fails. Fixed TAF profile $c = 1 - y$ and parameter values $\kappa = 0$, $\mu = 4.5$, $\beta = 0$, $\chi = 0.5$, $\lambda_1 = 0.5$, $\alpha = 5$, $B = 0.001$, $D_1 = 0.0025$, $D_2 = 0.5$.

illary vessels are not as distinct in comparison with the vessels in Figures 13 and 14. This becomes more evident in Figure 20 where we can see that the EC have not formed well-defined structures. We would expect the circulation of blood through such inferior vessels to be poor and hence, the angiogenic process has failed to produce a viable network of capillaries. Hence, we conclude that the compactness of the vessels shown in in Figures 13 and 14 as compared to Figures 19 and 20 is due to haptotaxis.

Finally, we conducted a parameter sensitivity analysis on this model and found that the following parameter changes had an equivalent effect on the resultant solutions. Increasing the proliferation rate, increasing the diffusion coefficient of the haptotactic chemical or decreasing the haptotaxis coefficient all resulted in a loss of definition (compactness) of the capillary sprouts (similar to the result is shown in Fig. 20). Furthermore, the same effect could be achieved by increasing or decreasing the secretion rate of the haptotactic chemical. This implies that there is some optimal level of fibronectin production, such that the EC cannot respond to too little chem-

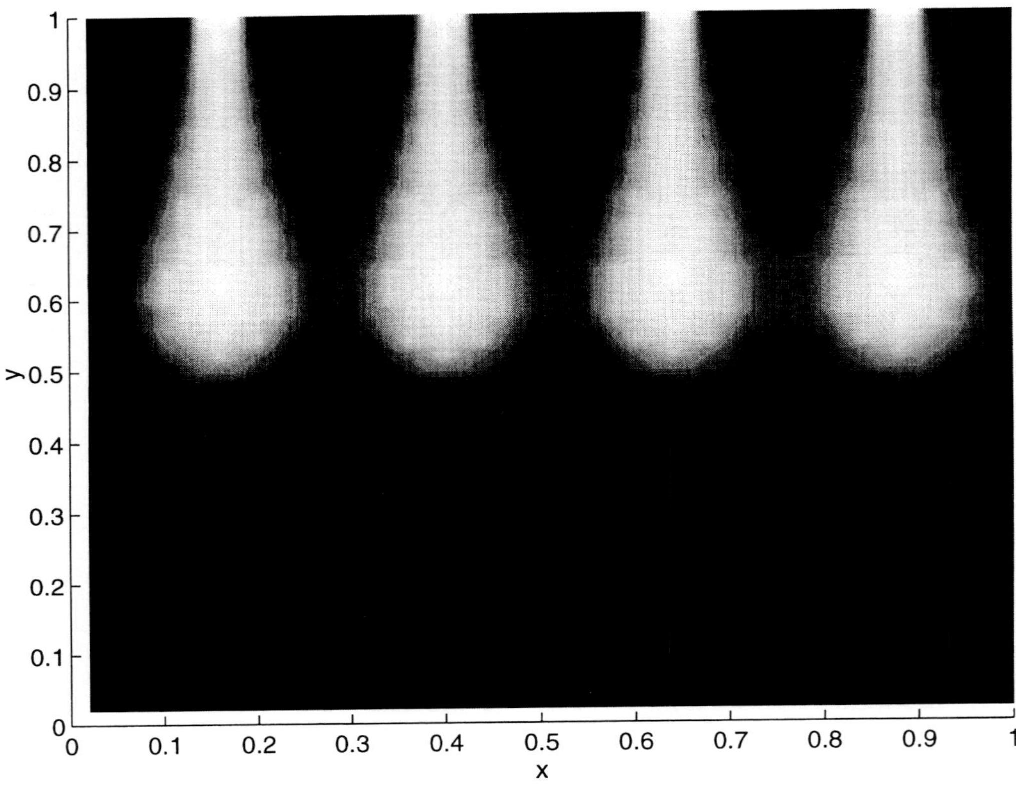

Fig. 19 Profile of EC density n at time t = 0.6 modeling the anti-angiogenesis strategy of preventing the EC from reacting to the haptotactic chemical, e.g., by affecting the receptors on the cell surface. Though the capillary sprouts continue to grow under the influence of chemotaxis, the sprouts are not as distinct as those in Figures 13 and 14. Fixed TAF profile $c = 1 - y$ and parameter values $\kappa = 0.65$, $\mu = 4.5$, $gb = 0$, $\chi = 0$, $\lambda_1 = 0.5$, $\alpha = 5$, $B = 0.001$, $D_1 = 0.0025$, $D_2 = 0.5$.

ical or become saturated when there is too much. A summary of the different numerical simulations conducted is given in Table 1.

Alternative Chemotaxis Model

We will now briefly consider an alternative mechanism for the formation of capillary sprouts during angiogenesis, which does not involve haptotaxis. We assume that the TAF produced by the tumor does not induce the secretion of fibronectin (or other such adhesive material) by the EC. Hence, by setting $p(x,y,t) = 0$ in the system (4.28)–(4.30), we obtain the submodel

$$\frac{\partial n}{\partial t} = D_1\left(\frac{\partial^2 n}{\partial x^2} + \frac{\partial^2 n}{\partial y^2}\right) - \kappa\left(\frac{\partial n}{\partial x}\frac{\partial c}{\partial x} + \frac{\partial n}{\partial y}\frac{\partial c}{\partial y} + n\left(\frac{\partial^2 c}{\partial x^2} + \frac{\partial^2 c}{\partial y^2}\right)\right)$$

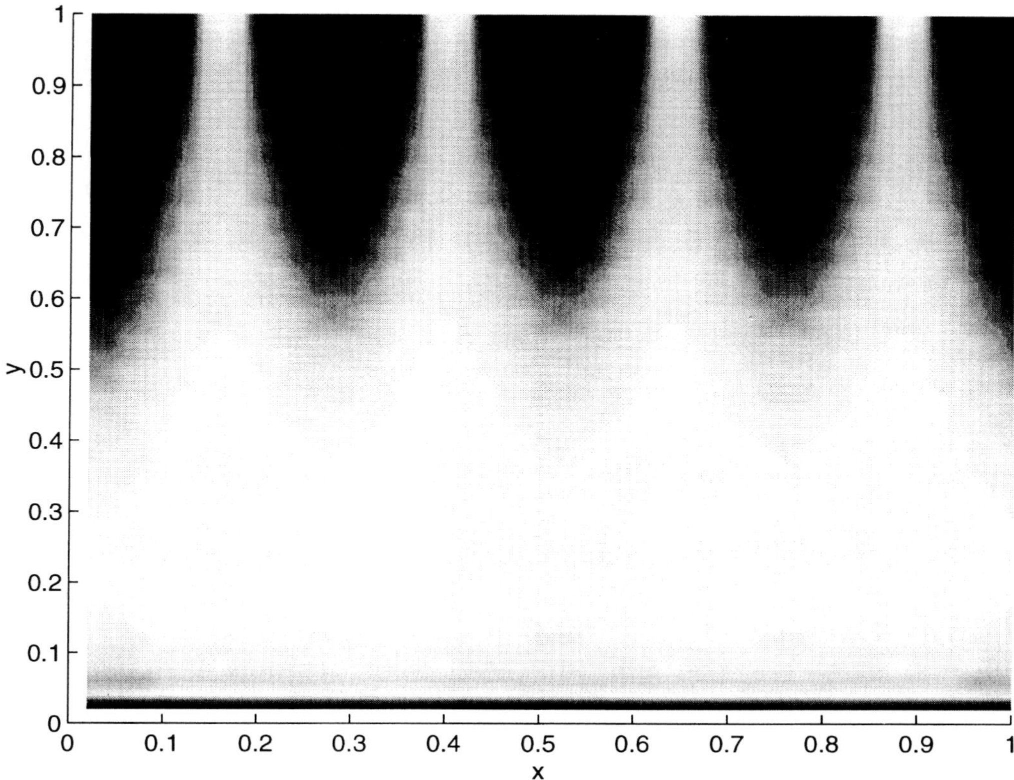

Fig. 20 Profile of the EC density n at time $t = 1.2$ modeling the anti-angiogenesis strategy of preventing the EC from reacting to the haptotactic chemical, e.g., by affecting the receptors on the cell surface. Though the strategy has not prevented the outgrowth of endothelial cells, the EC have not formed well-defined structures, and hence there will be a poor circulation of blood. Fixed TAF profile $c = 1 - y$ and parameter values $\kappa = 0.6$, $\mu = 5$, $\beta = 0$, $\chi = 0$, $\lambda_1 = 0.5$, $\alpha = 5$, $B = 0.001$, $D_1 = 0.0025$, $D_2 = 0.5$.

$$+\mu n(1 - n) - \beta n \tag{6.33}$$

$$\frac{\partial c}{\partial t} = D_3\left(\frac{\partial^2 c}{\partial x^2} + \frac{\partial^2 c}{\partial y^2}\right) - s_2 nc - \lambda_1 c \tag{6.34}$$

Here, the capillary vessels act as sinks, which absorb the TAF and hence, create local chemical gradients. This may provide an alternative mechanism for the formation of capillary branches and anastomoses. This is investigated in a preliminary numerical simulation of the system (6.33)–(6.34) with parameter values $\kappa = 0$, $\mu = 5$, $\beta = 0$, $\lambda_2 = 0.5$, $s_2 = 1$, $D_1 = 0.0025$, $D_3 = 0.5$. The resultant pattern of capillary growth is shown in Figure 21. We see the beginnings of capillary outgrowth toward the tumor located at $y = 0$. However, in the absence of haptotaxis, the ves-

Table 1. A summary of the different anti-angiogenesis strategies and a parameter sensitivity analysis and the effect on the solution as compared to Figures 13 and 14.

Anti-angiogenesis strategies

Action	*Change in parameter*	*Effect on solution*
Cytotoxic targeting of EC	$\beta = 50$	Complete regression of capillary sprouts
Inhibit EC mitosis	$\mu = 0$	Capillary sprouts stop growing
Anti-chemotaxis	$\kappa = 0$	Capillary sprouts stop growing
Anti-haptotaxis	$\chi = 0$	Loss of compactness

Parameter sensitivity analysis

Action	*Change in parameter*	*Effect on solution*
Increase proliferation	Increase μ	Loss of compactness
Increase diffusion of haptotactic chemical	Increase D_2	" "
Increase secretion of haptotactic chemical	Incresae α	" "
Decrease secretion of haptotactic chemical	Decrease α	" "

sels are not as well-defined as Figures 13 and 14. After a time $t = 0.8$, the vessels are almost indistinguishable, and such poor definition would result in poor circulation of blood. In this case, angiogenic process has failed to produce a viable capillary network.

Discussion

We have presented some mathematical models of tumor-induced angiogenesis. The initial mathematical model (in one space dimension) was able to capture, qualitatively, many of the experimental observations associated with angiogenesis. However, angiogenesis is essentially a two- or three-dimensional process and so the model was extended to two space dimensions in order to more faithfully replicate experiments performed in the cornea of test animals. The results from Figures 6–12 were clearly an improvement on the original model and many of the features of angiogenesis (e.g., anastomosis) were now captured explicitly. However, it was felt that because of the imposition of a circular source of TAF, these features may have been caused solely by the imposed geometry of the problem. In order to isolate the precise mechanisms responsible for some of the specific processes involved (e.g., tip-splitting) we finally developed the two-dimensional model to incorporate not only EC and TAF but also the possible effect of fibronectin. This model has indeed explicitly captured key features of angiogenesis, namely, the outgrowth of capillary sprouts, branching, and loop formation (anastomoses). In this model, anastomoses is brought about as a result of the haptotactic movement of the EC in response to an adhesive chemical (fibronectin) which the EC themselves secrete. This is therefore an improvement on the ini-

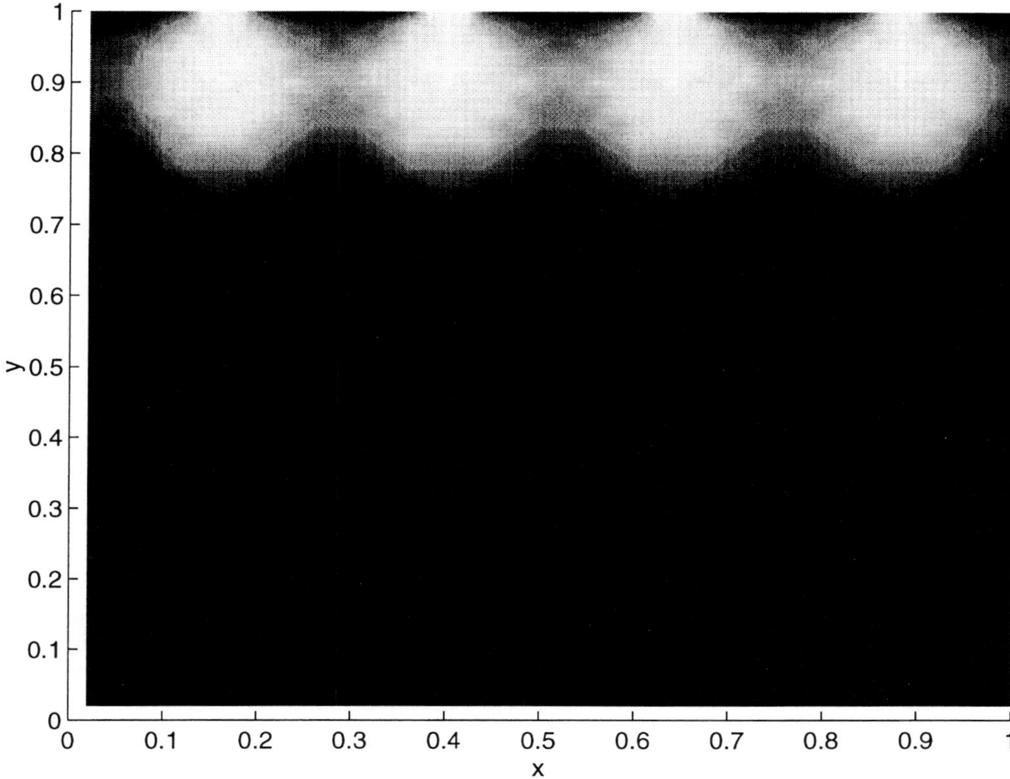

Fig. 21 Profile of the EC density n time t = 0.8 from the chemotaxis submodel. Though the capillary sprouts grow toward the tumor, the vessels are not well defined and as a result the flow of blood through the vasculature would be poor. In this case, we consider angiogenesis to have failed. Parameter values $\kappa = 0.6$, $\mu = 5$, $\beta = 0$, $\lambda_2 = 0.5$, $s_2 = 1$, $D_1 = 0.0025$, $D_3 = 0.05$.

tial two-dimensional model where the profile of the TAF was assumed to be circular and hence, anastomosis occurs as a consequence of the geometry.

The results from the final two-dimensional model are, in theory, experimentally testable. The initial and boundary conditions used in the model suggest an in vitro experiment, where a square petri-dish contains TAF suspended in gel and four clusters of EC located at the edge of the dish. Such an experiment has yet to be carried out.

The numerical simulations presented in the previous sections imply that both chemotaxis and haptotaxis are required for successful vascularization of the tumor, though they have different roles in the angiogenic process. Chemotaxis is the underlying mechanism which drives the outgrowth of the complete capillary network, whereas haptotaxis controls the finer structure and continuity of the vasculature. Though it has been shown that the proliferation of the EC is not required for the initial stages of capillary sprout formation (Sholley et al., 1984), it is essential for the successful completion of angiogenesis (Ausprunk and Folkman, 1977; Paweletz and Knierim, 1989).

We have also identified a number of ways by which the angiogenic process can be disrupted. Anti-angiogenesis has a lot of potential as an adjunctive therapy, whereby it is used alongside conventional treatments, or on its own in circumstances where other methods result in unacceptable damage of normal host tissue (Folkman, 1995; Harris et al., 1996). One problem for pathologists is the variety of metastatic patterns in patients with the same type of cancer (Frank et al., 1995; Nicolson, 1988). By establishing a range of different therapies, treatment can be tailored to meet the individual requirements of each patient. Furthermore, anti-angiogenesis can be used to control the growth of the metastases as well as the primary tumor.

We finally mention that work is currently in progress to extend the mathematical models above by incorporating a cellular automaton structure into them. We are at present devising models where the discretized partial differential equations (used in the numerical simulations) form the basis for the rules which govern the cellular automaton (see Fig. 22). In this way we hope to utilize the strengths of both continuum models and cellular automaton models (Chaplain and Anderson, 1997). Cellular automaton models have the advantage that they can be used to follow

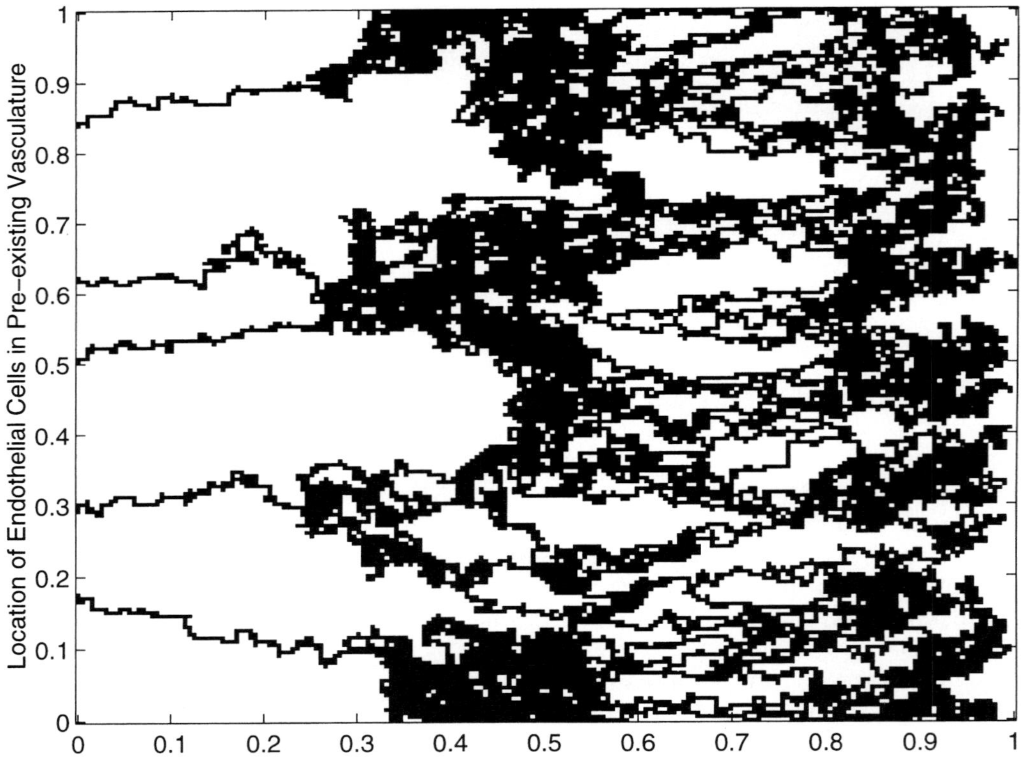

Fig. 22 Simulation of angiogenesis with explicit branching using a cellular automaton model based on a discretized version of the partial differential equations used in (4.30) and (4.31).

individual endothelial cells. We are therefore perhaps on the verge of genuine quantitative modeling and prediction.

References

Adam J.A. (1987): A mathematical model of tumor growth II: Effects of geometry and spatial nonuniformity on stability. Math. Biosci. 86:183–211.

Alessandri G., Raju K.S., Gullino P.M. (1986): Interaction of gangliosides with fibronectin in the mobilization of capillary endothelium. Invas. Metast. 6:145–165.

Ausprunk D.H., Folkman J. (1977): Migration and proliferation of endothelial cells in preformed and newly formed blood vessels during tumour angiogenesis. Microvasc. Res. 14: 53–65.

Balding D., McElwain D.L.S. (1985): A mathematical model of tumour-induced capillary growth. J. Theor. Biol. 114:53–73.

Blood CH, Zetter B.R. (1990): Tumor interactions with the vasculature: Angiogenesis and tumor metastasis. Biochim. Biophys. Acta 1032:89–118.

Bourrat-Floeck B., Groebe K., Mueller-Klieser W. (1991): Biological response of multicellular EMT6 spheroids to exogonous lactate. Int. J. Cancer 47:792–799.

Bowersox J.C., Sorgente N. (1982): Chemotaxis of aortic endothelial cells in response to fibronectin. Cancer Res. 42:2547–2551.

Bredel-Geissler A., Karbach U., Walenta S., Vollrath L., Mueller-Klieser W. (1992): Proliferation-associated oxygen consumption and morphology of tumor cells in monolayer and spheroid culture. J. Cell. Physiol. 153:44–52.

Brooks P.C., Montgomery A.M.P., Rosenfeld M., Reisfled R.A., Hu T., Klier G., Cheresh D.A. (1994): Integrin $\alpha v\beta_3$ antagonists promote tumor regression by inducing apoptosis of angiogenic blood vessels. Cell 79:1157–1164.

Britton N.F., Chaplain M.A.J. (1993): A qualitative analysis of some models of tissue growth. Math. Biosci. 113:77–89.

Brown J.M., Gaccia A.J. (1994): Tumour hypoxia: the picture has changed in the 1990s. Int. J. Radiat. Biol. 65:95–102.

Burton A.C. (1966): Rate of growth of solid tumours as a problem of diffusion. Growth 3:157–176.

Bussolino F., Di Renzo M.F., Ziche M., Bocchietto E., Olivero M., Naldini L., Gaudino G., Tamagnone L., Coffer A., Comoglio P.M. (1992): Hepatocyte growth factor is a potent angiogenic factor which stimulates endothelial cell motility and growth. JCB 119(9):629–641.

Byrne H.M., Chaplain M.A.J. (1995a): Mathematical models for tumour angiogenesis: Numerical simulations and nonlinear wave solutions. Bull. Math. Biol. 57:461–486.

Byrne H.M., Chaplain M.A.J. (1995b): Explicit solutions of a simplified model of capillary sprout growth during tumour angiogenesis. Appl. Math. Lett. 8:71–76.

Byrne H.M., Chaplain M.A.J. (1995c): Growth of non-necrotic tumours in the presence and absence of inhibitors. Math. Biosci. 130:151–181.

Byrne H.M., Chaplain M.A.J. (1996): Growth of necrotic tumours in the presence and absence of inhibitors. Math. Biosci. 135:187–216.

Carter S.B. (1965): Principles of cell motility: The direction of cell movement and cancer invasion. Nature 208:1183–1187.

Casciari J.J., Sotirchos S.V., Sutherland R.M. (1992): Mathematical modelling of microenvironment and growth in EMT6/Ro multicellular tumour spheroids. Cell Prolif. 25:1–22.

Chaplain M.A.J. (1995a): Reaction-diffusion pre-patterning and its potential role in tumour invasion. J. Biol. Syst. 3:929–936.

Chaplain M.A.J. (1995b): The mathematical modelling of tumour angiogenesis and invasion. Acta Biotheor. 43:387–402.

Chaplain M.A.J. (1996): Avascular growth, angiogenesis and vascular growth in solid tumours: The mathematical modelling of the stages of tumour development. Math. Comp. Modell. 23(6):47–87.

Chaplain M.A.J., Anderson A.R.A. (1997): The mathematical modelling, simulation and prediction of tumour-induced angiogenesis. Invas. Metast. 16:222–234.

Chaplain M.A.J., Britton N.F. (1993): On the concentration profile of a growth inhibitory factor in multicell spheroids. Math. Biosci. 115:233–245.

Chaplain M.A.J., Byrne H.M. (1996): The mathematical modelling of wound healing and tumour growth: Two sides of the same coin. WOUNDS 8:42–48.

Chaplain M.A.J., Orme M.E. (1996): Travelling waves arising in mathematical models of tumour angiogenesis and tumour invasion. FORMA 10:147–170.

Chaplain M.A.J., Sleeman B.D. (1990): A mathematical model for the production and secretion of tumour angiogenesis factor in tumours. IMA J. Math. Appl. Med. Biol. 7:93–108.

Chaplain M.A.J., Stuart A.M. (1991): A mathematical model for the diffusion of tumour angiogenesis factor into the surrounding host tissue. IMA J. Math. Appl. Med. Biol. 8:191–220.

Chaplain M.A.J., Stuart A.M. (1993): A model mechanism for the chemotactic response of endothelial cells to tumour angiogenesis factor. IMA J. Math. Appl. Med. Biol. 10:149–168.

Chaplain M.A.J., Benson D.L., Maini P.K. (1994): Nonlinear diffusion of a growth inhibitory factor in multicell spheroids Math. Biosci. 121:1–13.

Chaplain M.A.J., Giles S.M., Sleeman B.D., Jarvis R.J. (1995): A mathematical analysis of a model for tumour angiogenesis J. Math. Biol. 33:744–770.

Cliff W.J. (1963): Observations on healing tissue: A combined light and electron microscopic investigation. Phil. Trans. R. Soc. Lond. 246:305–325.

Degner F.L., Sutherland R.M. (1988): Mathematical modelling of oxygen supply and oxygenation in tumor tissues: Prognostic, therapeutic and experimental implications. Int. J. Radiat. Oncol. Biol. Phys. 15:391–397.

Durand R.E. (1990): Multicell spheroids as a model for cell kinetic studies. Cell Tissue Kinet. 23:141–159.

Ellis L.E., Fidler I.J. (1995): Angiogenesis and breast cancer metastasis. Lancet 346:388–389.

Folkman J. (1975): Tumor angiogenesis. Cancer 3:355–388.

Folkman J. (1976): The vascularization of tumors. Sci. Am. 234:58–73.

Folkman J. (1985): Tumor angiogenesis. Adv. Cancer Res. 43:175–203.

Folkman J. (1995): Angiogenesis in cancer, vascular, rheumatoid and other disease. Nature Med. 1:21–31.

Folkman J., Klagsbrun M. (1987): Angiogenic factors. Science 235:442–447.

Frank R.E., Saclarides T.J., Leurgans S., Speziale N.J., Drab E.A., Rubin D.B. (1995): Tumor angiogenesis as a predictor of recurrence and survival in patients with node-negative colon cancer. Ann. Surg. 222(6):695–699.

Gimbrone M.A., Cotran R.S., Leapman S.B., Folkman J. (1974): Tumor growth and neovascularization: An experimental model using the rabbit cornea. J. Natl. Cancer Inst. 52:413–427.

Greenspan H.P. (1972): Models for the growth of a solid tumor by diffusion. Stud. Appl. Math 51:317–340.

Greenspan H.P. (1974): On the self-inhibited growth of cell cultures. Growth 38:81–95.

Greenspan H.P. (1976): On the growth and stability of cell cultures and solid tumours. J. Theor. Biol. 56:229–242.

Groebe K., Mueller-Klieser W. (1991): Distributions of oxygen, nutrient and metabolic waste concentrations in multicellular spheroids and their dependence on spheroid parameters. Eur. Biophys. J. 19:169–181.

Harris A.L., Zhang H.T., Moghaddam A., Fox S., Scott P., Pattison A., Gatter, K., Stratford I., Bicknell R. (1996): Breast cancer angiogenesis—New approaches to therapy via anti-angiogenesis, hypoxic activated drugs, and vascular targeting. Breast Cancer Res. Treat. 38: 97–108.

Konerding M.A., van Ackern C., Steinberg F., Streffer C. (1992): The development of the tumour vascular system: 2-D and 3-D approaches to network formation in human xenografted tumours. In: Angiogenesis in Health and Disease, Maragoudakis M.E., Gullino P., Lelkes P.I., eds. New York: Plenum Press.

Landry J., Freyer, J.P., Sutherland R.M. (1982): A model for the growth of multicell spheroids. Cell Tissue Kinet. 15:585–594.

Liotta L.A., Saidel G.M., Kleinerman J. (1977): Diffusion model of tumor vascularization and growth. Bull. Math. Biol.39:117–129.

McCulloch P., Choy A., Martin L. (1995): Association between tumour angiogenesis and tumour cell shedding into effluent venous blood during breast cancer surgery. Lancet 346: 1334–1335.

McElwain D.L.S., Ponzo P.J. (1977): A model for the growth of a solid tumor with non-uniform oxygen consumption. Math. Biosci. 35:267–279.

Maggelakis S.A., Adam J.A. (1990): Mathematical model of prevascular growth of a spherical carcinoma. Math. Comp. Model. 13:23–38.

Mueller-Klieser W. (1987): Multicellular spheroids: A review on cellular aggregates in cancer research. J. Cancer Res. Clin. Oncol. 113:101–122.

Muthukkaruppan V.R., Kubai L., Auerbach R. (1982): Tumor-induced neovascularization in the mouse eye. J. Natl. Cancer Inst. 69:699–705.

Nicolson G.L. (1988): Cancer metastasis: Tumor cell and host organ properties important in metastasis to specific secondary sites. Biochim. Biophys. Acta 948:175–224.

Norton J.A. (1995): Tumor angiogenesis: The future is now. Ann. Surg. 222:693–694.

O'Reilly M.S., Holmgren L., Shing Y., Chen C., Rosenthal R.A., Moses M., Lane WS., Cao Y., Sage E.H., Folkman J. (1994): Angiostatin: A novel angiogenesis inhibitor that mediates the suppression of metastases by a Lewislung carcinoma. Cell 79:315–328.

Orme M.E., Chaplain M.A.J. (1996): A mathematical model of vascular tumour growth and invasion. Math. Comp. Model. 23(10):43–60.

Orme M.E., Chaplain M.A.J. (1996): A mathematical model of the first steps of tumour-related angiogenesis: Capillary sprout formation and secondary branching. IMA J. Math. Appl. Med. Biol. 13:73–98.

Paku S., Paweletz N. (1991): First steps of tumor-related angiogenesis. Lab. Invest. 65:334–346.

Paweletz N, Knierim M (1989): Tumor-related angiogenesis. Crit. Rev. Oncol. Hematol. 9:197–242.

Pettet G., Chaplain M.A.J., McElwain D.L.S., Byrne H.M. (1996): On the role of angiogenesis in wound healing. Proc. Roy. Soc. Lond. B 263:1487–1493.

Reidy M.A., Schwartz S.M. (1981): Endothelial regeneration III. Time course of intimal changes after small defined injury to rat aortic endothelium. Lab. Invest. 44:301–312.

Schoefl G.I. (1963): Studies on inflammation III. Growing capillaries: Their structure and permeability. Virchows Arch. Pathol. Anat. 337:97–141.

Sherratt J.A., Murray J.D. (1990): Models of epidermal wound healing. Proc. R. Soc. Lond. B 241:29–36.

Sholley M.M., Gimbrone M.A., Cotran R.S. (1977): Cellular migration and replication in endothelial regeneration. Lab. Invest. 36:18–25.

Sholley M.M., Ferguson G.P., Seibel H.R., Montour J.L., Wilson J.D. (1984): Mechanisms of neovascularization. Vascular sprouting can occur without proliferation of endothelial cells. Lab. Invest. 51:624–634.

Shymko R.M., Glass L. (1976): Cellular and geometric control of tissue growth and mitotic instability. J. Theor. Biol. 63:355–374.

Stokes C.L., Rupnick M.A., Williams S.K., Lauffenburger D.A. (1990): Chemotaxis of human microvessel endothelial cells in response to acidic fibroblast growth factor. Lab. Invest. 63:657–668.

Stokes C.L., Lauffenburger D.A. (1991): Analysis of the roles of microvessel endothelial cell random motility and chemotaxis in angiogenesis. J. Theor. Biol. 152:377–403.

Sutherland R.M. (1988): Cell and environment interactions in tumor microregions: The multicell spheroid model. Science 240:177–184 (1988).

Sutherland R.M., Durand R.E. (1984): Growth and cellular characteristics of multicell spheroids. Rec. Res. Cancer Res. 95:24–49.

Sutherland R.M., McCredie J.A., Inch W.R. (1971): Growth of multicell spheroids as a model of nodular carcinomas. J. Natl. Cancer Inst. 46:113–120.

Terranova V.P., Diflorio R., Lyall R.M., Hic S., Friesel R., Maciag T. (1985): Human endothelial cells are chemotactic to endothelial cell growth factor and heparin. J. Cell Biol. 101:2330–2334.

Ungari S., Katari R.S., Alessandri G., Gullino P.M. (1985): Cooperation between fibronectin and heparin in the mobilization of capillary endothelium. Invas. Metast. 5:193–205.

Vernon R.B., Angello J.C., Iruela-Arispe M.L., Lane T.F., Sage E.H. (1992): Reorganization of basement membrane matrices by cellular traction promotes the formation of cellular networks in vitro. Laboratory 66:536–547.

Warren B.A. (1966): The growth of the blood supply to melanoma transplants in the hamster cheek pouch. Lab. Invest. 15:464–473.

Yamada K.M., Olden K. (1978): Fibronectin-adhesive glycoproteins of cell surface and blood. Nature 275: 179–184.

Young W.C., Herman I.M. (1985): Extracellular matrix modulation of endothelial cell shape and motility following injury in vitro. J. Cell. Sci. 73:19–32.

3.5

Is the Fractal Nature of Intraorgan Spatial Flow Distributions Based on Vascular Network Growth or on Local Metabolic Needs?

James B. Bassingthwaighte, Daniel A. Beard, Zheng Li, and Tada Yipintsoi

Introduction

In this chapter we attempt to relate the coronary vascular anatomy, a system of parallel paths with differing transit times, to regional flow distributions, to transit time distribution, and to measures of regional physiological functions. To do this we explore some approaches to algorithmic vascular growth. "Growth" from embryonic beginnings through development of the heart is not what we treat here; rather we try to capture the essence of the adult form of the vascular network in algorithmic form in order to see how well its behavior matches that of the real system. This first part of the exercise, instead of following the dimensional and physiological changes that occur from embryonic to adult life, is an attempt to determine whether or not forming a network model from a set of statistically defined vessels, and positioning them within the contours of the adult heart, produces a network which shows appropriate physiological behavior.

Earlier explorations with simple dichotomous branching networks (Van Beek et al., 1989) illustrated that the fractal distributions of flows could be explained by a small degree of asymmetry at each branching. The four different methods that they used to create asymmetry at branch points gave similar results for the distributions of flows at the end points of the branching network. Whether the asymmetry was random around 50% or was set at a particular ratio at each bifurcation, the results were that there was a self-similar scaling between the variances in flows vs. the size of the regions in which the flow was measured. The lung model of Mandelbrot (1983) is in this class: a little asymmetry at each branch would result in a heterogeneity of resistances between the entrance and the periphery, and the higher the order the more the heterogeneity.

In the past, theoretical studies of the influence of network morphology on regional flow distribution have relied on simplified models of network anatomy. Conversely, Van Bavel and Spaan (1992) have simulated the hydraulics of blood flow in a more realistic representation of an arterial network based on reconstruction from a statistical morphometric study, but did not attempt to produce a three-dimensional representation of the network. The comprehensive set of morphometric data on the pig's coronary vasculature produced by Kassab et al. (1993) gave

Beard and Bassingthwaighte (1997, in review) the opportunity to create a detailed mathematical model of a whole-organ arterial network. Their reconstructed network exhibits physiological properties of the coronary vasculature remarkably well. Beard used a simple scheme for positioning the segments of a reconstructed network into a three-dimensional tissue space. This provided a framework for simulating a large-scale physiological system which allowed us to study the interaction between network geometry, regional perfusion, and tracer washout kinetics.

Normal flows in the myocardium are broadly heterogeneous and are found to be fractally correlated in space. The heterogeneity is characterized by the relative dispersion, RD_o, observed at a certain reference mass, m_o, and the fractal dimension, D, which describes how the relative dispersion scales for different observed tissue mass:

$$RD = RD_o(m/m_o)^{1-D} \tag{1}$$

Relative dispersion in baboon hearts is found to be about 13.5% for 1.0 gram tissue samples; the fractal dimensions were around 1.24. This translates into a predicted near-neighbor correlation coefficient, $r_1 = 0.43$, using the relationship $r_1 = 2^{3-2D} - 1$. A mathematical reconstruction of the coronary vascular network can be used to study the contribution of the network anatomy to the fractal correlation of regional flows in the myocardium. Measures of both the fractal dimension and the near-neighbor correlation coefficient can be calculated independently and serve as independent tests of any proposed vascular reconstruction algorithm.

This modeling effort helps in interpreting data from studies using external detection techniques such as positron emission tomography and nuclear magnetic resonance: if certain characteristics of the regional flow distributions or of tracer washout kinetics are universal, then they can be included in the basic models used to interpret data, reducing the numbers of free parameters in an optimization or fitting procedure and providing stronger estimates of the unknown, desired parameters such as the local oxygen consumption. Characterization of the vascular anatomy in three dimensions is necessary for modeling the dispersion through the tissue of highly-diffusible substances such as oxygen.

The downslope of the residue and outflow curves of tracer-labeled water from rabbit myocardium have been shown to be power law functions, following $t^{-\alpha}$ and $t^{-\alpha-1}$, respectively, with an α of about 2. The simulated washout of a flow-limited tracer through the whole-organ network model can be assessed to determine if it also shows a power–law relationship and if a physiologically reasonable value for α is obtained. This comparison gives another means of validation of the network model. When it comes to transit time distributions, not only the lengths of the vessels have an influence: we must also consider the other sources for longitudinal dispersion of material occurring during transit through the network. Substances carried in the blood are dispersed in their traversal of a vascular network by a number of processes: (1) erythrocyte rotation in blood augments molecular diffusion; (2) intravascular velocity profiles are dispersive because substances or particles located near the center of the tube travel faster than those near the wall; (3) axial diffusion contributes to spreading along a tube while radial diffusion reduces the degree of axial spreading due to radial velocity profiles; (4) pulsatile flow, eddies, and vortices in curved and branching vessels spread transit time distribution; (5) different parallel paths have different transit times, a result of heterogeneities in path lengths, vessel volumes, and flows.

Method of Network Reconstruction

To describe their observations, Kassab et al. (1993) used a "diameter-defined Strahler ordering," a classification scheme in which the arterial segments were classed into 11 diameter ranges without overlap from the largest artery, order 11, to the smallest precapillary arterioles, order 1. A vessel element of a given order may consist of a few segments. They defined segment length as the length of a vessel between two nearest bifurcations; the diameter was defined as the average lumen diameter of a segment. The data provided statistical summaries of numbers of segments, lengths, diameters, and the connectivities of the vessel segments. The connectivity is a probability that a daughter segment of a given order comes off a parent of a specific order.

Reconstructing the left anterior descending coronary artery, LAD, begins with a parent vessel of any order. The number of segments N of the parent element is chosen from a Gaussian distribution defined by Kassab et al. (1993). The segment lengths are chosen from another Gaussian distribution defined by them; lengths varied from 2.82 ± 1.96 mm for order 11 to 0.609 ± 0.48 mm for order 6, to 0.056 ± 0.038 mm for order 1. For this particular heart, the order 11 vessel, the LAD, was 4.79 cm long. The mean diameters were 3,176 μm for the LAD, 150 μm for order 6, and 9.2 μm for order 1. The N segments are connected in series, and branching vessels of lesser order are then placed at each bifurcation point. The connectivity matrix maps the relative probability of each parent vessel order giving rise to each possible daughter order. The order of each daughter vessel is chosen accordingly for its placement at the end of a segment. There is no correlation between the order of the daughter vessel and its position along the parent element. (This assumption is false but adequate as a starter; in a recent refinement of their data, Kassab et al., 1997a,b define a positional matrix giving the correlation between position and daughter order.) The process is iterated upon until the topology of the network is created from order down to the capillaries. This algorithm is used to create a mathematical model of the entire LAD arterial network topology, composed of approximately 10^6 vessel segments of known length, diameter, and connectivity.

Positioning of the Network in a Constrained Three-Dimensional Space

While the topology of the network is determined from the data of Kassab et al. (1993), data on the three-dimensional positions of all of the elements in a network were not reported by them. The key feature of the construction developed by Beard and Bassingthwaighte (1997) is an "avoidance" algorithm by which segments are placed to point toward open regions and away from previously positioned segments, and which tends to fill space. Consequently, this model is different than an angiogenic growth model that branches and grows as the tissue grows.

Self-Avoidance Algorithm. The first vessel segment in the network is positioned arbitrarily to start at an edge and lie along a surface. The direction of any daughter segment is determined by generating a vector calculated from the positions of other segments of higher order previously placed. Each existing vessel segment repels new vascular growth away from its position with a strength inversely related to the distance between its endpoint and the origin of the new vessel. This forces the network to tend to grow away from existing structures and toward more sparsely fed regions. An exponent ζ is an "avoidance exponent," governing the degree to which a new segment avoids previously positioned segments. Higher values of ζ means that vessels at greater distances are less influential than nearby vessels and creates more uniformly space-filling networks. Decreasing ζ increases the influence of the distant segments relative to that of nearby vessels, as illustrated in Figure 1. A low ζ produces wispy, hairlike structures; a high ζ produces cluttered, clumped structures. Realistic-appearing networks were generated with $\zeta \approx 2$.

Boundary Avoidance. This is accomplished in a similar fashion, generating vectors that force the positioning of a new segment in a direction such that it does not pierce the surface of the

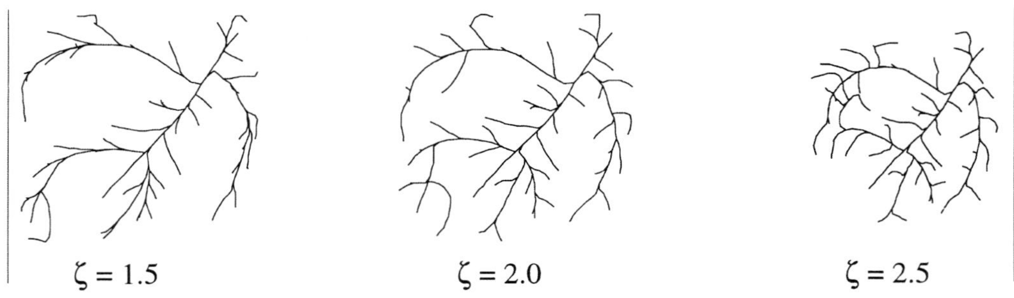

Fig. 1 Effect of avoidance strength ζ on network configuration. Stick-figure representations of large arteries (diameter greater than 0.7 mm) are shown for three values of ζ.

shell into which the vessel segments are being sequentially positioned (Beard and Bassingthwaighte, 1997).

Results of the Arterial Network Reconstruction

Figure 2 shows a stereo image of a network stemming from an epicardial ninth-order vessel with branches of lower order entering the myocardium.

A network reconstruction of the left anterior descending coronary artery, LAD, is shown in Figure 3. The first-order 11 segment is in the left lower corner and the main order 11 branch ascends toward the upper right. Larger vessels branch off from this element approximately perpendicular to the main branch.

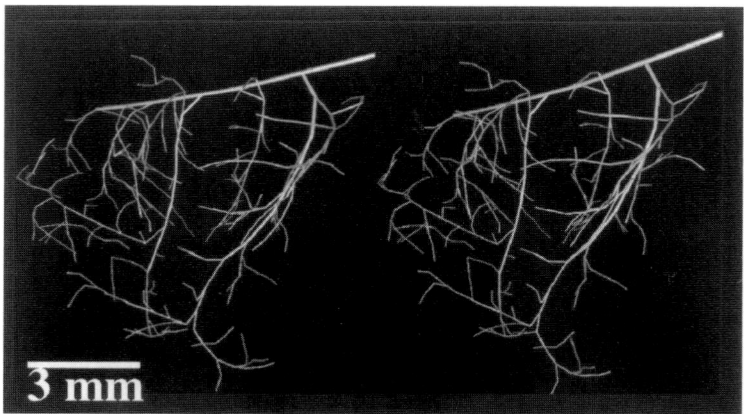

Fig. 2 Stereo image of a section of a reconstructed LAD network. An order 9 vessel lies on the surface of the myocardium and the smaller vessels of order 8 to 5 penetrate the tissue. ζ = 2.0. The three-dimensional representation is given by rotating the right image 10° from the left.

Fig. 3 Reconstruction of an LAD network using $\zeta = 2.0$ in the avoidance algorithm. Arterial vessels with diameters greater than 0.4 mm (order 8 to 11) are shown.

The smallest arterioles, of orders 3 to 1, form arterial groups from which the blood flows into the capillary swamp, a continuous syncytium. The spacing between these sources of nutrient looks quite sparse in Figure 4. Comparisons between the predictions (arterioles/mm^3) and anatomic measures have not been made.

Figure 5 shows order 9 to 11 vessels on the surface of a 1-cm-long cylindrical portion of the LV, smaller vessels (order 5 and larger are shown) penetrating into the tissue. The result is portrayed as if it were the X-ray image of a slice of the cylinder. The subendocardial region naturally has a larger fraction of smaller vessels, since the larger vessels are confined to the epicardial region.

Pressure Profiles Along the Network

Network Hemodynamics. Pressure profiles are dependent only on topology and not on spatial distribution, so long as capillary pressures are close to uniform. Assuming Poiseuille flow, stable laminar flow, the relative resistance of each vessel segment from the largest artery to the smallest arteriole (not including capillaries) was calculated from the viscosity times the length

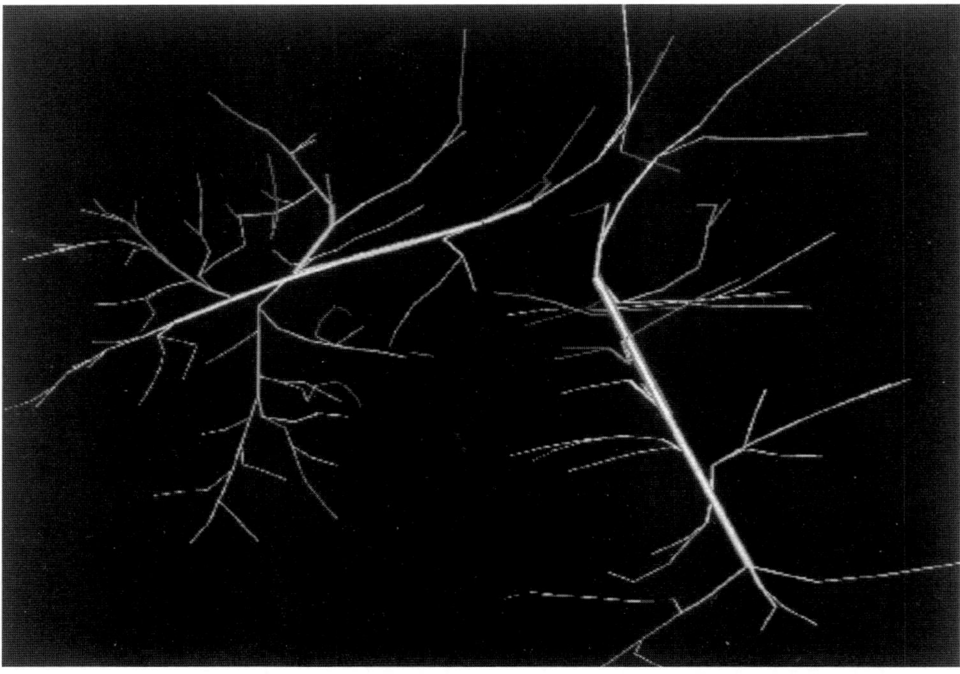

Fig. 4 Small arterioles of diameter less than 20 μm (orders 1, 2, and 3) in a block of tissue.

divided by the radius to the fourth power. The relative viscosity was calculated from the empirical viscosity law found by Pries et al. (1990), which gives the viscosity as a function of vessel radius and discharge hematocrit, and accounts for the Fahraeus effect, a reduction in apparent viscosity in small vessels. The hematocrit diminishes in smaller vessels, and we chose arbitrarily a linear diminution with vessel order. Since the hematocrit is fixed at each order, no phase-separation law is used and the flow simulation is soluble algebraically.

The variability in capillary length, e.g., 550 ± 203 μm (Kassab et al., 1993) was used to estimate the variability of capillary resistance, that is, a coefficient of variation of 203/550 or 37%, which might be an underestimate since it doesn't account for variation in capillary radius, but also may be an overestimate since the capillary network is a syncytium with many cross-connections to even out the resistances.

Aortic pressure was arbitrarily set to 100 mmHg, and the resistances and pressures calculated for steady flow assuming that the postcapillary venular pressure was 20 mmHg. The relative flows through this strictly bifurcating network of approximately 10^6 elements are determined in less than 10 minutes computation. The pressures are shown in Figure 6 as the mean and standard deviations of pressures at the upstream and downstream ends of the segments. Most of the pressure drop occurs in the smaller vessels. The coefficient of variation of the pressures increases to 0.30 at order 2.

Also plotted in Figure 6 are data obtained by Chilian (1991) from the endocardium and epicardium of pigs. The solid circle represents data from Tillmanns et al. (1981), who reported pressure in the small epicardial coronary arteries (25–100 μm diameter) of the rat heart. These

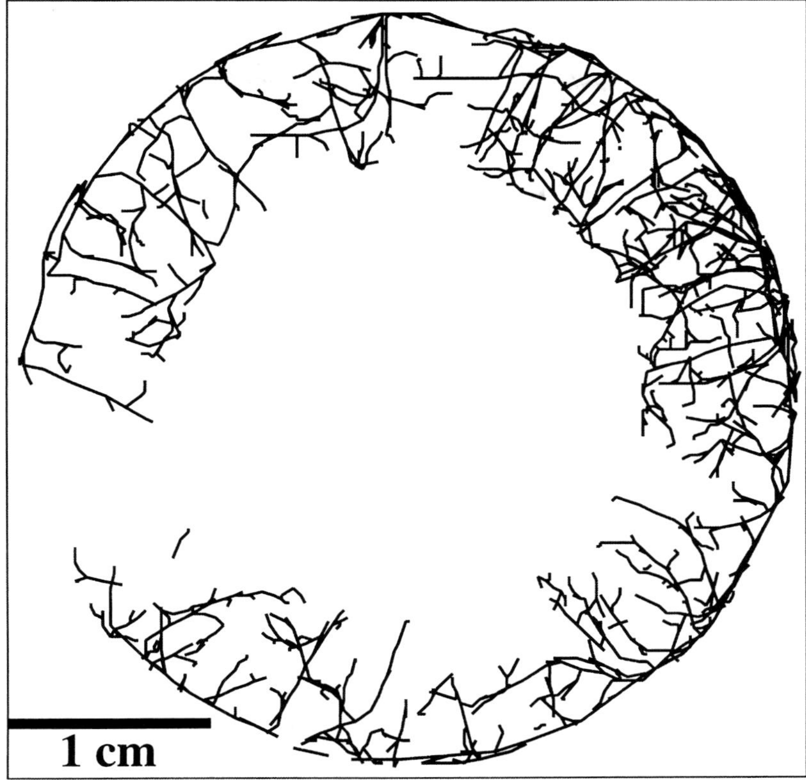

Fig. 5 Stick-figure representation of large arteries in a 3-mm thick slice through the cylindrical geometry is shown for vessels of order 5 and greater.

results are close to the data of Nellis et al. (1981), who measured a mean of about 70% of the aortic pressure occurring in 140-μm diameter arteries in rabbit heart.

Fractal Dispersion of Regional Myocardial Blood Flows

Myocardial blood flow heterogeneity is dependent on network topology, network rheology, and spatial positioning. The regional flow in a volume of tissue was calculated by summing the flows through all terminal arterioles within individual volume elements of a chosen size; this gives flow per ml of tissue. Spatial flow distributions are compared in Figure 7 for two different levels of resolution in the cylindrical geometry, for 0.0625 g voxels (left) and 0.015625 g voxels (right).

The regional flows were described statistically by the probability density functions of regional flows at different levels of resolution (Fig. 8, upper panels) in the network and in baboon studies (King et al., 1985). As element size is decreased, the density functions of local flow per ml broaden and the relative dispersion (RD) increases. This behavior shows that there is hetero-

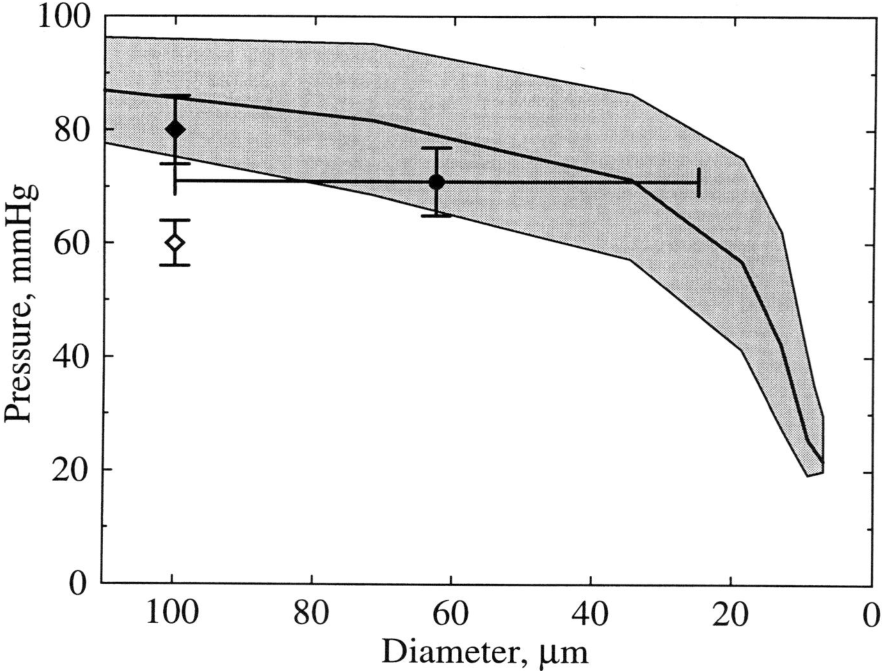

Fig. 6 Pressure in smaller diameter arterioles when the aortic mean pressure was set at 100 mm Hg. Solid line indicates mean of the pressures at the midpoints of segments, and shaded area covers ±1 SD for the vessels of a given diameter. Micropuncture pressures obtained by Chilian (1991) are plotted as a solid diamond (epicardium) and an open diamond (endocardium); error bars indicate one standard deviation. Data from Tillmanns et al. (1981) are plotted as a solid circle with vertical bars indicating one standard deviation. Horizontal bars on data from Tillmanns et al. indicate the range of diameters measured.

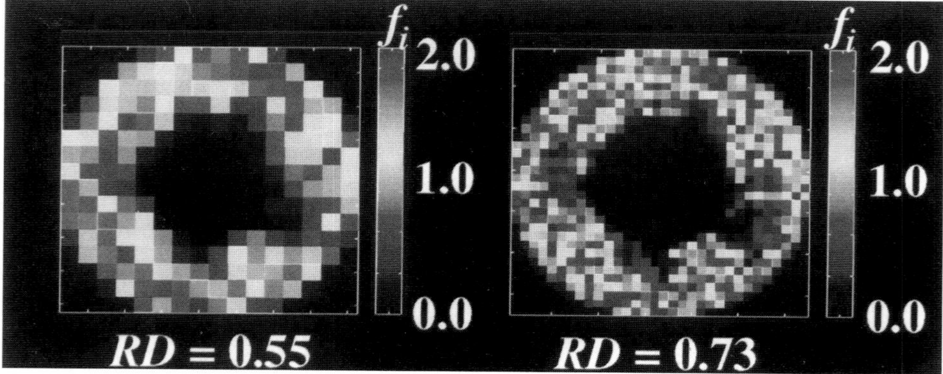

Fig. 7 Spatial distributions of regional flows for different levels of resolution are shown for a reconstruction in a 1-cm thick cylindrical wall. Voxel sizes are 0.0625 g and 0.016 g.

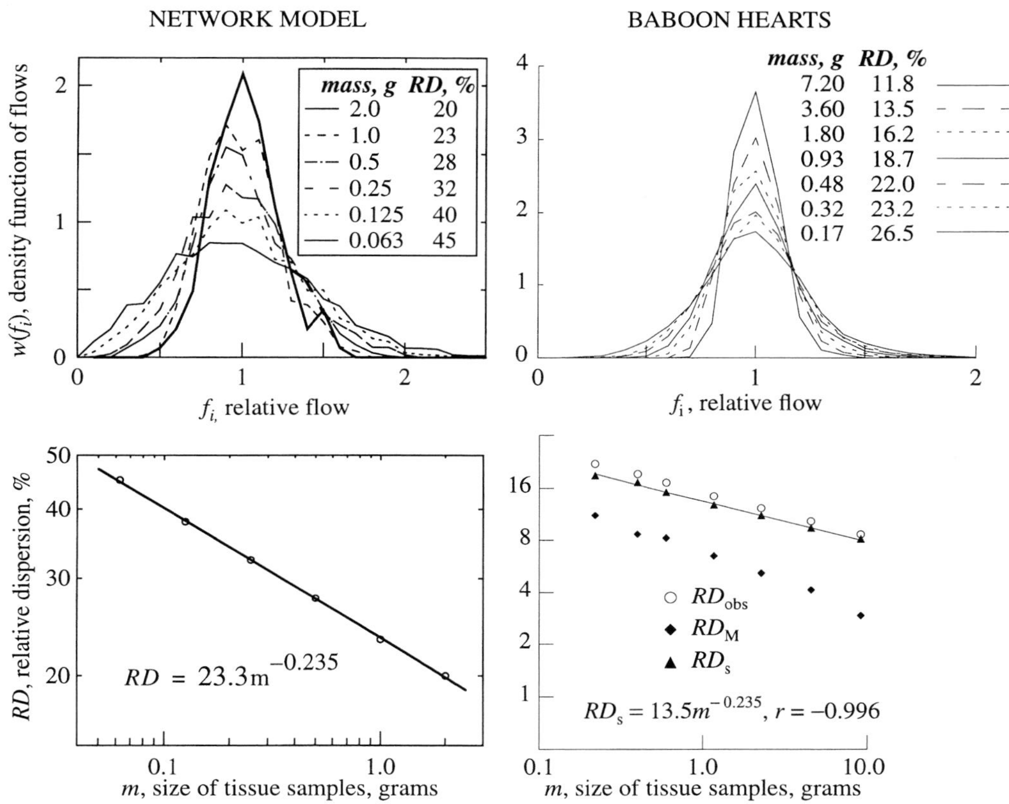

Fig. 8 Spatial flow heterogeneities. Upper panels: Probability density function of relative regional flows for six different levels of resolution for the network and seven levels for the baboons. Bin width = 0.1 times mean flow. The relative dispersion, RD, of the regional flows increases as resolution increases. Lower panels: Relative dispersion, RD, is a power-law function of voxel mass. The slope of logRD vs. logm is 1 – D = –0.235, or a fractal dimension D = 1.235 for the fractal network, left lower, and was by coincidence the same, 1.235 for the flows in a baboon heart, right lower. (For the baboon data corrected spatial relative dispersion, $RD_S^2 = (RD_{obs}^2 - RD_M^2)$, when the subscript obs refers to the observed data and subscript M refers to methodological variability.) The constants 23.3% and 13.5% are the RDs in pieces of mass 1 g.

geneity internally in volume elements of all sizes and therefore suggests looking for a fractal relationship between the measurement scale and the measured variability.

In Figure 8 (lower panels), the coefficient of variation or relative dispersion, RD, of the several density functions are plotted vs. voxel mass. The RD scales with voxel size in accord with the power–law relationship of Eq. (1). The slope of log RD vs. log voxel mass in a network realization in the left lower panel was –0.235 and the fractal dimension, D, is 1.235, implying positive correlation between flows in neighboring regions. The relationship in the right lower panel is for regional flows in the heart of an awake baboon; the fractal dimension was 1.235; that the fractal dimensions were identical is just coincidence, for the baboon hearts the estimates of D ranged from 1.18 to 1.28.

Near-Neighbor Correlation

In another assessment, the spatial autocorrelation function for regional flows was determined in volume elements of four different sizes in order to see whether or not there was self-similarity in correlation properties independent of scale, that is, whether the correlation function was also a fractal.

Inspection of the spatial flow patterns in Figure 8 (upper) suggests that high-flow regions tend to be adjacent to high-flow regions and low-flow regions tend to lie near low-flow regions. The autocorrelation function for spatial flows provides an independent measure of the fractal nature of the regional flow distribution and in theory should show the same fractal D. If the distribution is fractal the form of the autocorrelation function should be independent of the resolution of the measurement. The autocorrelation functions were calculated for four different voxel sizes. They approximately superimposed (Beard and Bassingthwaighte, 1997), which is taken as further evidence for the fractal nature of the correlation of regional flows. This result was similar to that illustrated by Bassingthwaighte and Beyer (1991) for sheep hearts at two levels of resolution. For the nearest neighbor correlation coefficient, r_1, the expression reduces to $r_1 = 2^{3-2D} - 1$. For the regional flows from the network in the slab geometry the calculated r_1 was 0.40 for 0.25 g voxels and 0.41 for 0.0625 g voxels, while the expected value from the fractal dimension of D = 1.235 was 0.44.

Transit Times and the Simulation of Tracer Washout

The transport function, h(t), of a tracer through an organ system is the probability density function of transit times and the normalized outflow concentration-time curve obtained following a flow-proportional tracer-labeled unit impulse input at time t = 0. This h(t) is the fraction per unit time of indicator or blood which traverses the system with transit time, t. The form of h(t) is dependent on topology and on the dispersion in fluid velocities within each segment of the network. Two different assumptions were used as extreme alternatives, either dispersionless, plug flow in each vessel segment, or dispersive Poiseuille flow, a parabolic velocity profile. In either case the mean transit through each segment, $\bar{t}_s$ is equal to the volume of the segment divided by the flow in the segment. A constant volume representing the venous vasculature is added to each pathway so that the sum of venous volumes equals the total venous volume measured by Kassab et al. (1997). The mean transit time through an individual pathway, $\bar{t}_p$, is calculated by summing the mean transit times of all segments in the pathway. The time axis for the washout curve is divided into 900 intervals of 0.1 seconds in order to accumulate the flow-weighted sum of all the pathways with transit times falling into a given interval.

Figure 9 shows the outflow dilution curve, h(t), or the frequency function or probability density function of transit times through the network in response to an impulse input into the model, showed that, for times greater than about 5 seconds, the different degrees of pathway dispersion had no effect on the shapes of the tails of the washout curves. While on a semilog plot the tails were concave upward, indicating that the data cannot be described by a monoexponential function, the data fell on a straight line on a log–log plot indicating that the tail of the washout is a power–law function, similar to the water washout curves observed by Bassingthwaighte and

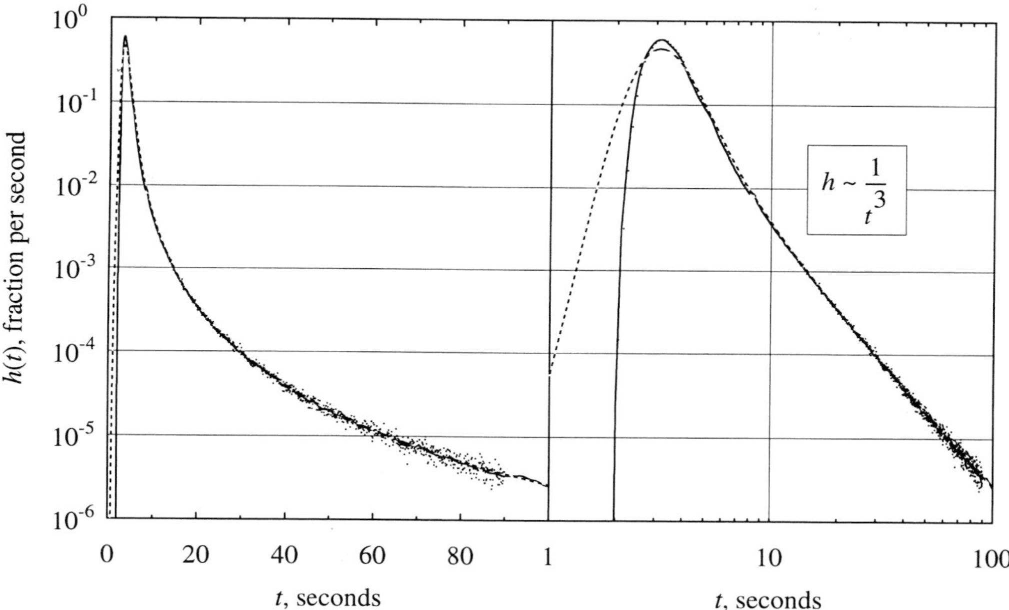

Fig. 9 Network impulse response is plotted against time on semilog scale (left) and on log–log scale (right). Continuous line is for a network with nearly plug flow pathways (RD = 0.01); the dashed line is for the same network with physiologically dispersive pathways (RD = 0.15); the dots are for a network with plug flow pathways. The tail of the washout curve fits a power-law function of the form h (t) = at^3 (right panel), and does not fit a monexponential washout, which would be straight on semilog plot (left).

Beard (1995). They found that the tails of flow-limited tracer-labeled water outflow curves from rabbit hearts follow $h(t) = a\,t^{-\alpha-1}$, with $\alpha = 2.1 \pm 0.3$. The tail of the outflow curve from the simulated network follows $h(t) = a\,t^{-3}$, or $\alpha = 2.0$. Since the form of the tail of the washout curve does not depend on the amount of dispersion occurring in individual pathways, it must depend only on the heterogeneity of transit times associated with the many parallel pathways in the network!

Perspective on Fractal Coronary Modeling

The coronary network we construct herein results from a simple algorithm for the anatomical arrangement of arteries, is based on the anatomical statistics, and is predictive, without any adjustment of parameters, of the anatomical arrangements and of five sets of physiological observations of primary importance:

(1) Anatomy: Since the network model was constructed based on the data reported by Kassab et al. (1993) on the pig coronary arterial system, it satisfies an extensive set of statistical morphometric data on the connectivity, vessel diameters, and vessel lengths from a coronary arterial network.

(2) Heterogeneity of flow: The heterogeneity of local blood flows per gram of tissue matches experimental data on the probability density functions of regional flows (Bassingthwaighte, King, and Roger, 1989).

(3) Fractal nature of the flow distribution: The measured degree of heterogeneity, or variance in flows, depends upon the mass of the tissue units observed, and in baboons, sheep and rabbits, scales as a power law function with a self-similarity fractal dimension D of 1.15 to 1.3; the network result was D = 1.24.

(4) Spatial autocorrelation decay: The correlation in local blood flows between neighboring pieces falls off with distance in accord with the fractal autocorrelation properties theorized by Bassingthwaighte and Beyer (1991); the shape of the autocorrelation function is independent of the unit size chosen, that is, it shows self-similarity independent of scale.

(5) Pressure distribution: The arterial pressures predicted from the resistances in the network are in accord with the means and variances observed for vessels of given diameters in physiological situations. The main resistance is in the smaller arterioles.

(6) Transit time distributions: The simulated time course of tracer washout following pulse injection into the inflow predicts the fractal washout process observed by Bassingthwaighte and Beard (1995).

The fractal nature of myocardial regional flow heterogeneity has been observed by all who have made measurements at different levels of resolution. The spatial autocorrelation is similar to values reported by Austin et al. (1990) and Matsumoto et al. (1996). However, the pressure distributions reported by Nellis et al. (1981) in the coronaries of the rabbit heart and by Chilian (1991) in the pig heart provide an independent measure, and again the results of the network construction appear to match the data adequately, given that Chilian's aortic pressures were lower than we used here.

Although the model remains incomplete, the capillary and venous networks are only crudely modeled, and the spatial reconstruction is imperfect, the model proves to be sufficient to accurately describe the gross behavior of coronary flow. Since mainly the arterial network structure is modeled here, this implies either that the anatomy of the arterial network is the most important factor in governing the fractal nature of the regional flow distribution and the kinetics of flow-limited tracer washout, or that the venous network has an effect which is similar.

Undoubtedly the venous network also plays some role in shaping the washout, for the venous volume is larger than the arterial volume. The general form of the venous network is similar to the arterial network, and many arteries and arterioles are accompanied by a pair of veins, a grouping which persists through several orders of branching. Therefore it is reasonable to suppose that the effects of the arterial network are mirrored by the venous network and either network by itself is sufficient to describe the gross characteristics of the system.

The anatomic measurements of arteries made by Kassab et al. (1993), upon which this reconstruction is based, were made under maximally dilated conditions. Therefore, the regional flow distributions that we have simulated might be expected to correspond to a maximally dilated state. Since relative regional flows do not correlate well between the dilated and autoregulated states, and spatial heterogeneity has been shown to decrease with decreasing coronary arterial oxygen tension, it will be important to include a vasoregulation mechanism in future versions of the model.

This algorithm for positioning segments and avoiding already positioned branches is unnatural. Normally vessels grow in growing tissues; invading capillaries bud and branch, soon enlarge and gain smooth muscle or remodel and sometimes even disappear. Will more statistical data on the relationship between the position of a branch along a parent element and the diameter of the daughter improve the approximation to reality? The angles between branches have not been reported by Kassab et al. (1993), but there are observations that angles tend toward a hemodynamic optimum. Is there any tendency for small branches to be at right angles and larger ones more nearly in the direction of the parent?

Distances between arterioles entering the capillary swamp are not accurately measured but appear to be about 500 to 1000 μm (Bassingthwaighte et al., 1974), larger than the distances between arterial and venular zones in left ventricle (mean capillary lengths, 512 ± 163 μm) reported by Kassab et al. (1993). The estimates of Batra and Rakusan (1992) for capillary lengths were about 800 μm, as were those of Bassingthwaighte et al. (1974), but the positions of the arteriolar entries into and the venous drainages from the capillary swamp need numerical assessment to provide greater precision before we can use them to evaluate our model network. The generally parallel arrangement of myocardial capillaries, about one per muscle fiber, allows capillaries to extend for many millimeters, but their functional lengths are the distances between the arteriolar inflow points and the venous drainage points. The cross-connections occurring about every 80 μm allow flow across fiber directions and help to equalize capillary pressures, but are probably not effective as collaterals for more than a millimeter.

Causation Versus Description

These observations of the efficacy of the fractal network model show that the fractal model is a good descriptor, but gives no proof that the process of vascular growth was fractal, natural though this would seem. Taking the point of view that it is the tissue that provides the signal for vascular growth, not the vessels, then one would look for the evidence in terms of the local metabolic needs.

One hint comes from the work of Caldwell et al. (1994) who observed that the transport capacities for fatty acid in different regions of the heart were vastly different, and that they were approximately proportional to the regional flows. Their studies were done in dogs running up a grade on a treadmill, heavy exercise for 5 minutes, so that the average coronary flows were several times the resting levels. Their results are shown in Figure 10; the main graph plots the deposition of tracer-labeled fatty acid analog vs. the local flow, and show the approximate proportionality. Such proportionality might occur if all of the tracer were taken up in a single passage through the heart, but this is not the explanation since less than half is taken up, and a smaller fraction would be taken up in high flow regions than low flow regions if their transport capacities were equal. The modeling analysis tested the hypotheses that the regional transport capacity could be the same everywhere, as seems to be the case for sucrose and other inert small solutes, or could be linearly increasing or decreasing with respect to higher flow regions. The only solution compatible with the data is shown by the curved line through the deposition data; this best fitting line is based on the relationship between regional transport capacity and regional flow shown in the insert panel: the transport capacity, PS for permeability surface area product, was found to be not merely higher in high flow regions, metabolically more active regions, but almost exactly proportional, i.e., PS/F is equal to a constant. The inference is that the transport capacity, which is believed to require expression of a special transmembrane protein or of an albumin receptor on the endothelial luminal surface, is driven by local metabolic needs.

The regional flows themselves infer that there is heterogeneity of metabolism. One cannot argue that regional oxygen requirements might be the same everywhere in a heart all of whose parts contract in synchrony: while high flow regions might have excess flow relative to needs for oxygen, it is not true that a low flow regions uses as much oxygen as the average for the heart, because there is simply not that much oxygen delivered to it. For example, if oxygen extraction

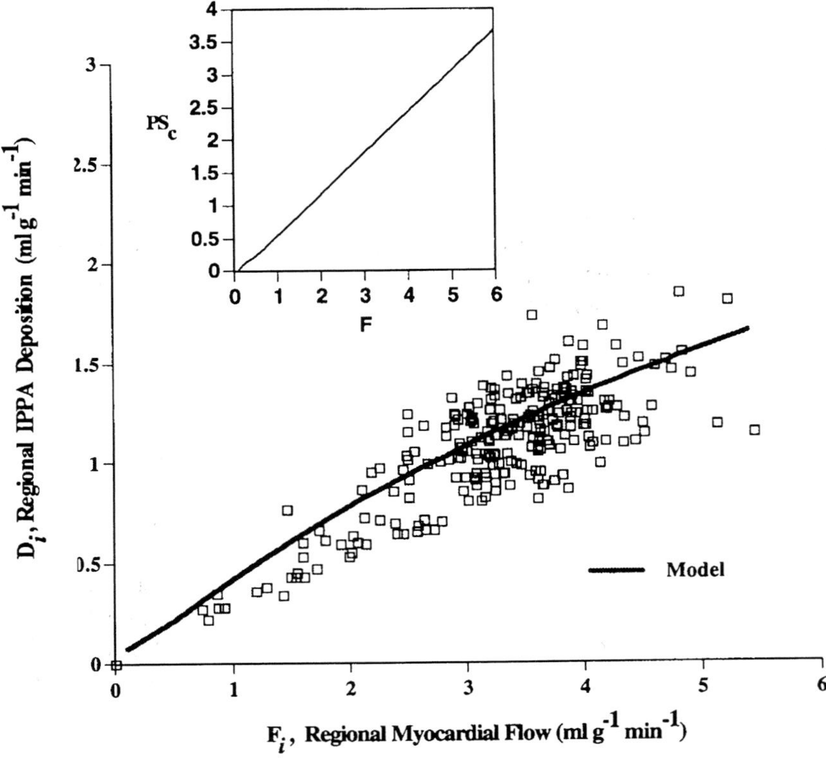

Fig. 10 Regional deposition of a fatty acid analog, iodophenylpentadecanoic acid, IPPA, in small myocardial regions after a pulse intravenous injection while the dog was running on a treadmill, plotted vs. the regional flow in the same tissue pieces. The relation providing the best fit of a model solution for deposition versus flow, the solid line in the main panel, was that the local fatty acid transport capacities, PS_c, ordinate of insert panel, is linearly proportional to the local flow, F, abscissa of insert panel. (From Caldwell et al., 1994, with permission.)

is 50%, and consumption is at a uniform rate, then a region with a flow of 25% of the mean flow does not receive enough oxygen to remain viable. Nor is the "twinkling" fluctuation of regional flows the answer, since the high and low flow regions are spatially stable (King and Bassingthwaighte, 1989). So the next question is "What is the direct evidence that metabolic requirements are higher in high flow regions and low in low flow regions?"

The answer comes from studies of regional myocardial oxygen utilization in the hearts of dogs studied using positron emission tomography. Using ^{15}O-oxygen single breath inhalation and analysis of the transient curve of the regional content, the residue function, of the sum of ^{15}O-oxygen and the metabolic product ^{15}O-water in small volume elements of the myocardium, Li et al. (1997a) used a nonlinear oxygen/water dual model to determine the regional oxygen metabolism. The analysis takes into account the very high membrane permeabilities for both oxygen and water, and the nonlinearities of oxygen binding to hemoglobin and myoglobin. The results illustrated that the local rate of oxygen transformation to water is approximately proportional to the local flow (Li et al., 1997b), as shown in Figure 11.

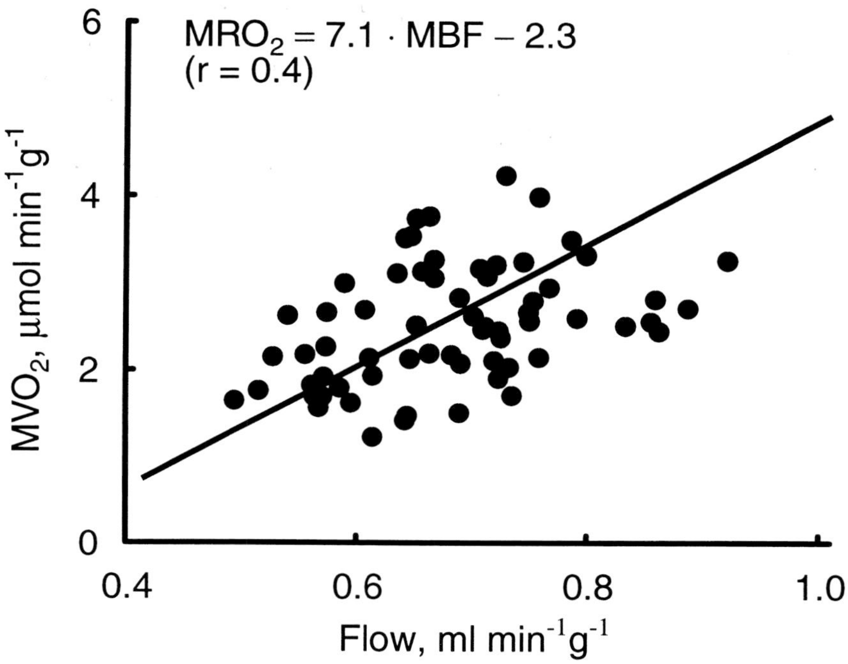

Fig. 11 The oxygen consumption, MVO_2, in small volume elements of dog myocardium plotted vs. the regional blood flow, MBF, in the individual voxels of size 0.5 ± 0.1 g. The experiments was in an open chest dog. The mean MBF was 0.68 (RD = 15%) ml $g^{-1}min^{-1}$, and the mean MVO_2 was 2.52 (RD = 27%) µmol. gm^{-1} min^{-1}. (Data from Li et al. 1997b.)

The next stages would be to link local oxygen consumption to local ATP turnover rates and to local ATP requirements for contractile work and for maintaining cellular homeostasis, particularly for ions. Technologies to collect such data are likely to evolve.

The implications of these observations with respect to how vascular growth occurred are stimulating. Leaving aside the question of why different regions of the heart have different levels of oxygen utilization (presumably because the regions have requirements proportional mainly to the local work of contraction), but appreciating that these regions of high or low flow and proportional oxygen consumption maintain their flows and oxygen consumptions relative to the average over a wide range of conditions, one may surmise that tissue requirements are the driving force for the growth. This is a reaffirmation of the ideas of Meinhardt (1982). See the chapter by Meinhardt in this book. His various schema of production of growth factors, diffusion into the matrix of cells, interstitium, and vasculature, washout by vascular convection, and uptake by receptors for growth have the essence of what must be happening. Such schema are compatible with the ideas on how vessels grow, by endothelial budding, invasion of the interstitium, and development of capillary paths which may later be enveloped by pericytes and smooth muscle cells (Hudlicka, 1984). There is thus a continuum in the perspective of how growth occurs all the way from the unnatural positioning of segments we have used here, through the convection-diffusion-reaction ideas to the mechanics of cellular growth and vascularization. See the chapters in this book by Sage and by Murray.

Our collaborator, C. Y. Wang of Michigan State University, has developed a simple recursive algorithm that makes the point that the production of a growth factor by growing cells with inadequate substrate supply can give rise to a vascular tree with fractal characteristics. An example in shown in Figure 12; the algorithm is to position a set of dots, or cells, within an arbitrarily shaped three-dimensional space, then to allow recursive development of the vascular tree. In this case the algorithm for the tree is to start growth from an arbitrary point on the periphery of the domain, extend a sprout in the direction of the center of mass of the cells by an arbitrary 1/3 of the distance toward that center of mass, divide the mass of tissue cells into two by a plane of separation along the line of the sprout, and starting at its end grow two daughter sprouts by the same rules toward the two centers of mass. Repetition of this set of rules until the tips of the vascular tree are within an arbitrarily small distance from the cells, or to an arbitrary number of generations, gives rise to a fractal network with a heterogeneity of path lengths, a heterogeneity of resistances and flows, a dichotomous branching topology, and a pattern which appears realistic. Such an algorithm can be applied to a growing tissue just as well as to a geometrically fixed structure, for example by augmenting the algorithm to allow remodeling of the diameters of higher order branches as the needs of the tissue at the periphery change over time as new cells are formed.

This kind of development sets the stage for understanding not only the growth of vessels in a particular organ, but in organs and organisms, as was suggested in the discussion of the original observations of fractal scaling of flow heterogeneity (Bassingthwaighte, 1988). The fractal branching appears to provide the basis for the scaling relationships found for respiratory and arte-

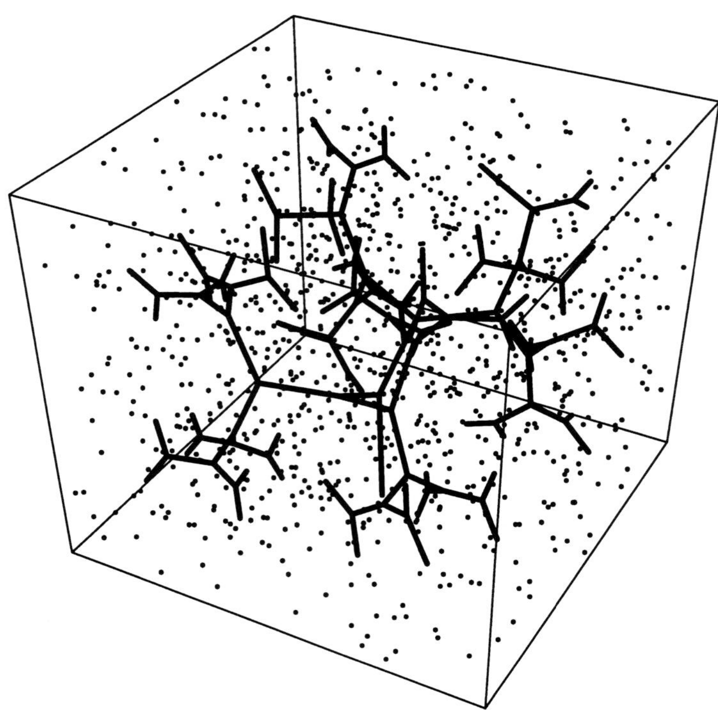

Fig. 12 First few generations of a three-dimensional arterial network resulting from C. Y. Wang's recursion algorithm for growth toward the centers of mass of cells.

rial branchings in animals from shrews to whales, according to West, Brown, and Enquist (1997). The 3/4 power scaling relationships appear to be driven by the fact that the metabolic needs in tissues of larger mammals are smaller per gram than for smaller mammals, presumably a consequence in part of the larger surface to volume ratios in small animals. West et al. argue, with mathematical elegance, albeit with some apologies for their approach being "zero order," that the fraction of large vessel volumes is no larger in large animals than in small animals, something which we feel is unlikely, but certainly the ratio of large to small vessel volumes is kept more nearly constant because the capillary volumes pre gram of tissue are less in more slowly metabolizing tissues. Their arguments are global, and, in assuming that there is perfect impedance matching at branch points, ignore the fact the there really are pressure wave reflections in the vascular system (e.g., Remington and Wood, 1956) that show that the branching area ratios are not exactly optimal in the sense that C. D. Murray (1926) or Lefèvre (1983) envisioned. Nevertheless, their derivation of nice relationships between fractal vascular structures and metabolic scaling behavior is a nice generalization.

Acknowledgments

The authors appreciate the work of James E. Lawson in the preparation of the manuscript and of James Ploger in the oxygen studies. This study was supported by National Institutes of Health grants HL-50238, HL-07403, and HL19139 from the National Heart Lung and Blood Institute and RR-1243 from the National Center for Research Resources.

References

Austin R.E., Jr., Aldea G.S., Coggins D. L., Flynn A.E., and Hoffman J.I.E. Profound spatial heterogeneity of coronary reserve: Discordance between patterns of resting and maximal myocardial blood flow. Circ. Res. 67:319–331, 1990.

Bassingthwaighte J.B., Yipintsoi T., and Harvey R.B. Microvasculature of the dog left ventricular myocardium. Microvasc. Res. 7:229–249, 1974.

Bassingthwaighte J.B. Physiological heterogeneity: Fractals link determinism and randomness in structures and functions. News Physiol. Sci. 3:5–10, 1988.

Bassingthwaighte J.B., King R.B., and Roger S.A. Fractal nature of regional myocardial blood flow heterogeneity. Circ. Res. 65:578–590, 1989.

Bassingthwaighte J.B., and Beyer R.P. Fractal correlation in heterogeneous systems. Physica D 53:71–84, 1991.

Bassingthwaighte J.B., and Beard D.A. Fractal ^{15}O-water washout from the heart. Circ. Res. 77:1212–1221, 1995.

Bassingthwaighte J.B., Beard D.A., and King R.B. Fractal regional myocardial blood flows: the anatomical basis. In: Fractals in Biology and Medicine, edited by Losa G., Weibel E., and Nonnenmacher T. Basel: Birkhauser, 1997, pp. 000–000.

Batra S., and Rakusan K. Capillary length, tortuosity, and spacing in rat myocardium during cardiac cycle. Am. J. Physiol. 263 (Heart Circ. Physiol. 32):H1369–H1376, 1992.

Beard D.A., and Bassingthwaighte J.B. Fractal nature of myocardial blood flow described by a whole-organ model of arterial network. Circ. Res. 1997.

Caldwell J.H., Martin G.V., Raymond G.M., and Bassingthwaighte J.B. Regional myocardial flow and capillary permeability-surface area products are nearly proportional. Am. J. Physiol. 267 (Heart Circ. Physiol. 36):H654–H666, 1994.

Chilian W.M. Microvascular pressures and resistances in the left ventricular subepicardium and subendocardium. Circ. Res. 69:561–570, 1991.

Hudlická O. Development of microcirculation: capillary growth and adaptation. In: Handbook of Physiology. Section 2: The Cardiovascular System Volume IV, edited by Renkin E.M. and Michel C.C. Bethesda, Maryland: American Physiological Society, 1984, pp. 165–216.

Kassab G.S., Rider C.A., Tang N.J., and Fung Y.B. Morphometry of pig coronary arterial trees. Am. J. Physiol. 265 (Heart Circ. Physiol. 34):H350–H365, 1993.

Kassab G.S., Berkley J., and Fung Y.C.B. Analysis of pig's coronary arterial blood flow with detailed anatomical data. Ann. Biomed. Eng. 25:204–217, 1997a.

Kassab G.S., Pallencaoe E., and Fung Y C. The longitudinal position matrix of the pig coronary artery and its hemodynamic implications. Ann. Biomed. Eng. 25, 1997b.

King R.B., Bassingthwaighte J.B., Hales J.R.S., and Rowell L.B. Stability of heterogeneity of myocardial blood flow in normal awake baboons. Circ. Res. 57:285–295, 1985.

King R.B., and Bassingthwaighte J.B. Temporal fluctuations in regional myocardial flows. Pflügers Arch. (Eur. J. Physiol.) 413/4:336–342, 1989.

Lefèvre J. Teleonomical optimization of a fractal model of the pulmonary arterial bed. J. Theor. Biol. 102:225–248, 1983.

Li Z., Yipintsoi T., and Bassingthwaighte J.B. Nonlinear model for capillary-tissue oxygen transport and metabolism. Ann. Biomed. Eng. 25:604–619, 1997.

Mandelbrot B.B. The Fractal Geometry of Nature. San Francisco: W.H. Freeman and Co., 1983, 468 pp.

Matsumoto T., Goto M., Tachibana H., Ogasawara Y., Tsujioka K., and Kajiya F. Microheterogeneity of myocardial blood flow in rabbit hearts during normoxic and hypoxic states. Am. J. Physiol. 270:H435–441, 1996.

Meinhardt H. Models of Biological Pattern Formation. New York: Academic Press, 1982.

Murray C.D. The physiological principle of minimum work applied to the angle of branching of arteries. J. Gen. Physiol. 9:835–841, 1926.

Nellis S.H., Liedtke A.J., and Whitesell L. Small coronary vessel pressure and diameter in an intact beating rabbit heart using fixed-position and free-motion techniques. Circ. Res. 49:342–353, 1981.

Pries A.R., Secomb T.W., Gaehtgens P., and Gross J.F. Blood flow in microvascular networks. Experiments and simulation. Circ. Res. 67:826–834, 1990.

Remington J.W., and Wood E H. Formation of peripheral pulse contour in man. J. Appl. Physiol. 9:433–442, 1956.

Tillmanns H., Steinhausen M., Leinberger H., Thederan H., and Kübler W. Pressure measurements in the terminal vascular bed of the epimyocardium of rats and cats. Circ. Res. 49:1202–1211, 1981.

van Bavel E., and Spaan J.A. Branching patterns in the porcine coronary arterial tree. Estimation of flow heterogeneity. Circ. Res. 71:1200–1212, 1992.

van Beek J.H.G.M., Bassingthwaighte J.B., and Roger S.A. Fractal networks explain regional myocardial flow heterogeneity. In: Oxygen Transport to Tissue XI. Adv. Exp. Med. Biol. 248, edited by Rakusan K. New York: Plenum Press, 1989, pp. 249–257.

West G.B., Brown J.H., and Enquist B.J. A general model for the origin of allometric scaling laws in biology. Science 276:122–126, 1997.

Meinhardt H. Models for the formation of netlike structures. In: Vascular Morphogenesis: In Vivo, In Vitro, In Mente, edited by Little C.D., Mironov V., and Sage E.H. Boston: Birkhauser, 1997 pp. 147–172.

Sage E.H. Introduction to Part II. In: Vascular Morphogenesis: In Vivo, In Vitro, In Mente, edited by Little C.D., Mironov V., and Sage E.H. Boston: Birkhauser, 1997 pp. 73–77.

Murray J.A., Manoussaki D., Lubkin S.R., and Vernon R. A mechanical theory of in vitro vascular network formation. In: Vascular Morphogenesis: In Vivo, In Vitro, In Mente, edited by Little C.D., Mironov V., and Sage E.H. Boston: Birkhauser, 1997 pp. 173–188.

Index